TEXTBOOK OF BIO-ORGANIC CHEMISTRY

TEXTBOOK OF BIO-ORGANIC CHEMISTRY

By
Dr. Syed Aftab Iqbal
M.Sc. , Ph.D., FICS
FICC, FIAEM, MNASc.
Professor
Department of Chemistry
Saifia Science College
Barkatullah University
Bhopal (India)

DISCOVERY PUBLISHING HOUSE PVT. LTD.
NEW DELHI-110 002

Published by:
Tilak Wasan

DISCOVERY PUBLISHING HOUSE PVT. LTD.
4383/4B, Ansari Road, Darya Ganj
New Delhi-110 002 (India)
Phone : +91-11-23279245, 43596064-65
Fax : +91-11-23253475
E-mail : discoverypublishinghouse@gmail.com
sales@discoverypublishinggroup.com
web : www.discoverypublishinggroup.com

***First Edition:* 2011**

***Reprinted:* 2017**

ISBN: 978-81-8356-832-6

Textbook of Bio-organic Chemistry

Printed at:
Infinity Imaging Systems
Delhi

Preface

The book "Text Book of Bio-Organic Chemistry" has been written the requirements of graduate students of all Indian Universities. It covers the complete syllabus of physics prescribed by Technical Universities. It covers the complete syllabus of electronics prescribed by Technical Universities. I have spared no pains in applying my long experience of degree class teaching to make the book useful to the students.

The subject matter of this book has been presented in this book has sufficient theory and maximum solved examples. One of salient course syllabi features of the present edition of the book is the inclusion of a large number of typical worked out problems which will elucidate various abstract principles and theories discussed in the text. The author in general hope.

The present book will be warmly received by the students and teachers. We shall indeed be very thankful to our colleague for their recommendations this book of for their students. Suggestion for improvement of the book will be gratefully acknowledged.

—*Author*

Contents

1

SYNTHETIC PENICILLINS AND PURINES

INTRODUCTION

It has been found that most strains of staphylococci are highly sensitive to penicillin, but after a time become resistant. This result has been shown to be due to the fact that these resistant strains produce the enzyme penicillinase which converts penicillin into the inactive penicilloic acid.

Of all the natural penicillins, benzylpenicillin (penicillin G) is still the best. It has been recently found that different types of penicillin are produced by Penicillium chrysogenum when the cultural conditions are changed. Batchelor *et al.* (1959) isolated pure 6-aminopenicillanic acid from fermentation liquors to which no precursors had been added. This acid had already been synthesised by Sheehan. It has also been shown that (I) is the site of action of the enzyme penicillin amidase (Rolinson *et al.*, 1960; Claridge *et al.*, 1960) and, as mentioned above, (2) is the site of action of penicillinase.

Many 'synthetic' penicillins have now been prepared. α-Aminobenzylpenicillin (Rolinson *et al.*, 1961) has been synthesised and shows considerable activity against many organisms against which benzylpenicillin is not very effective. 6-Aminopenicillanic acid itself has also been used as the starting point of many new penicillins by acylation of the acid (Doyle *et al.*, 1963).

ANALYSIS OF THE PENICILLINS

The penicillins are all strong monobasic acids, *e.g.*. they form salts. They are hydrolysed by hot dilute inorganic acids: one carbon atom is eliminated as carbon dioxide and two products are obtained in equimolecular amounts, one being an amine, penicillamine, and the other an aldehyde, penilloaldehyde. All the penicillins give the same amine, but different aldehydes; it is the latter which contain the R group.

$$C_9H_{11}N_2O_4SR + 2H_2O \xrightarrow{HCl} CO_2 + C_5H_{11}NO_2S + C_3H_4NO_2R$$

D-Penieillamine, $C_5H_{11}NO_2S$. This compound gave colour reactions with sodium nitroprusside and ferric chloride which were characteristic of the thiol group (SH). Electrometric titration showed three pKa values: 1.8, 7.9, and 10.5. These correspond to carboxyl, α-amino, and thiol groups, respectively. Since penicillamine combined with acetone to give an isopropylidene derivative which no longer contained a free amino or free thiol group and was reconverted into penicillamine on hydrolysis, this suggested that these two groups were attached to adjacent carbon atoms.

Oxidation of penicillamine with bromine water gave a sulphonic acid (this reaction is characteristic of a thiol). The Kuhn-Roth determination of methyl side-chains gave a very low value (~0.2 molecules). This suggested that the amine contained an isopropyl end-group and not a methyl endgroup. It was therefore proposed on the foregoing evidence that penicillamine was β,β-dimethylcysteine, and this was confirmed by synthesis, *e.g.*, Penicillin is the name given to the mixture of natural compounds having the molecular formula $C_9H_{11}N_2O_4SR$, and differing only in the nature of R. There are at least six natural penicillins. Commercial preparations of penicillin contain one or more of the penicillins in varying proportions. It has been found that the addition to the culture medium of various compounds containing a benzyl group, *e.g.*, phenylacetic acid, phenylacetamide, etc., increases the total yield of penicillin, and also the proportion of benzylpenicillin.

Similarly the addition of compounds containing the p-hydroxybenzyl group to the culture medium increases the proportion of p-hydroxybenzylpenicillin. On the other hand, by adding various compounds to the culture medium, a number of, 'unnatural' penicillins have been prepared.

Some of the steps leading to the azlactone and from this to the thiazoline derivative are uncertain; a possible sequence is:

The racemic amine was resolved as follows: the amine was converted into the formyl derivative, which was then resolved by means of brucine.

D-Penicillamine was obtained after removal of the formyl group by hydrolysis.

$$\underset{\text{DL-form}}{\underset{\text{SH}\ \ \text{NH}_2}{(CH_3)_2C—CHCO_2H}} \xrightarrow{HCO_2H} \underset{\text{DL-form}}{\underset{\text{SH}\ \ \text{NHCHO}}{(CH_3)_2C—CHCO_2H}} \xrightarrow[\text{(ii) HCl, (iii) pyridine}]{\text{(i) brucine}} \underset{\text{D-penicillamine}}{\underset{\text{SH}\ \ \text{NH}_2}{(CH_3)_2C—CHCO_2H}}$$

This was found to be identical with the natural penicillamine.

When treated with diazomethane, penicillin is converted into its methyl ester and this, on treatment with an aqueous solution of mercuric chloride, gives the methyl ester of penicillamine. Thus the carboxyl group in penicillamine is the carboxyl group in penicillin itself. Penilloaldehyde. On vigorous hydrolysis, all the penilloaldehydes give a substituted acetic acid and aminoacetaldehyde. Thus the penilloaldehydes are acylated derivatives of aminoacetaldehyde.

$$RCONHCH_2CHO + H_2O \rightarrow RCO_2H + NH_2CH_2CHO$$

This structure has been confirmed by synthesis:

$$RCOCl + NH_2CH_2CH(OC_2H_5)_2 \rightarrow RCONHCH_2CH(OC_2H_5)_2 \xrightarrow{HCl} RCONHCH_2CHO$$

As pointed out above, the acid hydrolysis of penicillin gives penicillamine, penilloaldehyde and carbon dioxide. The formation of this molecule of carbon dioxide gave rise to the belief that it is formed by the ready decarboxylation of an unstable acid. Such an acid is a β-keto-acid, and so a possible explanation is that penilloaldehyde carboxylic acid (penaldic acid) is formed as an intermediate in the hydrolysis of penicillin.

$$\underset{\text{penaldic acid}}{\underset{CO_2H}{RCONHCHCHO}} \longrightarrow CO_2 + RCONHCH_2CHO$$

The problem now is: How are the two fragments, penicillamine and penilloaldehyde, combined in penicillin? The hydrolysis of penicillin with dilute alkali or with the enzyme penicillinase produces penicilloic acid (a dicarboxylic acid), which readily eliminates a molecule of carbon dioxide to form penilloic acid. This suggests that a carboxyl group is in the β-position with respect to an electronattracting group (cf. above). Penilloic acid, on hydrolysis with aqueous mercuric chloride, gives penicillamine and penilloaldehyde. This hydrolysis is characteristic of compounds containing a thiazolidine ring. Thus penilloic acid could be (I) since this structure would give the required products.

Hence, if (I) is penilloic acid, then penicilloic acid would be (II).

```
                           S
                          / \
RCONHCH ——— HC       C(CH3)2
   |             |         |          ——> CO2 + (I)
  CO2H         HN ——— CHCO2H
                 (II)
```

Structure (II) is supported by the fact that the treatment of penicillin with methanol gives methyl penicilloate which, on hydrolysis with aqueous mercuric chloride, gives methyl penaldate (see also above) and penicillamine.

```
                                  S                                             HS
                                 / \                                              \
           CH3OH    RCONHCH ——— HC     C(CH3)2   H2O     RCONHCHCHO            C(CH3)2
Penicillin ——————>     |         |       |      ——————>     |                   |
                    CH3O2C      HN ——— CHCO2H   HgCl2     CH3CH2            H2NCHCO2H
```

On the basis of the foregoing evidence, two structures are possible for penicillin, viz. (III) and (IV). It was not possible to decide between them on chemical evidence alone, since penicillin readily undergoes molecular rearrangements, e.q., on treatment with dilute acid, penicillin rearranges to penillic acid. It was therefore desirable to examine the molecule by physical methods (thereby leaving the molecule intact).

(i) The infrared spectra of many penicillins were examined and a correlation between various bands and functional groups was carried out by examining the spectra of synthetic model compounds which contained different parts of structures (III) and (IV) that had been proposed on chemical evidence. This may be illustrated v-ith the methyl ester and sodium salt of benzylpenicillin, which showed the following maxima (characteristi' of all the penicillins in these regions).

Methyl ester: 3333, 1770, 1748, 1684, 1506 cm^{-1}

Sodium salt: 3333, 1770, 1613, 1681, 1515 cm^{-1}

The band at 3333 cm^{-1} in both compounds was assigned to the NH group (str.), and the 1748 cm^{-1} band of the ester and the 1613 cm^{-1} band of the salt were assigned to the carbonyl group (str.) in the carboxyl group (as ester or salt). Then model oxazolones were studied; these showed two characteristic bands, one at 1825 cm^{-1} for the carbonyl group, and one at 1675 cm^{-1} for the C=N group. The absence of the first but possible presence of the second in the benzylpenicillin derivatives would not permit a decision to be reached between (III) and (IV). When a large number of thiazolidines were examined in the double bond region down to 1 470 cm^{-1}, only the carbonyl band was found to be present (~ 1748 and 1613 cm^{-1}). A large number of amides were now examined.

All three types had a band close to 1670 cm^{-1}, which can be attributed to the carbonyl group, but with the primary amides there was also a band near 1613 cm^{-1}, and with the secondary amides the band was close to 1515 cm^{-1}. These results suggest that penicillins have the secondary amide structure (*i.e.*, (IV)). since the secondary amide band at 1670 cm^{-1} ≡ 1684 and 1681 cm^{-1}, and the band at 1515 cm^{-1} ≡ 1506 and 1515 cm^{-1}. Thus, four of the five bands have been accounted for. Finally, a number of simple β-lactams and fused thiazolidine-β-lactams were examined. The former did not show a band near 1770 cm^{-1}, but all the latter were found to have a band at 1770 cm^{-1}. This accounts for the fifth band, and so it follows that (IV) is the structure of the penicillins.

(ii) The X-ray analysis of the sodium, potassium and rubidium salts of benzylpenicil lin showed the presence of a fl-lactam ring; thus (IV) is the structure of penicillin.

Sheehan's later approach (1957, 1959) was the ring-closure of natural penicilloic acids. Because of the sensitivity of the β-lactam ring to acids in particular, ring-closure was effected by means of dicyclohexylcarbodi-imide; this isa mild reagent for forming the amide bond (Sheehan *et al.*, 1955), and was carried out at the last step of the synthesis. The amine was protected by the phthalimido group and one of the carboxyl groups (the one involved in ring-closure) was protected by t-butyl ester formation. Removal of the phthalimido group was carried out by means of hydrazine since this left the ester group intact.

The final step of cyclisation was carried out on the potassium salt of the penicilloic acid (IV). Purification of the potassium salt of penicillin V by means of counter-current distribution between isobutyl methyl ketone and two successive phosphate buffers, gave a yield of 5.4 per cent pure crystalline potassium salt.

The synthesis of penicillins was improved by preparing 6-aminopenicillanic acid and acylating this directly to a penicillin (Sheehan *et al.*, 1959, 1962). In this route, the β-lactam ring was formed prior to the removal of the protecting groups; no side reaction to form oxazolone was now possible. Also, the final step should be noted; this had to be carried out under strictly controlled conditions. The formation of the aldehydate results in the introduction of one chiral centre and so the product is the (±)-form. Condensation of this with D-penicillamine produces a new chiral centre in (V).

(V) therefore contains three chiral centres, but since D-penicillamine was used as one of the starting materials, all possible optical isomers are derived from D-penicilloates. However, in the formation of (V), ring-closure has occurred to give the thiazolidine ring.

Hence, (V) can theoretically exist as two geometrical isomers, (VI) and (VII), and each of these can theoretically give rise to four optical isomers (two diastereoisomers). Because of steric effects, it can be anticipated that the $CHZCO_2$ But group would take up preferentially the trans position with respect to the carboxyl group already in the ring, *i.e.*, (VII) is the anticipated product (exclusive or predominant). Of the four possible diastereoisomers only two in fact appeared to have been formed, α and γ, corresponding to structure (VII). The minor one was the required α-isomer; this was the form which had the same stereochemistry as that of the corresponding product obtained by degradation of natural penicillins. The two isomers were separated by fractional crystallisation (α more soluble than γ).

Since the γ-isomer underwent epimerisation (in pyridine solution in an atmosphere of hydrogen) to give an equilibrium mixture containing about 25 per cent of the α-isomer, it was therefore possible to increase the amount of the required α-isomer for completion of the synthesis. Presumably, epimerisation is possible at both chiral centres, the $CHZCO_2Bu^t$ and the ring CO_2H group (via enolisation of CH with the adjacent carboxyl group). The $CHZCO_2Bu^t$ group cannot change its geometrical configuration with respect to the ring, and if the ring carboxyl group underwent epimerisation, this would bring it into the cis-configuration (VI).

Hence, it can be anticipated that only the $\overset{*}{C}$ of $\overset{*}{C}HZCO_2Bu^t$ is epimerised. As mentioned above, both (VI) and (VII) can each give rise to two diastereoisomers, *i.e.*, four diastereoisomers of the D-penicilloates are possible. All have been prepared and are designated as D-α-, D-β-, D-γ-, and D-δ-isomers. As was also mentioned above, the natural compounds are the D-a-isomer, the configuration of which has been established by the X-ray analysis of the penicillins.

$$HOCH_3-C\equiv C-C\equiv C-C\equiv C-CONH_2$$

agrocybin

Chloramphenicol (Chloromycetin)

Chloramphenicol is a laevorotatory compound that is produced by Streptomyces venezuelae (Carter *et al.*, 1948); it is very effective in the treatment of typhoid fever, etc.

The molecular formula of chloramphenicol is $C_{11}H_{13}C_{12}N_2O_5$, and its absorption spectrum is similar to that of nitrobenzene. The presence of a nitro-group was confirmed by the reduction of chloramphenicol with tin and hydrochloric acid, followed by diazotisation and then coupling to give an orange-red precipitate with 2-naphthol (Rebstock *et al.*, 1949). When catalytically reduced (palladium), chloramphenicol gives a product which has an absorption spectrum similar to that of p-toluidine, and the solution contains ionic chlorine.

The hydrolysis of chloramphenicol with acid or alkali produces dichloroacetic acid and an optically active base, $C_9H_{12}N_2O_4$. This base was shown to contain a primary amino-group, and when treated with methyl dichloroacetate, the base reformed chloramphenicol (Rebstock *et al.*, 1949).

Chloramphenicol is converted into a diacetyl derivative on treatment with acetic anhydride in pyridine; the base obtained from chloramphenicol forms a triacetyl derivative on similar treatment.

Thus chloramphenicol probably contains two hydroxyl groups. When the base is treated with periodic acid, two molecules of the latter are consumed with the formation of one molecule each of ammonia, formaldehyde and p-nitrobenzaldehyde. These products may be accounted for if the base is assumed to be 2-amino-l-p-nitrophenylpropane-1,3-diol (Rebstock *et al.*, 1949).

Reduction of (I) with aluminium isopropoxide gave predominantly the threo-compound (II) together with a small amount of the erythro-isomer. The threo-racemate was separated from the erythro-racemate by fractional crystallisation. (II), on hydrolysis, gave threo-(III), which was resolved by means of (+)-camphorsulphonic acid. D-threo-(III) was converted into (–)-chloramphenicol with methyl dichloroacetate. This structure has also been confirmed by crystallographic studies (Dunitz, 1952).

Chloramphenicol and the base contain two chiral centres; thus there are two possible pairs of enantiomers. Comparison of the properties of the base with those of norephedrine and nor-ψ-ephedrine showed that the configuration of the base was similar to that of nor-ψ-ephedrine (Rebstock *et al.*, 1949). Thus chloramphenicol is D-(–)-threo-2-dichloroacetamino-l -p-nitrophenylpropane-1,3 - diol. It is interesting to note that chloramphenicol is the first natural compound found to contain a nitro-group; the presence of the $CHCl_2$ group is also most unusual.

Lavandulol, $C_{10}H_{18}O$, b.p. 94-95°C/13 mm.

This occurs in the free state and as esters in French lavender oil. It is a particularly interesting acyclic monoterpenoid in that it does not obey the special isoprene rule, *i.e.*, two isoprene units are not joined head to tail.

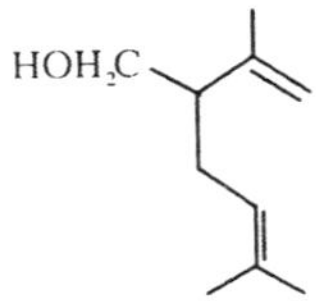

HOH_2C

Citronellal, $C_{10}H_{18}O$

This is an optically active compound which occurs in citronella oil. Citronellal is an aldehyde; reduction with sodium amalgam converts it into the alcohol citronellol, $C_{10}H_{20}O$, and oxidation gives citronellic acid, $C_{10}H_{18}O_2$. Oxidation of citronellal with chromic acid gives 3-methyladipic acid and acetone.

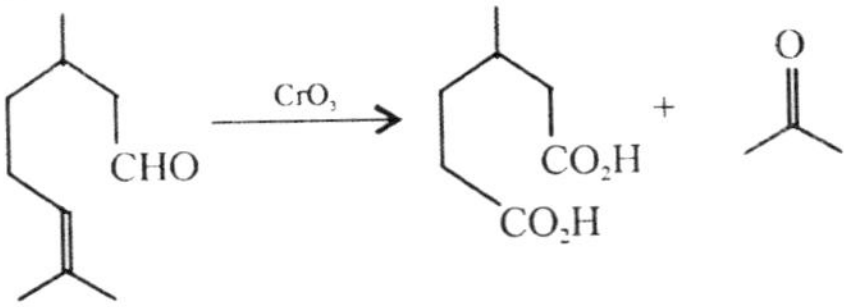

CHO $\xrightarrow{CrO_3}$ CO_2H CO_2H + O

The isopropenyl isomer was named rhodinal, but it is no longer believed to be present in natural citronellal.

Citronellol, $C_{10}H_{20}O$, b.p. 103°/5 mm

This occurs in the (–)-form in rose and geranium oils. Its structure was determined by oxidative degradation (to give acetone and 3-methyladipic acid), and by the following sequence of reactions:

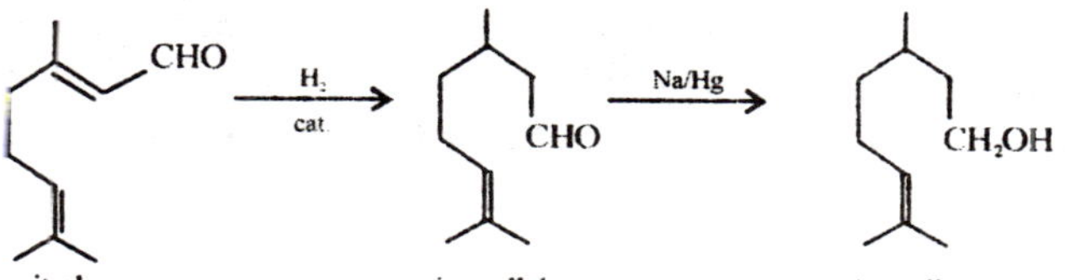

CHO $\xrightarrow[\text{cat.}]{H_2}$ CHO $\xrightarrow{\text{Na/Hg}}$ CH_2OH

citral citronellal citronello

Fig. 1

The isopropenyl isomer was named rhodinol, but its presence in natural citronellol is no longer accepted. It has been synthesised.

Arsenical Drugs

A particularly important use of arsenical drugs is in the treatment of syphilis.

Arsphenamine (Salvursan, '606') was first introduced by Ehrlich (1909); it is 3,3′-diamino-4,4′-dihydroxyarsenobenzene, and may be prepared as follows: Arsphenamine is an unstable compound; it is stable as its dihydrochloride which, however cannot be used as such but must be converted into the soluble sodium salt. Ehrlich (1912) overcame this difficulty by preparing neoarsphenamine (Neosalrarsan), a soluble compound, which may he produced by condensing arsphenamine with sodium formaldehydesulphoxylate, $HOCH_2SO_2Na$.

Atoxyl is the sodium salt of β-arsanilic acid (p-aminophenylarsonic acid): it is used in the treatment of sleeping sickness. p-Arsanilic acid may be prepared by heating aniline with arsenic acid at 200°C.

Tryparsamide is the sodium salt of N-phenylglycineamide-p-arsonic acid; it is less toxic than Atoxyl, and may be prepared by refluxing the latter with chloroacetamide.

Patulin

This has been obtained from various moulds. It is an optically inactive solid, and it inhibits Staphylococci and coliforms. It has, however, never become important as an antibiotic because it has bad side-effects, *e.g.*, it slows down the healing process, although limiting the infection. Nevertheless, it is discussed here because the elucidation of its structure is a very good example of the use of ultraviolet and infrared spectroscopy. The molecular formula of patulin is $C_7H_6O_4$; it is a neutral substance and forms a monoacetate. Hydrolysis of patulin with acid produces one molecule of formic acid and a small yield (10 per cent) of tetrahydro-γ-pyrone-2-carboxylic acid (I). Catalytic reduction followed by further reduction with hydrogen iodide and red phosphorus gives3-methylhexoic acid (II) and the lactone of 4-hydroxy-3-methylhexoic acid (III) [Raistrick *et al.*, 1943]. On the basis of these results, Raistrick proposed (VI) as the structure of patulin.

On the other hand, Bergel *et al.* (1944) obtained an acid from patulin and formulated it as 3-methyltetrahydro-g-pyrone-2-carboxylic acid (IV) and so Bergel supported formula (VI), but at the same time believed that patulin was a tautomeric compound of four forms, (VI) and three others, with (VII) being present in a large amount. (VII) was synthesised by Puetzer *et al.* (1945) and the product was found to be different from patulin. (VII) behaved as a simple lactone, and it was concluded that (VI) and (VII) were not patulin. It was then found that (IV) was not the structure of the acid isolated by Bergel; the structure was shown to be tetrahydro-ypyrone-3-acetic acid (V) and this led Plattner *et al.* (1949) to propose structure (VIII) for patulin.

Woodward *et al.* (1950) examined the ultraviolet and infrared spectra of patulin, as well as its chemical properties. The ultraviolet spectrum showed a maximum at 276 nm (e 16600), which suggests the presence of a keto group conjugated with more than one double bond (the chromophore C=C–C=O absorbs around 220 nm and may be raised to ~ 250 nm by substituents on the a- and b-carbon atoms). Partly on the basis of this fact, Woodward rejected structure (VIII) and proposed (IX). Furthermore, Woodward synthesised (VIII) and showed it was different from patulin ((VIII) is now known as allopatulin). Patulin is readily acetylated and readily converted into an ether. It therefore appears that patulin contains a hydroxyl group. This was supported by the fact that the infrared spectrum showed a band at 3660 cm^{-1}, a region typical of a free hydroxyl group.

Furthermore, when patulin was acetylated, this band disappeared, but most of the other bands were still present, in particular, the bands at 1792 (v.s.), 1685 (s), and 1636 cm^{-1}(s). Now, the band corresponding to a keto group is always very strong (usually being the most intense band in the infrared spectrum), and so the maximum at 1792 cm^{-1} was assigned to a keto group. This frequency is higher than the usual range for acyclic (1725–1700 cm^{-1}) and 5- and 6-ring ketones (1750–1700 cm^{-1}), and α, β-unsaturated ketones (1690-1660 cm^{-1}). This information, taken in conjunction with the ultraviolet spectrum, thus suggests the presence of a g-lactone (1800-1760 cm^{-1}). Wocdward synthesised (X) [as a model compound] and found that its infrared spectrum was essentially identical with that of patulin in the 1600-1800 cm^{-1} region, and then finally synthesised patulin itself as follows: The monoacetate (obtained above) was shown to be identical with that obtained from patulin.

The Macrolide Group of Antibiotics

These are macrocyclic lactones and are also known as the erythromycin group because the most useful macrolide is erythromycin. The lactone ring is joined by glycosidic links to one or more amino-sugars; in some cases the sugar may not be of the amino-type. Erythromycin A, $C_{37}H_{67}NO_{13}$, has structure (I). It contains a fourteen-membered lactone ring which is joined to desosamine (A) and to cladinose (B). Erythromycin B has structure (I) but there is no hydroxyl group at C-12.

Methymycin, $C_{25}H_{43}NO_7$, is (II); the sugar is desosamine. Kinumaki *et al.* (1972) isolated a new antibiotic, m.p. 68-70°C, from cultures producing methymycin (II). The mass spectrum of this new compound showed its molecular formula to be $C_{25}H_{43}NO_6$ (M^+ 453.3048). The ultraviolet spectrum (ethanol) showed the following bands, which were assigned as indicated: λ_{max} 225-226 and 285 nm; an α, β-unsaturated ketone. The infrared spectrum data and their assignments were: V_{max} 3420 cm^{-1} (OH); 1730 cm^{-1} (lactone and carbonyl); 1695 cm^{-1} (conjugated ketone); and 1635 cm^{-1} (conjugated double bond). These spectroscopic data suggested a structural similarity to methymycin (II). The NMR spectrum of the new antibiotic showed the presence of six C-methyl groups and one NMe_2 group.

There were also the following signals: τ 3.6 (d, J 16 Hz; 8-H) and τ 3.18 (dd, J 16. 4-5 Hz; 9-H). The assignments were made as shown, and because of the shift to lower field (than the usual r-values for olefinic protons, τ 4.8), the authors suggested these values indicated the presence of the grouping –CO–CH=CH–CH <. There was also a broad triplet at τ 5.02. and this was assigned to the methine proton at C-11 (III).

POLYPEPTIDE ANTIBIOTICS

Many antibiotics formed by micro-organisms of the genus Bacillus and the genus Streptomyces are polypeptides containing six to twelve amino-acid residues. Their structures have been elucidated mainly by the methods used in protein chemistry. All contain rings and D-amino-acids are often present, *e.g.*, bacitracin A (note the presence of the thiazoline ring).

POLYACETYLENE

Many of this group are known, some are highly toxic and apparently none has been used clinically. Their structures are unbranched carbon chains which, in addition to acetylenic bonds, may contain ethylenic bonds and functional groups, *e.g.*, hydroxyl, carboxyl, carbonamido; *e.g.*, mycomycin and agrocybin.

Pyrimidine and Purine Bases Spectra

Infrared spectra of these bases have been determined and their main use, so far, has been in settling the problem of keto-enol tautomerism. On the other hand, ultraviolet spectra have been of great value in the determination of the structure of these bases and in the study of nucleic acid chemistry. In the earlier work, changes in the absorption maxima with pH were believed to be due to changes in the keto-enol equilibria, but now it has been shown that these changes are due to ionisation into different species. Some absorption maxima at different pH values (water as solvent). Tinoco. Jr. *et al.* (1965) have correlated a number of the absorption bands (250-300 nm) in pyrimidine and purine bases; these are believed to be $\pi \rightarrow \pi^*$ transitions.

It is possible to identify various bases, and it has been found that the ultraviolet maxima of nucleosides generally lie fairly close to those of the bases that they contain. This is the case at the lower pH values, but at pH 13 there is a considerable difference, *e.g.*, uridine : λ_{max} 262 nm (4.0) at pH 1-7 and 263 nm (3.87) at pH 13; adenosine: 257 nm (418) at pH 2, 259 nm (4.19) at pH 7 and 259 rim. (4.19) at pH 11.

The ultraviolet spectra of most nucleotides are very similar to those of their corresponding nucleosides, but the ultraviolet spectra of the nucleic acids generally show about 40 per cent lower absorbance than the equivalent solution of the component nucleotides. This result, known as hypochromism, is believed to be partly due to the hydrogen bonds between the base residues in the double helix system since, when this double helix is broken, *e.g.*, by heating above 80°C the absorbance increases by about 40 per cent. Another contributing factor to hypochromism is believed to be the change in the resonance effect in the bases when they are chemically bound in the polynucleotide.

Sugars: Two sugars have been isolated from the hydrolysates of nucleic acids; both are pentoses: D(-)-ribose ad 2-deoxy-D-(–)-ribose

$$\underset{\text{ribose}}{OHC(HOH)_3CH_2OH} \quad \underset{\text{2-deoxyribose}}{OHCCH_2(CHOH)_2CH_2OH}$$

The nucleic acids are classified according to the nature of the sugar present: the ribonucleic acids (RNA), and the deoxyribonucleic acids (DNA). Ribonucleoproteins are found mainly in the cytoplasm of the cell, whereas deoxyribonucleoproteins are found mainly in the cell nucleus. D(–)-Ribose is the pentose of yeast, liver and pancreas RNAs; 2-deoxy-D(–)-ribose occurs in thymus DNA. Nucleic acids also occur in plant and animal viruses.

It has now been found that some RNAs contain minute amounts of 2′-O-methvlribose.

Bases: There are two types of bases which occur in nucleic acids: purines and pyrimidines. The most common purine bases are adenine and guanine. Many other purines have been isolated, *e.g.*, 1-, 2-, and 3-methyladenine, 6-methylaminopurine, 3-methylguanine, etc. The most common pyrimidine bases are uracil, thymine, and cytosine.

Other pyrimidines have been isolated, *e.g.*, 5-methylcytosine and 5-hydroxymethylcytosine. Both types of nucleic acids (RNA and DNA) contain adenine and guanine.

On the other hand, RNAs also contain uracil and cytosine, whereas DNAs contain thymine and cytosine. This distribution of pyrimidines, however, is not rigid, *e.g.*, uracil has been found in certain DNAs.

Angell (1961) has shown, from infrared studies, that in the solid state and in ribose and deoxyribose nucleosides derived from these bases, adenine exists in the amino form, cytosine and guanine exist in the keto-amino form and uracil in the diketo form. Furthermore, X-ray analysis of the various bases has shown that all are planar. Combination of a base (either a purine or pyrimidine) with a sugar (ribose or deoxyribose) gives rise to a nucleoside, *e.g.*, adenosine (ribose + adenine), guanosine (ribose + guanine), cytidine (ribose + cytosine), uridine (ribose + uracil), thymidine (deoxyribose + thymine). The nucleoside derived from hypoxanthine and ribose is named inosine.

Combination of a nucleoside with phosphoric acid produces a nucleotide, *i.e.*, nucleotides are nucleoside phosphates, *e.g.*, adenylic, guanylic, cytidylic, inosinic, and uridylic acids. It might be noted here that the term nucleotide is now used to embrace a large group of compounds composed of the phosphates of N-glycosides of heterocyclic bases, and the pyrophosphates and polyphosphates containing one or more nucleosides.

Nucleotides Structure

When nucleotides are carefully hydrolysed, ribose monophosphate may be isolated from the products; thus the phosphoric acid is attached to the sugar residue in nucleotides. Examination of the nucleoside structures shows that the point of attachment may be 2′, 3′ or 5′ in the ribose molecule, and 3′ or 5′ in the deoxyribose molecule. On reduction with hydrogen in the presence of platinum, ribose phosphate is converted into an optically inactive phosphoribitol (Levene *et al.*, 1932, 1933).

This product can be optically inactive only if the phosphate residue is attached to the centre hydroxyl group of the ribose molecule, *i.e.*, at the 3′-position.

Later work, however, resulted in the isolation of 2′-, 3′-, and 5′-phosphates. Enzymic hydrolysis of nucleic acids can give rise to either 3′- or 5′-phosphates; alkaline hydrolysis gives a mixture of 2′- (RNAs only) and 3′-phosphates. The mixture of 2′- and 3′-phosphates (from RNAs) has now been shown to arise only in alkaline hydrolysates and has been explained on the basis of the formation of an intermediate 2′,3′-cyclic phosphate (Todd *et al.*, 1953-1955). The mechanism proposed is:

Cyclic 2′,3′- and 3′,5′-phosphates have been prepared by synthesis.

Inspection of the above formulae shows that only a ribonucleotide 5′-phosphate has an adjacent pair of cis-hydroxyl groups (2′ and 3′). Hence, the 5′-phosphate can be readily distinguished from its isomers-2′- and 3′-phosphates-by means of the periodic oxidation.

This method, however, cannot differentiate between deoxyribonucleoside 3′- and 5′-phosphates since neither of these contains a 2′-hydroxyl group. The determination of the position of the phosphate residue (other than by the periodate oxidation) has been carried out by synthesis, X-ray analysis, etc.

Nucleoside di- and triphosphates are also known, *e.g.*, ADP and ATP. These, however, do not occur in nucleic acids.

Nucleotides have been synthesised in various way, *e.g.*, Levene et al. (1937) synthesised adenosine-5'phosphate from 2′,3′-O isopropylideneadenosine. This was phosphorylated with phosphoryl chloride in pyridine, followed by careful hydrolysis with acid to remove the isopropylidene residue. 2′- and 3′-phosphates are more difficult to synthesise because of their ready interconversion.

Todd *et al.* (1954) synthesised adenosine-2′-phosphate by phosphorylating 3′,5′-di-O-acetyladenosine in the 2′-position with dibenzylphosphochloridate [$(PhCH_2O)_2POCl$] and removing the benzyl groups (as toluene) by hydrogenation (Pd), and finally removing the acetyl groups by treatment with alkali. Under these conditions no phosphate migration is possible.

Biosynthesis of Pyrimidines and Purines

An interesting point about the biosynthesis of these bases is that they are not formed in the free state as an intermediate; they are formed as nucleotides. Also, deoxyribose nucleotides are formed directly from ribose nucleotides, *i.e.*, the glycosidic bond is not broken in the conversion.

Anthocyanidins Structure

The anthocyanin is first hydrolysed with hydrochloric acid and the anthocyanidin is isolated as the chloride. The usual analytical methods are applied to determine the number of hydroxyl and methoxyl groups present in the molecule. The structure of the anthocyanidin is ascertained by the nature of the products obtained by fusing the anthocyanidin with potassium hydroxide (Willstatter *et al.*, 1915); phloroglucinol or a methylated phloroglucinol and a phenolic acid are always obtained, *e.g.*, cyanidin chloride gives phloroglucinol and protocatechuic acid.

This method suffers from the disadvantage that the fusion (or boiling with concentrated potassium hydroxide solution) not only degrades the anthocyanidin, but also often demethylates it at the same time. Thus the positions of the methoxyl groups in the original compound are now rendered uncertain. This difficulty was overcome by Karrer *et al.* (1927), who degraded the anthocyanidin with a 10 per cent solution of barium hydroxide or sodium hydroxide in an atmosphere of hydrogen; in this way, the methoxyl groups are left intact.

The next problem is to ascertain the positions of the sugar residues. After hydrolysis (of the anthocyanin), the sugar is identified by the usual methods of sugar chemistry, and this includes the use of paper chromatography. If two or more monosaccharide molecules are obtained for each molecule of anthocyanin, it is necessary to determine whether they were present as such, or as a disaccharide (or trisaccharide) which was hydrolysed. One method is first to methylate the anthocyanin and then hydrolyse with suitable enzymes. In this way, methylated disaccharides may be isolated intact.

(i) Karrer *et al.* (1927) methylated the anthocyanin, removed the sugar residues by hydrolysis (hydrochloric acid), and finally hydrolysed with barium hydroxide solution in an atmosphere of hydrogen; the positions of the free hydroxyl indicate the points of attachment of the sugar residues. In some cases, however, interpretation of the results is uncertain, *e.g.* (G represents a sugar residue): The problem is: Which of the two hydroxyl groups in monomethylphloroglucinol was originally attached to G? The above results do not lead to a definite answer, since had the structure of the anthocyanin been (IV) instead of (I), (III) would still have been obtained:

(ii) Hydrogen peroxide (15 per cent) attacks anthocyanins as follows (Karrer *et al.*, 1927): If the anthocyanin (V) has a glucose residue in the 3-position, then this glucose residue in (VI) is readily hydrolysed by dilute ammonia. If the glucose residue in (V) is in either the 5- or 7-position, then this glucose residue in (VI) is removed only by heating with dilute hydrochloric acid. Thus position 3 can be distinguished from positions 5 or 7, but the latter two cannot be distinguished from each other.

(iii) Anthocyanins with a free hydroxyl group in the 3-position are very readily oxidised by ferric chloride; the anthocyanins are rapidly decolorised in this oxidation (Robinson *et al.*, 1931).

The final problem is to determine whether the sugar linkage (to the anthocyanidin) is α or β. This is ascertained by hydrolysis with the enzymes maltase (α-linkage) and emulsin (β-linkage).

Conclusive evidence for the positions and nature of the linkages of the sugar residues is afforded by the synthesis of the anthocyanins (see, *e.g.*, cyanin). In general, it has been found that glucose residues are linked at positions 3 or 3,5 and that the linkage is usually β. Now that a large number of structures have been elucidated, it has been possible to correlate, to a large extent, physical data with structure. Thus, the flavonoid is fairly easily assigned to its class by its absorption spectrum (visible and ultraviolet) and by colour reactions.

It has already been pointed out that the known anthocyanidins belong to three main types: cyanidin (λ_{max} 535 nm), pelargonidin (λ_{max} 520 nm), and delphinidin λ_{max} 564 nm) [all measured in McOH-HCl solution]. Thus these may be distinguished from each other. Furthermore, the introduction of sugar residues into these anthocyanidins shifts the maxima towards the shorter wavelengths, the shifts being characteristic for sugar residues in the 3- or 3,5-positions, and for the 5-position. Also, since 3-glycoside spectra show a pronounced shoulder (440-460 nm region), it is therefore possible to obtain information about the positions of sugar residues in anthocyanins.

Examination of the absorption spectra of anthocyanidins in the presence of certain reagents also gives information about structure. Thus aluminium chloride and sodium ethoxide shift the maxima to the longer wavelengths, the actual shift depending on the positions of the hydroxyl groups in the molecule, *e.g.*, cyanidin (λ_{max} 535 nm) has its λ_{max} at 553 nm in the presence of aluminium chloride. In this way, it may be possible to obtain some indication of the relationship of the compound under investigation to a flavonoid of known structure, thereby simplifying further structure determination. Also, since the R_F values of the known anthocyanins have been determined (Harborne, 1959), these compounds are quite readily identified.

GENERAL METHODS OF SYNTHESISING THE ANTHOCYANIDINS

(i) Willstatter (1914) synthesised anthocyanidins starting from coumarin.

(ii) Robinson has introduced a number of methods whereby all anthocyanidins can be prepared. The basic reaction of these methods is the condensation between o-hydroxybenzaldehyde and acetophenone in ethyl acetate solution which is then saturated with hydrogen chloride.

The original method of Robinson (1924) resulted in the formation of a product in which the substituent groups were either all hydroxyl groups or all methoxyl groups, *e.g.*, Robinson (1928, 1931) then modified this method so that the product could have both hydroxyl and methoxyl substituent groups, *e.g.*, The following is a brief account of the methods used by Robinson and his co-workers for preparing the substituted acetophenones and substituted benzaldehydes.

Flavonol (3-hydroxyflavone), $C_{15}H_{10}O_3$

Flavonol is widely distributed in the plant kingdom, usually in the form of glycosides. Flavonols show two absorption bands: Band I, 350-390 nm, and Band II, 250-270 nm.

These, taken in conjunction with their specific colour reactions, make it possible to identify this group of compounds. When boiled with an ethanolic solution of potassium hydroxide, flavonol gives o-hydroxybenzoylmethanol and benzoic acid. This suggests that flavonol is 3-hydroxyflavone (3-hydroxy-2phenyl-γ-chromone).

Selenium dioxide may also be used (Venkataraman *et al.*, 1935, 1936), but not if the molecule contains free hydroxyl groups. If these are present, the dehydrogenation can be carried out via the acetylated derivative (Seshadri *et al.*, 1954). On the other hand, flavonol may be prepared by the Algar-Flymr Oyamada reaction (1934). 2′-Hydroxychalcone, on treatment with alkaline hydrogen peroxide, is first converted into 3-hydroxyflavanone and then into flavonol.

This reaction affords a very good general synthesis, of flavonoids. An alternative general method for preparing flavones based on the flavonol synthesis is as follows (Kostanecki *et al.*, 1898): This synthesis has been simplified by Wheeler *et al.* (1955), who condensed ω-chloro-o-hydroxyacetophenones with aromatic aldehydes in the presence of ethanolic sodium hydroxide, *e.g.*,

2,5-Dihydroxy-7-methoxyflavanone: Chadenson *et al.* suggested that a compound, $C_{16}H_{15}O_5$, m.p. 170-172°C, isolated from Populus nigra buds had structure (I) on the basis that it readily gave the flavone (II) on cyclodehydration with acid.

When Chadenson *et al.* (1972) treated (II) with sodium peroxide in pyridine in the expectation of regenerating (I), the experiment failed (cf. flavone). On the other hand, when (II) was heated with anhydrous potassium hydroxide in pyridine, the product was identical with the natural compound (believed to be (I)).

A synthesis of benzoyl-(2,6-dihydroxy-4-methoxybenzoyl)methane (I) was then attempted by the Baker-Venkataraman method as shown. The product (Z) was identical with the natural compound, but their physical data were not consistent with those expected of structure (I). The i.r. spectrum of (Z) showed only one carbonyl band (1640 cm^{-1}) instead of the two (1640 and 1680 cm^{-1}) which have been observed for the colourless forms of o-hydroxyJibenzoylmethanes (type (I) structure). The mass spectrum of (Z) showed a strong peak at M-17. This corresponds to (111), which is expected to be derived from structure (IV) by comparison with the known fragmentation pattern of flavanones.

Furthermore, this cyclic structure, (IV)≡(Z), was supported by the NMR spectrum (100 MHz; $(CD_3)_2CO$; –60°C): τ 7.21 (d, J_{gem} 17 Hz), 3-H (eq); 6.70 (q, J_{gem} 17 Hz, J 2 Hz), 3-H-2-OH, 3-H (ax); 6.20 (s), OMe; 2.91 (s), 6-and 8-H; 2.74 (d, J 2 Hz), 2-OH (ax); 25 (3′-, 4′-, and 5′-H); 2.2 (2′- and 6′-H); –2.40 (s), 5-OH.

The equatorial position of the 2-phenyl group is indicated by the long-range coupling (2 Hz) between 2-OH and 3-H. Also, the positions of the signals of the 3-H are similar to those of flavanones.

Quercetin, $C_{15}H_{10}O_7$

This occurs as the glycoside quercitrin in the bark of Quercus tinctoria; quercitrin appears to be the most widely distributed natural pigment. On hydrolysis with acid, quercitrin forms quercetin and one molecule of rhamnose.

$$C_{21}H_{20}O_{11} + H_2O \xrightarrow{HCl} C_{15}H_{10}O_7 + CH_3(CHOH)_4CHO$$

Quercetin contains five hydroxyl groups; no methoxyl groups are present; on fusion with potassium hydroxide, phloroglucinol and protocatechuic acid are obtained (cf. cyanidin).

Also, when quercetin is methylated and the product, pentamethylquercetin, boiled with an ethanolic solution of potassium hydroxide, 6-hydroxy-ω,2,4-trimethoxyacetophenone and veratric acid are obtained. These results suggest that quercetin is 3,3′,4′,5,7-pentahydroxyflavone.

This structure has been confirmed by synthesis, *e.g.*, Kostanecki *et al.* (1904). Another synthesis is that of Robinson *et al.* (1926); it is a general method for flavonols (cf. flavone): w-methoxyphloroacetophenone is condensed with veratric anhydride in the presence of the potassium salt of veratric acid (3,4-dimethoxybenzoic acid; this has been written as $ArCO_2H$). The position of the rhamnose residue in quercitrin has been shown to be 3 (Herzig *et al.*, 1912).

Before leaving this problem of quercetin, let us consider its relationship to cyanidin. As we have seen, the relationship between the two compounds is suggested by the fact that both give the same products when fused with potassium hydroxide. Willstatter *et al.* (1914) reduced quercetin with magnesium in hydrochloric acid containing mercury, and thereby obtained a small amount of cyanidin chloride.

Bauer *et al.* (1954) have converted the penta-acetate of quercetin into cyanidin chloride by means of lithium aluminium hydride.

King *et al.* (1957) have shown that the reductive acetylation of a flavonol, followed by the action of hot hydrochloric acid, gives the corresponding anthocyanidin; thus:

$$\text{quercetin} \xrightarrow[\text{(ii) HCl}]{\text{(i) Zn - AcONa; Ac}_2\text{O}} \text{cyanidin chloride}$$

Isoflavones

These are hydroxylated derivatives of isoflavone (3-phenyl-4-chromone) which may be partially alkylated. The isoflavones occur naturally, but are not so widespread as the flavones; they occur either in the free state or as glycosides. The general method of ascertaining the structure of isoflavones is similar to that used for the flavones. Thus fusion with potassium hydroxide breaks down the molecule into two fragments, and hydrolysis with ethanolic potassium hydroxide permits the isolation of intermediates. This may be illustrated with daidzein (Walz, 1931):

Monoterpenoids

The monoterpenoids may be subdivided into three groups: acyclic, monocyclic and bicyclic. This classification affords a convenient means of study of the monoterpenoids.

Myrcene, $C_{10}H_{16}$, b.p. 166-168°C

This is an acyclic monoterpenoid hydrocarbon (*i.e.*, is a terpene) which occurs in verbena and bay oils. Catalytic hydrogenation (platinum) converts myrcene into a decane, $C_{10}H_{22}$; thus myrcene contains three double bonds, and is an open-chain compound. Furthermore, since myrcene forms an adduct with maleic anhydride, two of the double bonds are conjugated (Diels *et al.*, 1929; see the Diels-Alder reaction). This conjugation is supported by the fact that myrcene shows optical exaltation. These facts, *i.e.*, that myrcene contains three double bonds, two of which are in conjugation, had been established by earlier investigators (*e.g.*, Semmler, 1901). Ozonolysis of myrcene produces acetone, formaldehyde and a ketodialdehyde, $C_5H_6O_3$, and the latter, on oxidation with chromic acid, gives succinic acid and carbon dioxide (Ruzicka *et al.*, 1924). These results can be explained by assigning structure (I) to myrcene. In terpenoid chemistry it has become customary to use conventional formulae rather than those of the type (I). In these conventional formulae only lines are used; carbon atoms are at the junctions of pairs of lines or at the end of a line, and unsaturation is indicated by double bonds. Inspection of (I) shows that the structure of myrcene is based on the 2,6-dimethyloctane skeleton.

This would normally be drawn in a zig-zag fashion, but it is common practice in terpenoid chemistry to draw the carbon skeleton in a ring fashion (the 'open' cyclohexane ring), since this representation usually clearly shows the relationships between various classes of terpenoids. Even so, these 'ring' structures have been, and still are, written differently, *e.g.*, (II), (III), and (IV), but (IV) is now the one that is recommended.

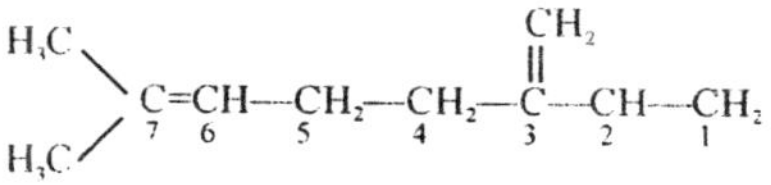

$$(H_3C)_2\overset{7}{C}{=}\overset{6}{C}H{-}\overset{5}{C}H_2{-}\overset{4}{C}H_2{-}\overset{3}{C}({=}CH_2){-}\overset{2}{C}H{=}\overset{1}{C}H_2$$

(I)

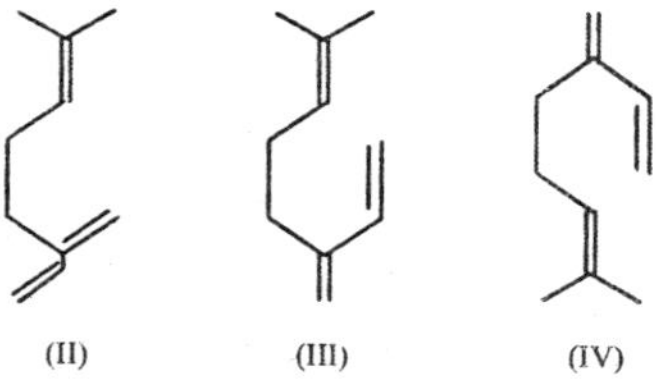

(II) (III) (IV)

Fig. 2

The systematic name of the compound is obtained by use of the rule for acyclic polyenes. Thus, myrcene is 7-methyl-3-methyleneocta-1,6-diene.

We can now represent the process of ozonolysis and oxidation of the ketoaldehyde as shown.

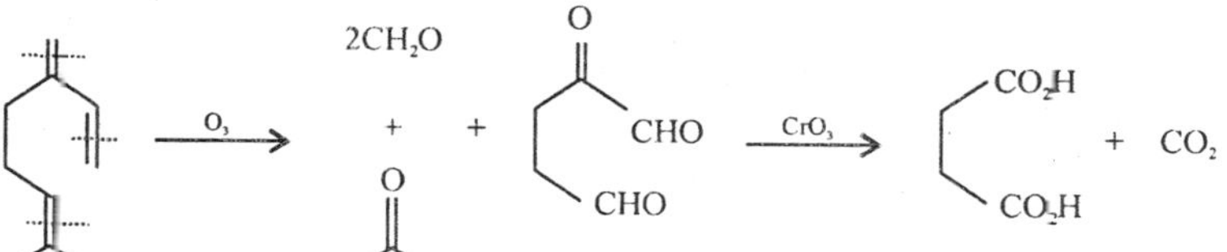

$\xrightarrow{O_3}$ $2CH_2O$ + $(CH_3)_2C{=}O$ + $OHC{-}CH_2{-}CH_2{-}CO{-}CHO$ $\xrightarrow{CrO_3}$ $HO_2C{-}CH_2{-}CH_2{-}CO_2H$ + CO_2

This structure for myrcene is supported by the fact that on hydration (under the influence of sulphuric acid), myrcene forms an alcohol which, on oxidation, gives citral. The structure of this compound is known and its formation is in accord with the structure given to myrcene.

Myrcene has λ_{max} 224 (ε 14 600) nm (calc. value is 214 + 5 = 219 nm) and, according to Sutherland *et al.* (1950), the absence of the band at 890 cm^{-1} shows the complete absence of the isopropenyl.

Ocimene, $C_{10}H_{16}$, b.p. 81°/30 mm

It occurs in the leaves of Ocimum basilicum. When catalytically hydrogenated, ocimene adds on three molecules of hydrogen to form a decane. Thus ocimene is an acyclic compound which contains three double bonds. Furthermore, since ocimene forms an adduct with maleic anhydride, two of the double bonds are conjugated. On ozonolysis, ocimene produces formaldehyde, methylglyoxal, laevulaldehyde, acetic and malonic acids, and acetone. All of these products are accounted for by structure (I) for ocimene (this has an isopropenyl end-group), and also by structure (II) (this has the isopropylidene end-group; Dupont *et al.*, 1938).

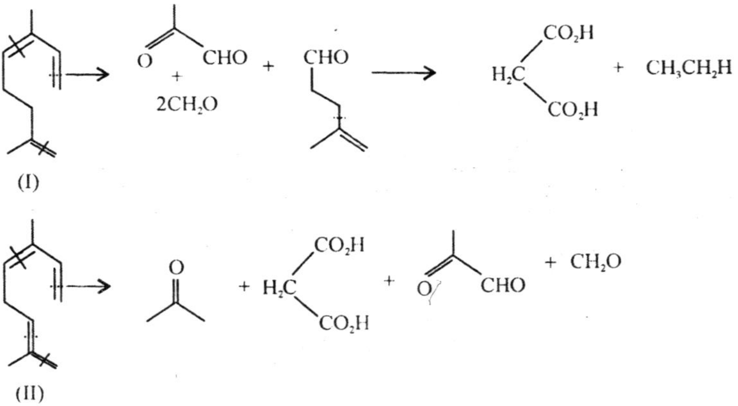

(I) → CHO + $2CH_2O$ + CHO → $H_2C(CO_2H)_2$ + CH_3CH_2H

(II) → O + $H_2C(CO_2H)_2$ + CHO + CH_2O

Fig. 3

From the relative amounts of formaldehyde and acetone obtained, the authors believed that (II) was the major constituent in the mixture. More recent work has cast some doubt on these results. There is also evidence that ocimene is a mixture of geometrical isomers, α- and β-ocimene (structure II). Ocimene is an unstable compound, so much so that it has not yet been obtained in a pure form. When heated, it readily isomerises to allo-ocimene (III), in which the three double bonds are conjugated (λ_{max} 275 nm). This structure has been confirmed by synthesis.

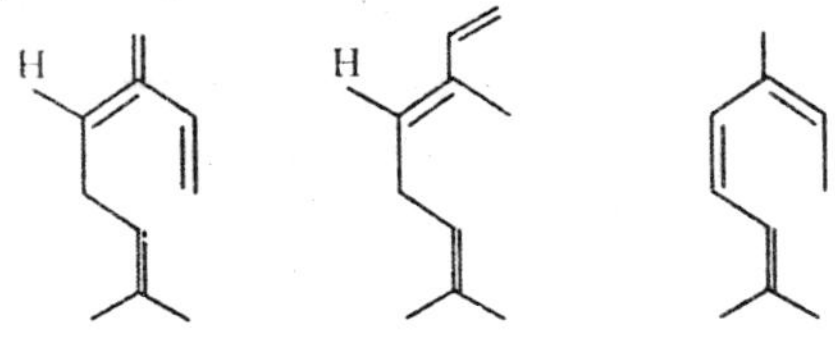

α- β- (III)

Fig. 4

The λ_{max} of ocimene is 237 (ε 40 000) nm; this is 13 nm longer than that for myrcene, and indicates more substitution in the diene system; the calculated value is 214 + 5 + 5 = 224 nm.

Mass Spectrometry: Since isoprene is the building unit of the terpenoids, its mass spectrum is given here. The following include the peaks which are also usually observed in the spectra of terpenoids in general: 68 ($C_5H_8^+$, M^+), 67 ($C_5H_7^+$, B.P.), 53 ($C_4H_5^+$), 51 ($C_4H_3^+$), 41 ($C_3H_5^+$), 39 ($C_3H_3^+$), 29 ($C_2H_5^+$), 27 ($C_2H_3^+$). Paths that account for some of these are:

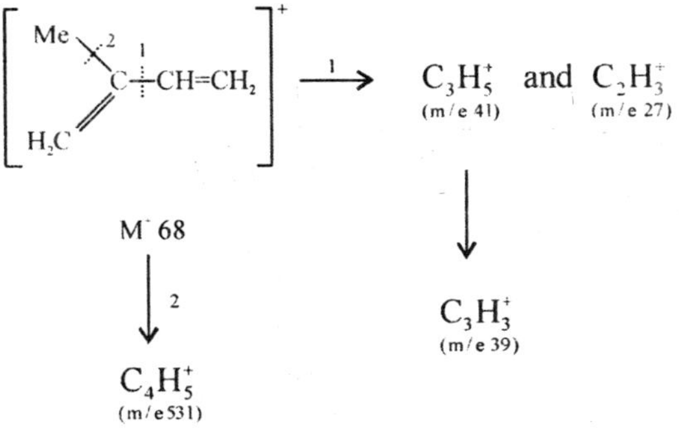

$[Me-C(=CH_2)-CH=CH_2]^{+}$ $\xrightarrow{1}$ $C_3H_5^+$ (m/e 41) and $C_2H_3^+$ (m/e 27)

M^+ 68 $\xrightarrow{2}$ $C_4H_5^+$ (m/e 531)

$C_3H_5^+$ (m/e 41) → $C_3H_3^+$ (m/e 39)

Fig. 5

The mass spectra of myrcene and allo-ocimene have been examined. The former shows a weak molecular-ion peak (M^+ 136), whereas the latter shows a strong one (M^+ 136). This difference is attributed to the stability of the extended conjugated system in allo-ocimene. In myrcene, there are two allylic systems, and allylic fission at the bond common to both can therefore be expected to occur readily. This accounts for the extremely strong peak m/e 69 $\left(C_5H_9^+\right)$ and the weak peak at m/e 67 $\left(C_5H_7^+\right)$. The base peak for myrcene is m/e 41 $\left(C_3H_5^+\right)$; its for-mation can only be explained by rearrangement. The base peak for allo-ocimene is m/e 121, corresponding to a loss of a methyl free radical (M-15 = 121). There are also peaks m/e 27 $\left(C_2H_3^+\right)$ and m/e 91 ($C_7H_7^+$; tropylium ion) present in both spectra.

Citral, $C_{10}H_{16}O$

This is the most important member of the acyclic monoterpenoids, since the structures of most of the other compounds in this group are based on that of citral. Citral is widely distributed and occurs to an extent of 60-80 per cent in lemon grass oil. Citral is a liquid which has the smell of lemons.

Citral was shown to contain an oxo group, *e.g.*, it forms an oxime, etc. On heating with potassium hydrogen sulphate, citral forms p-cymene (II) (Semmler, 1891). This reaction was used by Semmler to determine the positions of the methyl and isopropyl groups in citral; Semmler realised that the citral molecule was acyclic, and gave it the skeleton

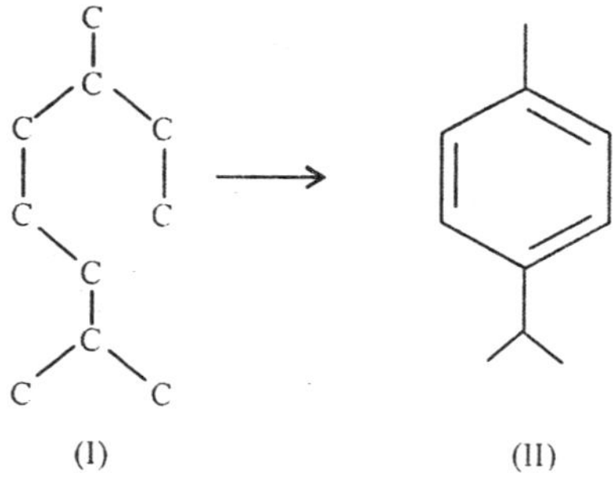

Fig. 6

structure (I) (two isoprene units joined head to tail). Citral can be reduced by sodium amalgam to an alcohol, geraniol, $C_{10}H_{18}O$, and is oxidised by silver oxide to geranic acid, $C_{10}H_{16}O_2$; since there is no loss of carbon on oxidation to the acid, the oxo group in citral is therefore an aldehyde group (Semmler, 1890). Oxidation of citral with alkaline permanganate, followed by chromic acid, gives acetone, oxalic and laevulic acids (Tiemann and Semmler, 1895). Thus, if citral has structure (III), the formation of these oxidation products may be accounted for. This structure

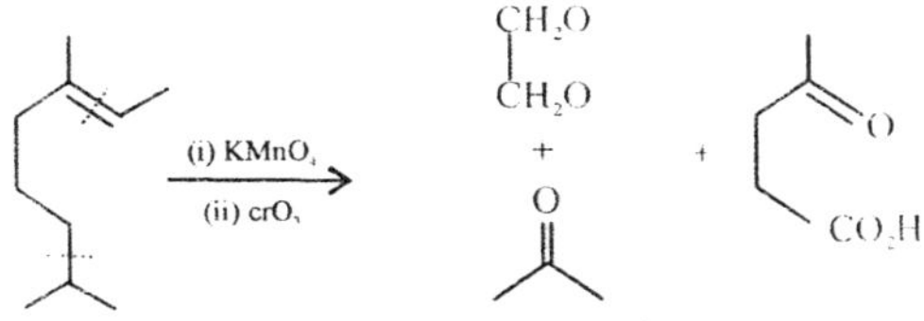

Fig. 7

is supported by the work of Verley (1897), who found that aqueous potassium carbonate converted citral into 6-methylhept-5-en-2-one (IV) and acetaldehyde. The formation of these products is readily explained by assuming (III) undergoes cleavage at the α, β-double bond, this cleavage by alkaline reagents is a general reaction of

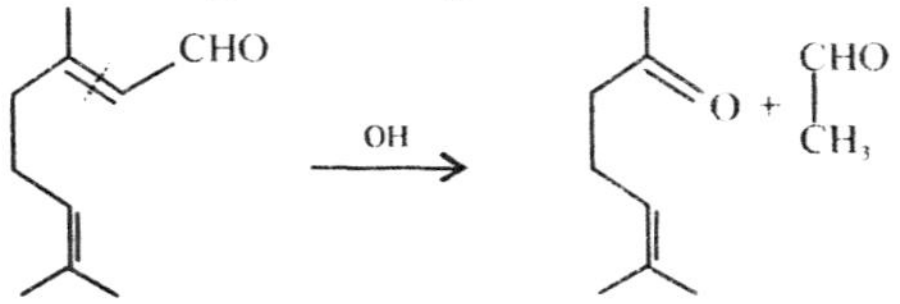

α, β-unsaturated oxo compounds. Furthermore, methylheptenone itself is also oxidised to acetone and laevulic acid; this is again in accord with structure (III). The structure of citral was confirmed by the synthesis of methylheptenone (IV), the conversion of this into geranic ester (Barbier *et al.*, 1896), which was then converted into citral by heating a mixture of the calcium salts of geranic and formic acids (Tiemann, 1898). It should be noted that an allylic rearrangement occurs in both parts of this synthesis. Ethoxyacetylenemagnesium bromide may conveniently be prepared from chloroacetaldehyde diethyl acetal as follows (Jones *et al.*, 1954):

$$ClCH_2CH(OC_2H_5)_2 \xrightarrow{NaNH_2} CH \equiv COC_2H_5 \xrightarrow{RMgBr} BrMgC \equiv COC_2H_5$$

Examination of the formula of citral shows that two geometrical isomers are possible. The functional group (aldehyde) is trans or cis with respect to the methylene group of the main chain. Both isomers occur in natural citral, *e.g.*, two semicarbazones are formed by citral; both forms of citral itself have also been obtained: citral-a (also known as geranial) has a b.p. 118-119 C/20 mm., and citral-b (also known as neral) has a b.p. 117-118°C/20 mm. The configurations of these two forms have been determined from a consideration of the ring closures of the

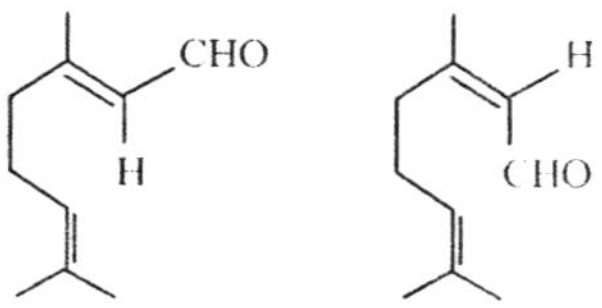

trans- (or E-) form citral-a: geranial

cis- (or Z-) form citral-b, neral

corresponding alcohols (see geraniol, 7). These assignments have been confirmed by the examination of the NMR spectra of citral-a and citral-b (in $CDCl_3$; Ohtsuru *et al.*, 1967). Thus, for example, the τ-values of CH_2 (a) and CH_3 (b) are different due to the different magnetic shielding effects of the carbonyl double bond (in CHO).

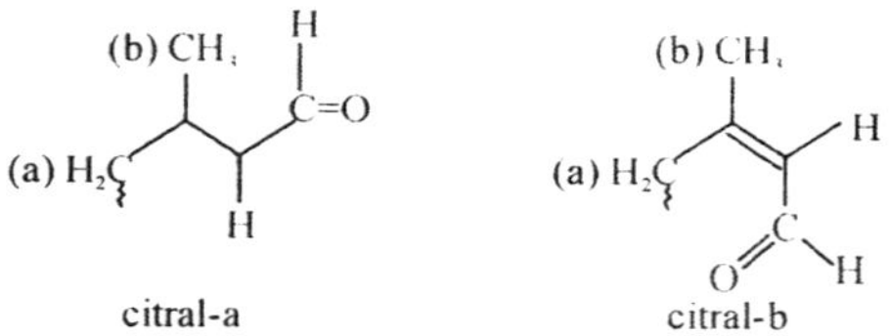

citral-a

citral-b

	CH_2 (a)	CH_3 (b)
citral-a	τ 7.76	τ 7.84
citral-b	τ 7.42	τ 8.02

Ozonolysis was very much used by the classical workers in the determination of structures of terpenoids. In most cases, this method produced two types of products, one arising from the terminal isopropylidene group, $Me_2C=$ (to give acetone), and the other arising from the terminal isopropenyl group, $CH_2=CMe-$ (to give formaldehyde). Because of this, it was originally believed that many acyclic monoterpenoids were mixtures of both structures. However, infrared spectroscopic studies showed the presence of exclusively (or almost exclusively) the isopropylidene group (Barnard *et al.*, 1950). In particular, a detailed study of the infrared spectrum of citronellol showed a maximum at 890 cm^{-1}. This corresponds to some isopropenyl structure (the absorption region of $R_2C=CH_2$ is 895-885 cm^{-1}, whereas that of the isopropylidene structure, $R_2C=CHR$, is 850-790 cm^{-1}). Also, on the basis of the intensities of the bands, the authors calculated that there was about 2-3 per cent of isopropenyl form present. According to the authors, during oxidative degradation, partial rearrangement from the isopropylidene to the isopropenyl structure occurs, and so this method of determining fine structure is unreliable.

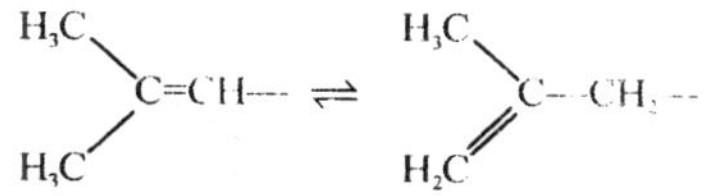

$$\begin{matrix} H_3C \\ \quad\diagdown \\ \quad C{=}CH{-}{-}{-} \\ \quad\diagup \\ H_3C \end{matrix} \rightleftharpoons \begin{matrix} H_3C \\ \quad\diagdown \\ \quad C{-}{-}CH_2{-}{-} \\ \quad/\!/ \\ H_2C \end{matrix}$$

All recent work appears to support this and so the compounds are considered to have isopropylidene structures.

It might also be noted that the presence of the α, β-unsaturated carbonyl system is shown from the ultraviolet absorption spectrum of citral; λ_{max} is 238 (ε 13 500) rim, but this does not distinguish between the isopropylidene and isopropenyl forms.

BIOSYNTHESIS OF THE FLAVONOIDS

Robinson (1936) considered the C, s skeleton of flavonoids to be composed of two parts, C6 and C9: Oxidation with alkaline hydrogen peroxide may also be used in degrading isoflavones; recognisable fragments are not usually obtained by this method, but sometimes information may be obtained about the substituents in the 3-phenyl nucleus, *e.g.*, genistein (4′,5,7-trihydroxyisoflavone) gives p-hydroxybenzoic acid. The final proof of the structure of an isoflavone lies in its synthesis. A general method of synthesising isoflavones is that of Spath *et al.* (1930); *e.g.*, isoflavone itself may be synthesised from benzyl o-hydroxyphenyl ketone and ethyl formate: By using substituted ketones, various isoflavones may be synthesised, *e.g.*, daidzein from 2,4dihydroxyphenyl p-hydroxybenzyl ketone (Wessely *et al.*, 1933): Another method of preparing isoflavones is the Baker-Ollis synthesis (1953). Benzyl o-hydroxyphenyl ketones react at room temperature with ethoxalyl chloride in pyridine, and the products, on alkaline hydrolysis followed by acidification and heating produce.

BIOSYNTHESIS OF THE FLAVONOIDS

Robinson (1936) considered the C_{15} skeleton of flavonoids to be composed of two parts, C_6 and C_9: Not very much is yet known about the large number of individual steps through which the biosynthesis proceeds, but it is well established that rings A and B are formed by different routes. Ring A is produced by the acetate pathway. This was proposed by Birch *et al.* (1953), and was shown to be correct by feeding experiments with labelled acetate, *e.g.*, Grisebach (1957) fed $^{14}CH_3CO_2H$ ($\dot{C}$) and $CH_3{}^{14}CO_2H$ ($\dot{C}$) to red cabbage plants and obtained cyanidin chloride labelled as shown. This is in keeping with a head-to-tail condensation between acetyl-coenzyme A units. However, by analogy with the biosynthesis of fatty acids, it has been assumed that malonyl-coenzyme A rather than acetyl-coenzyme A is the intermediate in flavonoid biosynthesis. Feeding experiments in the study of the biosynthesis of fatty acids showed that malonate was an excellent precursor (Lynen *et al.*, 1961) Experiments using labelled sodium hydrogen carbonate ($NaH\dot{C}O_3$) showed that this was not incorporated with labelled acetate. Hence, a possible pathway for the biosynthesis of the C_6-polyketide is: Ring B, *i.e.*, the C_6-C_3 unit, arises from the shikimic acid pathway.

Shikimic acid → prephenic acid → phenylpyruvic acid → phenylalanine → cinnamic acid. This pathway is indicated by the fact that shikimic acid, phenylalanine and p-hydroxycinnamic acid are good precursors for quercetin. Underhill *et al.* (1957), using labelled compounds, showed the following distributions in the quercetin produced:

Feeding experiments with cinnamic acid labelled at the carboxyl group have shown that this carbon atom is retained at position 4 in quercetin. As shown above, quercetin (a flavone) and cyanidin (an anthocyanidin) are produced from the chalcone by independent pathways. A possible sequence appears to be:

$$\text{Isoflavone} \longleftarrow \text{Chalcone} \underset{}{\overset{\text{isomn}}{\rightleftharpoons}} \text{Flavonone} \xrightarrow{-2H} \text{Flavone}$$

Flavonone → Flavonol; Flavonone → Anthocyanidin

Feeding experiments with labelled phenylalanine afford strong evidence that isoflavone is produced by migration of the aryl group (Grisebach, 1965). Whether the migration occurs in the chalcone or after ring-closure, or by modification of the C_9 precursor, appears to be a matter of debate. A point of interest in this connection is that 2-hydroxychalcones undergo rearrangement with methanolic thallic nitrate to form hydroxyacetals which, on treatment with acid, give isoflavones (Ollis *et al.*, 1970;

Depsides

Phenolic acids, by the interaction of the carboxyl group of one molecule with the hydroxyl group of another, give rise to depsides: If n is zero, then the molecule is a didepside; if n is l, then a tridepside; etc. The main sources of the depsides are the lichens.

In order to synthesise depsides in a known fashion, it is necessary to protect hydroxyl groups. Fischer (1919) carried this out by means of acetylation (acetic anhydride) or by introducing a carbomethoxyl group (with methyl chloroformate); two hydroxyl groups in the ortho-position may he protected by means of carbonyl chloride, *e.g.*, gallic acid forms the following compound.

(I) may be hydrolysed to the didepside by means of cold alkali. By using different phenolic acids, it is possible to synthesise a large variety of depsides. When the hydroxyl group is mesa or pare to the carboxyl group. the phenolic acid is readily carboxymethylated, but ortho-hydroxyl groups are very resistant under the same conditions (steric effect). Reaction can, however, be brought about by condensing o-hydroxyacids with methyl chloroformate in the presence of a base, *e.g.*, dimethylaniline. There is also the further difficulty that ortho-hydroxyl groups do not react with acid chlorides (steric effect). This has been overcome by condensing an acid chloride with an o-phenolic aldehyde, *e.g.*,

Tannins

These are widely distributed in plants; many are glycosides. One of the best sources of tannin is nutgall. The tannins are colourless non-crystalline substances which form colloidal solutions in water; these solutions have an astringent taste.

Tannins precipitate proteins from solution, and they form a bluish-black colour with ferric salts, a property which is used in the manufacture of ink. Tannins also precipitate many alkaloids from their solutions. The name tannins is derived from their ability to tan leather, and is not based on a class of compounds with a common basic structure. There are two groups of tannins: the hydrolysable tannins, which are esters of gallic acid and also glycosides of these esters; and the condensed tannins, which are polymers derived from various flavonoids.

PYRIMIDINE BIOSYNTHESIS

By means of labelled precursors (^{13}C and ^{15}N), the origin of the pyrimidine skeleton has been shown to be derived from carbon dioxide, ammonia, and aspartic acid. Ammonia and carbon dioxide react to form carbamyl phosphate (I), which combines with aspartic acid to form, in turn, N-carbamylaspartic acid (II), dihydro-orotic acid (III), and orotic acid (IV). (IV) condenses with 5′-phosphoribosyl 1-pyrophosphate (PRPP) to give orotidine 5′-phosphate (V) which, on loss of carbon dioxide, produces uridine 5′-phosphate (uridylic acid; (VI)).

Uridylic acid (VI) appears to be the ribonucleotide from which all other pyrimidine nucleotides are biosynthesised. Thus, cytidylic acid (VII) is believed to be produced via uridine triphosphate.

PURINE BIOSYNTHESIS

By means of labelled precursors (^{13}C, ^{15}N), it has been found that the purine skeleton arises from the compounds . The key purine. ribonucleotide is inosinic acid (hypoxanthine is the purine base), *i.e.*, all other purine ribonucleotides are derived from this. The biosynthetic pathway of inosinic acid is believed to be (note the formation of the imidazole ring first). Inosinic acid is converted into other purine nucleotides as shown (note the aminations with different amino-acids).

The series of reactions leading to the formation of inosinic acid from common cell constituents is known as the de novo synthesis of the pyrimidine and purine nucleotides. This name is given to distinguish the pathways whereby an organism utilises preformed pyrimidines and purines.

Chemical and Enzymic Syntheses of the Polynucleotides

The chemical synthesis of nucleotides has been described in, and their biosynthesis in. Several methods have been used to prepare oligonucleotides of known sequence and also for the polymerisation of these to give polynucleotides. One method widely used is that of Khorana *et al.* (1961-1967). p-Methoxytrityl chloride (p-$McOC_6H_4CPh_2C1$; written as MPh_3CCl) is used as a protecting reagent for the CH_2OH group (primary alcohol) and dicyclohexylcarbodi-imide (DCC) as the condensing reagent, *e.g.*, for deoxyribonucleotides; OH^- removes the 3′-Ac group; H^+ removes the MPh_3C group. (III) may be converted into the 3′,5′-dinucleotide 5′-phosphate as follows (via the use of dibenzylphosphochloridate).

By starting with (III), removing the acetyl group (by action of OH^- only), and condensing the product with (II) by means of DCC, a trinucleotide is formed; and so on. It should be noted that in these syntheses, involving bases containing an amino-group (A, G, C, etc.), this group is protected by, *e.g.*, benzoylation. In these cases, concentrated ammonium hydroxide removes the acetyl and the benzoyl group; acid is used to remove the MPh_3C group. Also, reagents besides DCC were used as condensing reagents, *e.g.*, aromatic sulphonyl chlorides.

Kornberg *et al.* (1961) have prepared biosynthetic DNA by the polymerisation of deoxyribonucleoside triphosphates (the four types in natural DNAs) by means of the enzyme DNApolymerase.

This enzymic synthesis, however, must be carried out in the presence of bivalent magnesium cations (Mg^{2+}) and a primer DNA (*i.e.*, some natural DNA which 'initiates' the polymerisation).

The biosynthetic DNA closely resembles the natural DNA primer, and can itself behave as a primer for the enzymic synthesis of DNA from the deoxyribonucleoside triphosphates. It has also been shown that whatever are the relative proportions of the triphosphates in the mixture, the DNA produced is a replica of the primer DNA.

The most important difference between biosynthetic DNA and its natural primer is that the former possesses no biological activity. Further work showed that DNA polymerase acts on the strand of the double helix from the 3′-end to the 5′-end. Since the two strands run in opposite directions, the 3′-end of one is opposite the 5′-end of the other. Hence, continuous synthesis of the complementary DNA takes place only on one strand (the 3′ to the 5′ direction). Kornberg (1967) therefore proposed that synthesis occurs in a backwards direction on the other strand (5′ to 3′ direction). Thus, when some of the double helix has unwound, synthesis has begun from the 3′-end strand and from '3′-point' in the 5′-end strand.

When the latter synthesis reaches its end (5′-), some more of the double helix has unwound and so synthesis starts again at a '3′-point', and so on. The net result is that continuous synthesis occurs on the 3′-end strand, but discontinuous short lengths are built up backwards on the 5′-end strand. In 1967 an enzyme was isolated that was capable of joining together short lengths of DNA. When Kornberg (1968) used this enzyme DNA lipase--ir. The enzymic synthesis of RNAs is believed to proceed by a method analogous to that of replication of DNAs. The appropriate ribonucleoside triphosphates are polymerised in the presence of RNA polymerase, and copy the pattern of the DNA molecule on which it is built.

SOME STRUCTURE OF THE PURINES

Introduction : Purine is the parent substance of a group of cyclic diureides and was used by E. Fischer to name systematically the naturally occurring derivatives. Its structure consists of a pyrimidine ring fused to an imidazole ring. Purine can exist in four tautomeric forms in which the hydrogen atom is joined to the different nitrogen atoms: N-1, N-3, N-7, and N-9. In the first two the aromaticity of the pyrimidine ring is lost, the ring now being virtually equivalent to the far less stable (and more reactive) ortho-quinonoid structure. In practice, purine appears to behave completely as the tautomers of N-7 H and N-9 H.

These formulae are now written as A or B. In this book, formula A is used (B is A turned upside down; there is no change in numbering, and so the reader can readily translate A into B).

Uric Acid

Guano (birds' excrement found on islands near the western coast of South America) contains up to about 25 per cent uric acid; about 90 per cent of snakes' excrement is ammonium urate. Small amounts of uric acid art present in human urine; it was first discovered by Scheele (1776) in urinary calculi.

Liebig and Wohler (1834) showed that the molecular formula of uric acid is $C_5H_4N_4O_3$. These authors also found, in 1838, that the oxidation of uric acid with nitric acid gives alloxan and urea in equimolecular proportions.

$$C_5H_4O_3 + H_2O + [O] \xrightarrow{HNO_3} C_4H_2N_2O_4 + NH_2CONH_2$$

Structure of Alloxan, $C_4H_2N_2O_4$

When hydrolysed with alkali, alloxan produces one molecule of urea and one molecule of mesoxalic acid.

$$C_5H_2N_2O_4 + 2H_2O \xrightarrow{KOH} NH_2CONH_2 + HO_2CCOCO_2H$$

Since alloxan contains no free amino or carboxyl groups, the products of hydrolysis suggest that alloxan is mesoxalylurea; this cyclic structure has been confirmed by the direct union of urea and mesoxalic acid to give alloxan (Liebig and Wohler, 1838).

Alloxan, as its monohydrate, is conveniently prepared from barbituric acid as follows: Alloxan is a strongly acidic compound; it crystallises with four molecules of water of crystallisation. Three of these are readily lost on warming, but the fourth is lost only when the monohydrate is heated to 150°C. Because of this, it is believed that the fourth molecule of water is not water of crystallisation but water of constitution.

Alloxan stains the skin purple (due to the formation of murexide). The 5-oxime of alloxan is violuric acid, and when reduced with zinc and hydrochloric acid, alloxan forms dialuric acid. When alloxan is reduced with hydrogen sulphide, the product is alloxantin. According to Tipson *et al.* (1951), however, if excess of hydrogen sulphide is used, the product is dialuric acid only. Alloxantin is produced by reducing alloxan (one molecule) with half a molecule of hydrogen sulphide, or by mixing aqueous solutions of alloxan and dialuric acid. When heated with ammonia in ethanolic solution, alloxantin forms murexide, which is the ammonium salt of purpuric acid (an unstable compound).

Murexide is soluble in water, giving a purple solution which turns blue on the addition of alkali. Purpuric acid slowly hydrolyses in solution to form alloxan and uramil. When uric acid is oxidised with an aqueous suspension of lead dioxide, the products are allantoin and carbon dioxide (Liebig and Wohler, 1838). These products are obtained in quantitative yield if the oxidation is carried out with alkaline permanganate (Behrend, 1904).

$$C_5H_4N_4O_3 + H_2O + [O] \longrightarrow C_4H_6N_4O_3 + CO_2$$

Structure of Allantoin, $C_4H_6N_4O_3$ (Baeyer, 1861-1864)

When hydrolysed with alkali, allantoin forms two molecules of urea and one molecule of glyoxylic acid.

$$C_4H_6N_4O_3 + 2H_2O \rightarrow 2NH_2CONH_2 + OHCCO_2H$$

The formation of these hydrolytic products suggests that allantoin is the diureide of glyoxylic acid.

On oxidation with nitric acid, allantoin forms urea and parabanic acid in equimolecular proportions.

$$C_4H_6N_4O_3 + [O] \xrightarrow{HNO_3} NH_2CONH_2 + C_3H_2N_2O_3$$

Parabanic acid, on hydrolysis, gives urea and oxalic acid, and since there are no free amino or carboxyl groups present in the molecule, this suggests that parabanic acid is oxalylurea.

This structure has been confirmed by synthesis, *e.g.*, oxalyl chloride condenses with urea to form parabanic acid (Bornwater, 1912). Thus, from the above facts, it can be seen that allantoin contains the parabanic acid nucleus joined to a molecule of urea. The point of the attachment is deduced from the following experimental evidence. When reduced with concentrated hydriodic acid at 100°C, allantoin forms urea and hydantoin.

$$C_4H_6N_4O_3 + 2[H] \xrightarrow{HI} NH_2CONH_2 + C_3H_4N_2O_3$$

Hydantoin, on controlled hydrolysis, gives hydantoic acid (ureidoacetic acid) and this, on further hydrolysis, gives glycine, ammonia and carbon dioxide. These results suggest that hydantoin is glycollylurea. In the formation of allantoin from uric acid by oxidation, one carbon atom is lost from the latter as carbon dioxide. The problem, then, is to fit this carbon atom into the allantoin structure. At the same time, the structure thus given to uric acid must also include the alloxan skeleton in order to account for the formation of this compound. Two structures that were proposed which both agreed with the facts known at the time were by Medicus (1875) and by Fittig (1878).

Fischer (1884) prepared two isomeric monomethyluric acids; one gave methylalloxan and urea on oxidation with nitric acid, and the other gave alloxan and methylurea. Fittig's formula, which is symmetrical, can give rise to only one monomethyluric acid; hence this structure is untenable. On the other hand, the Medicus formula satisfies the existence of at least two isomeric monomethyl derivatives: one methyl group in the pyrimidine nucleus (at position 1 or 3) would produce methylalloxan and urea, and a methyl group in the imidazole nucleus (at position 7 or 9) would produce alloxan and methylurea (Fischer showed that the two monomethyluric acids were the 3- and 9-derivatives).

Examination of the Medicus formula shows that it admits the possibility of four monomethyl, six dimethyl and four trimethyl derivatives. All of these have been prepared by Fischer and his co-workers, thus giving powerful support to the Medicus formula. Proof of the Medicus formula lies in the synthesis of uric acid; three syntheses are given here. In this reduction, some of the aminouracil is converted into hydroxyuracil. The mechanism of this change is not certain, but a possibility is as follows:

Which of these forms is the one that gives the disodium salt still appears to be uncertain. Fischer thought that the dianion is the 2,6-. Evidence that may be quoted to support this is that in this arrangement the pyrimidine ring will be 'aromatic' and so stabilised by resonance. Further evidence for the 2,6-form is afforded by the fact that the ultraviolet spectra of purine derivatives and pyrimidine derivatives show basic similarities. An interesting point about uric acid is that X-ray analysis of its 1,3,7,9-tetramethyl derivative has shown that there is hydrogen bonding between hydrogen (of the methyl group) and oxygen (Sutor, 1963).

It is also interesting to consider the path followed in the oxidation of uric acid to allantoin. Behrend (1904) suggested that the alkaline permanganate oxidation of uric acid (I) gives allantoin (IIIa and b) via the symmetrical intermediate (II).

Cavalieri *et al.* (1948) have carried out this oxidation using uric acid labelled with 15N at N-I and N-3, and found that the allantoin produced had this isotopic nitrogen distributed uniformly among all the four nitrogen atoms. This is in keeping with the intermediate formation of(II).

PURINE DERIVATIVES

When uric acid is treated with phosphoryl chloride, 2,6,8-trichloropurine is obtained. This trichloro compound is a very important intermediate in the synthesis of purine derivatives, and a point worth noting is that the reactivities of the chlorine atoms towards nucleophilic reagents are 6 > 2 > 8. Purine, m.p. 217°C, may be prepared from uric acid as follows: Catalytic reduction of trichloropurine (H_2—Pd in aqueous sodium acetate) gives purine (Brodereck *et al.*, 1962). Purine behaves as a weak monoacid base (pK_a 8.96). Diazomethane or dimethyl sulphate and alkali methylate purine in the 9-position. Purine has been found to occur naturally as its 9-D-ribofuranoside, nebularine. The NMR spectrum of purine shows three singlets: τ 1.34 (6-H), 1.47 (2-H), and 1.72 (8-H). These values are in the aromatic region. The mass spectrum of purine shows fragments arising from the consecutive losses of hydrogen cyanide molecules. This often occurs in heterocyclic compounds containing two nitrogen atoms in one ring.

STRUCTURE OF THE ANTHOCYANIDINS

The anthocyanin is first hydrolysed with hydrochloric acid and the anthocyanidin is isolated as the chloride. The usual analytical methods are applied to determine the number of hydroxyl and methoxyl groups present in the molecule. The structure of the anthocyanidin is ascertained by the nature of the products obtained by fusing the anthocyanidin with potassium hydroxide (Willstatter *et al.*, 1915); phloroglucinol or a methylated phloroglucinol and a phenolic acid are always obtained, *e.g.*, cyanidin chloride gives phloroglucinol and protocatechuic acid. This method suffers from the disadvantage that the fusion (or boiling with concentrated potassium hydroxide solution) not only degrades the anthocyanidin, but also often demethylates it at the same time. Thus the positions of the methoxyl groups in the original compound are now rendered uncertain. This difficulty was overcome by Karrer *et al.* (1927), who degraded the anthocyanidin with a 10 per cent solution of barium hydroxide or sodium hydroxide in an atmosphere of hydrogen; in this way, the methoxyl groups are left intact.

The next problem is to ascertain the positions of the sugar residues. After hydrolysis (of the anthocyanin), the sugar is identified by the usual methods of sugar chemistry, and this includes the use of paper chromatography. If two or more monosaccharide molecules are obtained for each molecule of anthocyanin, it is necessary to determine whether they were present as such, or as a disaccharide (or trisaccharide) which was hydrolysed. One method is first to methylate the anthocyanin and then hydrolyse with suitable enzymes. In this way, methylated disaccharides may be isolated intact.

(i) Karrer *et al.* (1927) methylated the anthocyanin, removed the sugar residues by hydrolysis (hydrochloric acid), and finally hydrolysed with barium hydroxide solution in an atmosphere of hydrogen; the positions of the free hydroxyl indicate the points of attachment of the sugar residues. In some cases, however, interpretation of the results is uncertain, *e.g.* (G represents a sugar residue): The problem is: Which of the two hydroxyl groups in monomethylphloroglucinol was originally attached to G? The above results do not lead to a definite answer, since had the structure of the anthocyanin been (IV) instead of (I), (III) would still have been obtained:

(ii) Hydrogen peroxide (15 per cent) attacks anthocyanins as follows (Karrer *et al.*, 1927): If the anthocyanin (V) has a glucose residue in the 3-position, then this glucose residue in (VI) is readily hydrolysed by dilute ammonia. If the glucose residue in (V) is in either the 5- or 7-position, then this glucose residue in (VI) is removed only by heating with dilute hydrochloric acid. Thus position 3 can be distinguished from positions 5 or 7, but the latter two cannot be distinguished from each other.

(iii) Anthocyanins with a free hydroxyl group in the 3-position are very readily oxidised by ferric chloride; the anthocyanins are rapidly decolorised in this oxidation (Robinson *et al.*, 1931).

The final problem is to determine whether the sugar linkage (to the anthocyanidin) is α or β. This is ascertained by hydrolysis with the enzymes maltase (α-linkage) and emulsin (β-linkage).

Conclusive evidence for the positions and nature of the linkages of the sugar residues is afforded by the synthesis of the anthocyanins (see, *e.g.*, cyanin). In general, it has been found that glucose residues are linked at positions 3 or 3,5 and that the linkage is usually β. Now that a large number of structures have been elucidated, it has been possible to correlate, to a large extent, physical data with structure. Thus, the flavonoid is fairly easily assigned to its class by its absorption spectrum (visible and ultraviolet) and by colour reactions.

It has already been pointed out that the known anthocyanidins belong to three main types: cyanidin (λ_{max} 535 nm), pelargonidin (λ_{max} 520 nm), and delphinidin λ_{max} 564 nm) [all measured in McOH–HCl solution]. Thus these may be distinguished from each other. Furthermore, the introduction of sugar residues into these anthocyanidins shifts the maxima towards the shorter wavelengths, the shifts being characteristic for sugar residues in the 3- or 3,5-positions, and for the 5-position. Also, since 3-glycoside spectra show a pronounced shoulder (440-460 nm region), it is therefore possible to obtain information about the positions of sugar residues in anthocyanins.

Examination of the absorption spectra of anthocyanidins in the presence of certain reagents also gives information about structure. Thus aluminium chloride and sodium ethoxide shift the maxima to the longer wavelengths, the actual shift depending on the positions of the hydroxyl groups in the molecule, *e.g.*, cyanidin (λ_{max} 535 nm) has its λ_{max} at 553 nm in the presence of aluminium chloride. In this way, it may be possible to obtain some indication of the relationship of the compound under investigation to a flavonoid of known structure, thereby simplifying further structure determination. Also, since the R_F values of the known anthocyanins have been determined (Harborne, 1959), these compounds are quite readily identified.

General Methods of Synthesising the Anthocyanidins

(i) Willstatter (1914) synthesised anthocyanidins starting from coumarin.

(ii) Robinson has introduced a number of methods whereby all anthocyanidins can be prepared. The basic reaction of these methods is the condensation between o-hydroxybenzaldehyde and acetophenone in ethyl acetate solution which is then saturated with hydrogen chloride.

The original method of Robinson (1924) resulted in the formation of a product in which the substituent groups were either all hydroxyl groups or all methoxyl groups, *e.g.*,

Robinson (1928, 1931) then modified this method so that the product could have both hydroxyl and methoxyl substituent groups, *e.g.*,

The following is a brief account of the methods used by Robinson and his co-workers for preparing the substituted acetophenones and substituted benzaldehydes.

Cyanidin Chloride, $C_{15}H_{11}ClO_6$

Cyanin chloride, on hydrolysis with hydrochloric acid, gives cyanidin chloride and two molecules of D-glucose.

$$C_{27}H_{31}ClO_{16} + 2H_2O \xrightarrow{HCl} C_{15}H_{11}ClO_6 + 2C_6H_{12}O_6$$

Since cyanidin chloride forms a penta-acetate, the molecule therefore contains five hydroxyl groups. No methoxyl groups are present, and so the potassium hydroxide fusion may be used to degrade this compound; this gives phloroglucinol and protocatechuic acid. The formation of phloroglucinol and protocatechuic acid by the alkaline fusion of cyanidin chloride suggests a relationship to quercetin, since the latter also gives the same fusion products.

Cyanidin chloride is a red salt which is insoluble in water but is very soluble in ethanol. The colour of the salt, however, varies with the pH of the solution. In aqueous sodium acetate solution (pH 8), the solution is violet due to the formation of the anhydrobase. On standing, the solution becomes colourless by conversion of the anhydrobase (quinonoid structure) into the colourless pseudobase (in which the quinonoid structure has been lost).

When this colourless solution is made alkaline with sodium hydroxide (pH 12), the colour changes to blue due to the formation of the anion of the anhydrobase. When this solution is made acid (pH 4), the colour turns red because of the regeneration of cyanidin chloride. On the other hand, on standing in alkaline solution, all of these compounds are converted into the yellow chalcone. On the basis of these ionic structures (positive for oxonium salts and negative for salts of the colour bases), anthocyanins should migrate in an electric field. Markakis (1960) has shown that various anthocyanins, when placed within an electric field applied across filter paper, move to the anode or cathode according to the pH of the solution. Markakis also showed that isoelectric point and the pH of minimum colour display coincide.

On the acidic side of the isoelectric point, the oxonium salt-form predominates; and when the pH is higher than that of the isoelectric point, the salt of the colour base predominates, and according to Markakis, it is the pseudobase which probably predominates at the isoelectric point. Cyanin was the first anthocyanin to be isolated and its structure determined. It has been synthesised by Robinson *et al.* (1932). Phloroglucinaldehyde (I) was condensed with tetra-acetyl-α-bromoglucose (II), in acetone solution to which has been added aqueous potassium hydroxide; the product was 2-O-tetra-acetyl-β-glucosidyl-phloroglucinaldehyde (III).

ω-Hydroxy-3,4-diacetoxyaceto-phenone (IV) was also condensed with tetra-acetyl-α-bromoglucose (II) in benzene solution to give ω-O-tetra-acetyl-β-glucosidoxy-3,4-diacetoxyacetophenone (V). Compounds (III) and (V) were then dissolved in ethyl acetate and the solution saturated with hydrogen chloride; the product (VI) was treated first with cold aqueous potassium hydroxide and then with hydrochloric acid, whereby cyanin chloride (VII) was produced.

Delphinidin Chloride, $C_{15}H_{11}ClO_7$

This is obtained, together with two molecules of glucose and two molecules of p-hydroxybenzoic acid, when delphinin chloride is hydrolysed with hydrochloric acid.

Delphinidin chloride contains six hydroxyl groups, and no methoxyl groups; on fusion with potassium hydroxide, the products are phloroglucinol and gallic acid.

This structure has been confirmed by synthesis, starting from 2-benzoylphloroglucinaldehyde and ω,3,4,5-tetra-acetoxyacetophenone (Robinson *et al.*, 1930). Delphin chloride, $C_{27}H_{31}ClO_{17}$, is the 3,5-diglucoside of delphinidin chloride (no p-hydroxybenzoic acid is present).

Structure of the Phthalocyanines

Analysis showed that the phthalocyanines had an empirical formula $C_{32}H_{16}N_8M$, where M is a bivalent metal, *e.g.*, copper, magnesium, etc.

The molecular weight determination of magnesium phthalocyanine by the ebullioscopic method with naphthalene as solvent showed that the empirical formula was also the molecular formula (Linstead *et al.*, 1934). This has been confirmed by means of X-ray measurements (Robertson, Linstead *et al.*, 1935).

Linstead showed that the phthalocyanines can be obtained by reaction between a metal and phthalonitrile (I), o-cyanobenzamide (II), phthalamide (III), but not with, for example, terephthalonitrile (IV), homophthalonitrile (V), or o-xylylene dicyanide (VI). It is therefore reasonable to infer that in the formation of phthalocyanines, the two nitrile groups involved must be in the ortho-position. Thus there are probably four $C_8H_4N_2$ units, each having an isoindole structure (VII) or a phthalazine structure (VIII). (VIII) was shown to be untenable since no phthalocyanine could be prepared from compounds containing this skeleton.

The oxidation of phthalocyanines with hot nitric acid, cold acid permanganate or ceric sulphate produces phthalimide and ammonium salts, the amount of phthalimide being that which would correspond to the presence of four isoindole units. The problem then is: How are these units joined together? The treatment of magnesium phthalocyanine with sulphuric acid replaces the magnesium atom by two hydrogen atoms.

$$(C_8H_4N_4)_4Mg \xrightarrow{H_2SO_4} (C_8H_4N_2)_4H_2$$

This suggests that in metal phthalocyanines, the metal has replaced two imino hydrogen atoms. A reasonable structure for phthalocyanine is one in which the four isoindole units are joined through nitrogen atoms to form a cyclic structure (IX). On the other hand, an open-chain structure could also be produced by joining four isoindole units through nitrogen atoms (X); in this case the molecular formula would be $(C_8H_4N_2)_4H_4$. It seems unlikely that (X) could be rejected on these grounds alone, since in a large molecule of this type it is difficult to estimate the hydrogen with certainty ((IX) contains approximately 3.5 per cent hydrogen, and (X) 3.9 per cent). (X), however, is unlikely, since phthalocyanine is a very stable substance; the presence of an imino group at the end of the molecule could be expected to render the compound unstable to, *e.g.*, acid reagents. Furthermore, the oxidation of phthalocyanine with ceric sulphate in dilute sulphuric acid proceeds according to the following equation (over 90 per cent of the phthalimide has been isolated).

$$(C_8H_4N_2)_4H_2 + 7H_2O + [O] \rightarrow 4C_8H_5NO_2 + 4NH_3$$

This agrees with (IX), but had the structure been (X) then the molecule would have required two atoms of oxygen.

$$(C_8H_4N_2)_4H_4 + 6H_2O + 2[O] \rightarrow 4C_8H_5NO_2 + 4NH_3$$

Thus (IX) represents best the known properties of phthalocyanine. The two imino hydrogen atoms are replaceable by a bivalent metal, and the remaining two nitrogen atoms form co-ordinate links (see formula (III)).

The most common metals in phthalocyanines are in the bivalent state, *e.g.*, Cu^{2+}, Fe^{2+}, Ni^{2+}, etc. It is also possible for the central metal atom to have a valency of three or four, and in this case one or two anions are also attached to the metal, *e.g.*, bromoaluminium and dichlorostannic phthalocyanines (cf. haem and haemin). Phthalocyanines containing two alkali metal ions, *e.g.*, Na, K (each metal atom joined to an adjacent pair of pyrrole N atoms) have been prepared, but these are readily hydrolysed by dilute acids to the metal-free phthalocyanine. Finally, the metal may have a zero valency. Thus, copper phthalocyanines are known in which the copper is copper (0), copper (I), or copper (II).

In metal phthalocyanines resonance is possible, and so all four nitrogen atoms linked to the metal atom would be equivalent (18n-electrons; 4n + 2; see (I) and (II)). Phthalocyanines (with and without a central metal atom) have been examined by means of X-ray analysis (Robertson, 1936), and the results shows that these compounds are large flat molecules with a centre of symmetry. The bond lengths of the C–N bonds indicate resonance, ds do those of the benzene ring (all the lengths are equal). Robertson also showed that for nickel phthalocyanine, if the radius of the nickel atom be assumed, then the positions of the other atoms in the molecule are exactly those obtained by chemical evidence.

Phthalocyanines are known to exist in at least three polymorphic forms : α, β and γ. The β-form is the most stable one, and is the one produced by sublimation of the phthalocyanine. Kendall (1953) has distinguished between the three forms by means of their infrared spectra, and Kahn *et al.* (1965) have shown the difference in the polymorphic structures between α- and β-cobalt phthalocyanines by means of their ESR spectra.

Phthalocyanines can act as catalysts, *e.g.*, they catalyse the combination of hydrogen and oxygen to form water, the decomposition of hydrogen peroxide, the isomerisation of dimethyl maleate to dimethyl fumarate, etc.

TETRACYCLINE ANTIBIOTICS

Aureomycin was isolated from cultures of Streptomyces aureofaciens, and is used in the treatment of typhoid fever, etc. Terramycin was isolated from cultures of Streptomcyes rimosus, and is very effective in the treatment of trachoma. The structures of these antibiotics are:

These compounds are classified as tetracyclines, tetracycline itself being obtained by replacement of the chlorine atom (R^1 = Cl) in aureomycin by hydrogen (R^1 = H). This conversion is readily carried out by the catalytic hydrogenation of aureomycin. The stereochemistry of aureomycin had been partly established from the chemical work, but it was completely determined by the X-ray analysis of its hydrochloride (Pepinsky *et al*., 1959), and Shemyakin *et al*. (1962) have established the absolute configuration of the tetracyclines by means of optical rotatory dispersion studies.

The structure of terramycin was the first of this group to be elucidated (Woodward *et al*., 1953). The molecular formula was shown to be $C_{22}H_{24}N_2O_9$ and the compound was found to contain eight active hydrogen atoms. Also present were a dimethylamino and a carbonamido group, and a C-methyl group. Diazomethane formed a dimethyl ether; this showed the presence of two acidic hydroxyl groups; the other two hydroxyl groups were shown to be alcoholic. On the evidence so far, the structure of terramycin may be written as:

(I) is terracinoic acid (formed by a naphthacene rearrangement); (II) is terranaphthol; and (III) is naphthacene. On the basis of the isolation of the hydroxybenzoic acids, compounds (I) and (II), the compound with structure (IV) was synthesised as a model for ultraviolet spectroscopic examination.

The spectrum of (IV) resembled that of terramycin itself and this led to the proposal of (V) as the partial structure of the antibiotic. The isolation of (III) and further work led to the structure given above for terramycin (5-oxytetracycline). Once the structure of terramycin had been elucidated, structure determination of the tetracyclines was facilitated by the use of spectroscopic methods. Total syntheses have also been carried out.

Antibiotics

Many micro-organisms produce within themselves chemical substances which, when excreted, interfere with the growth or metabolism of other micro-organisms. Such compounds are known as antibiotics, and need be present only in low concentration to bring about this antibiotic action. Antibiotics are thus chemotherapeutic agents. In 1929, Fleming discovered a mould of the Penicillium species which inhibited the growth of certain bacteria. This observation was investigated later by a number of workers and culminated in the isolation of the active principle penicillin. At the same time, research along this line led to the isolation of many other antibiotics.

The antibiotics cover a wide range of compounds of different chemical structures. A rational classification is very difficult, and many schemes have been suggested, *e.g.*, classification on their chemical structures or according to the nature of their activity.

MONOCYCLIC MONOTERPENOIDS

For the purposes of nomenclature of the monocyclic monoterpenoids, the fully saturated compound p-methylisopropylcyclohexane, hexahydro-p-cymene or p-menthane, $C_{10}H_{20}$, is used as the parent substance; it is a synthetic compound, b.p. 170°C. p-Menthane is (I), and (II) is a conventional method of drawing formula (I). The positions of substituents and double bonds are indicated by numbers, the method of numbering being shown in (I) and (II).

When a compound derived from p-menthane contains one or more double bonds, ambiguity may arise as to the position of a double bond when this is indicated in the usual way by a number which locates the first carbon atom joined by the double bond.

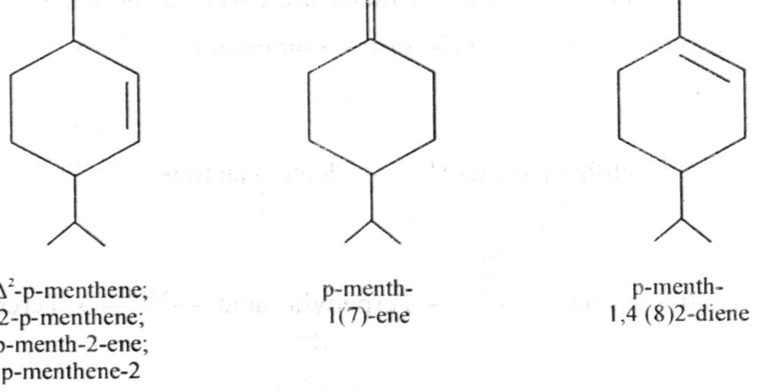

Fig. 8

To prevent ambiguity, the second carbon atom joined to the double bond is also shown, but is placed in parentheses. The examples illustrate the method of nomenclature in the first example, all the types of methods of nomenclature have been given; in the second and third examples, only the nomenclature that will be used in this book is given.

α-Terpineol

This is an optically active monoterpenoid that occurs naturally in the (+)-, (–)- and (±)-forms; it is a solid, m.p. (of the racemic modification) 35°C. The molecular formula of α-terpineol is $C_{10}H_{18}O$, and the oxygen atom is present as a tertiary alcoholic group (as shown by the reactions of α-terpineol). Since a-terpineol adds on two bromine atoms, it therefore contains one double bond. Thus the parent (saturated) hydrocarbon of a-terpineol has the molecular formula $C_{10}H_{20}$. This corresponds to C_nH_{2n}, the general formula of the (monocyclic) cycloalkanes, and so it follows that α-terpineol is a monocyclic compound.

When heated with sulphuric acid, α-terpineol forms some p-cymene. Taking this in conjunction with the tentative proposal that α-terpineol is monocyclic, it is reasonable to infer that α-terpineol contains the p-cymene skeleton. Thus we may conclude that α-terpineol is probably p-menthane with one double bond and a tertiary alcoholic group. The positions of these functional groups were ascertained by Wallach (1893, 1895) by means of graded oxidation. The following chart gives the results of Wallach's work; only the carbon content is indicated to show the fate of these carbon atoms (the formulae are given in the text).

$$\underset{\substack{C_{10}\\(I)}}{\alpha-\text{Terpineol}} \xrightarrow[\text{KMnO}_4]{1\%\text{ alk}} \underset{\substack{C_{10}\\(II)}}{\text{Trihydroxy compound}} \xrightarrow{\text{CrO}_3}$$

$$\underset{\substack{C_{10}\\(III)}}{[\text{Ketohydroxyacid}]} \longrightarrow \underset{\substack{C_{10}\\(IV)}}{\text{Keto}-\text{lactone}} \xrightarrow{\substack{\text{warm}\\\text{alk.}\\\text{KMnO}_4}}$$

$$\underset{\substack{C_{10}\\(IV)}}{\text{Keto}-\text{lactone}} \xrightarrow{\substack{\text{warm}\\\text{alk.}\\\text{KMnO}_4}} \underset{\substack{C_8\\(V)\\+\\CH_3CO_2H}}{\text{Terpenylic acid}} \xrightarrow{\text{KMnO}_4} \underset{\substack{C_7\\(VI)}}{\text{Terebic acid}}$$

Oxidation of α-terpineol (I) with 1 per cent alkaline potassium permanganate hydroxylates the double bond to produce the trihydroxy compound (II), $C_{10}H_{20}O_3$. This, on oxidation with chromic acid (chromium trioxide in acetic acid), produces a compound with the molecular formula $C_{10}H_{16}O_3$ (IV).

This compound was shown to contain a ketonic group, and that it was neutral, *e.g.*, it gave no reaction with sodium carbonate solution. When, however, (IV) was refluxed with excess of standard sodium hydroxide solution, and then back titrated, it was found that alkali had been consumed, the amount corresponding to the presence of one carboxyl group.

Thus compound (IV) appears to be the lactone of a monocarboxylic acid. Furthermore, since it is the lactone that is isolated and not the hydroxy-acid, this spontaneous lactonisation may be interpreted as being produced from a γ-hydroxy-acid, *i.e.*, (IV) is a γ-lactone, and therefore (III) is a y-hydroxy-acid. It is possible, however, for δ-hydroxy-acids to spontaneously lactonise, and so whether (IV) is a γ- or δ-lactone is uncertain at this stage of the evidence. Now, since (IV) is formed from (II) by scission of the glycol bond, and since there is no loss of carbon atoms in the process, the double bond must therefore be in the ring in (I).

On warming with alkaline permanganate, (IV) gave acetic acid and a compound $C_8H_{12}O_4$ (V). The formation of acetic acid suggests that (IV) is a methyl ketone, *i.e.*, a CH_3CO group is present. Thus (IV) is a methyl ketone and a lactone; it is known as homoterpenyl methyl ketone, and the structure assigned to it has been confirmed by synthesis (Simonsen *et al.*, 1932).

A study of the properties of terpenylic acid (V) showed that it was the lactone of a monohydroxydicarboxylic acid. Further oxidation of terpenylic acid gives terebic acid, $C_7H_{10}O_4$ (VI), which is also the lactone of a monohydroxydicarboxylic acid. The above reactions can be formulated as shown, assuming (I) (p-menth-I-en-8-ol) as the structure of α-terpineol. These reactions were formulated by Wallach, who adopted formula (I) which had been proposed by Wagner (1894). The structure of terpenylic (V) and terebic (VI) acids were established by synthesis, *e.g.*, those of Simonsen (1907). It is of interest to note here that Sandberg (1957) has prepared the β-acetotricarballylate in one step from acetoacetic ester and ethyl bromoacetate in the presence of sodium hydride (in benzene solution). These syntheses strengthen the evidence for the structure assigned to α-terpineol. A synthesis of α-terpineol itself has been carried out by Perkin, junior (1904), and by Perkin, junior, with Meldrum and Fisher (1908).

Only the second synthesis is given here; this starts with p-toluic acid.

Compound (VII) was also resolved with strychnine, each enantiomer treated as shown above (esterified, etc.), and thereby resulted in the formation of (+)- and (–)-terpineol. It should be noted that in the above synthesis the removal of a molecule of hydrogen bromide from 3-bromo-4-methylcyclohexane-l-carboxylic acid to give (VII) is an ambiguous step; instead of (VII), compound (VIII) could have been formed. That (VII) and not (VIII) is formed rests on the analytical evidence for the position of this double bond; (VIII) cannot give the products of oxidation that are actually obtained from a-terpineol. A much simpler synthesis of a-terpineol has been carried out by Alder and Vogt (1949); this makes use of the Diels-Alder reaction, using isoprene and methyl vinyl ketone as the starting materials.

Carvone, $C_{10}H_{14}O$. b.p. 230°C/755 mm

This occurs in various essential oils, *e.g.*, spearmint and caraway oils, in optically active forms and also as the racemic modification. Carvone behaves as a ketone and, since it adds on four bromine atoms, it therefore contains two double bonds. Thus the parent hydrocarbon is $C_{10}H_{20}$, and since this corresponds to the general formula C_nH_{2n}, carvone is monocyclic. When heated with phosphoric acid, carvone forms carvacrol; this suggests that carvone probably contains the p-cymene structure, and that the keto group is in the ring in the ortho-position with respect to the methyl group.

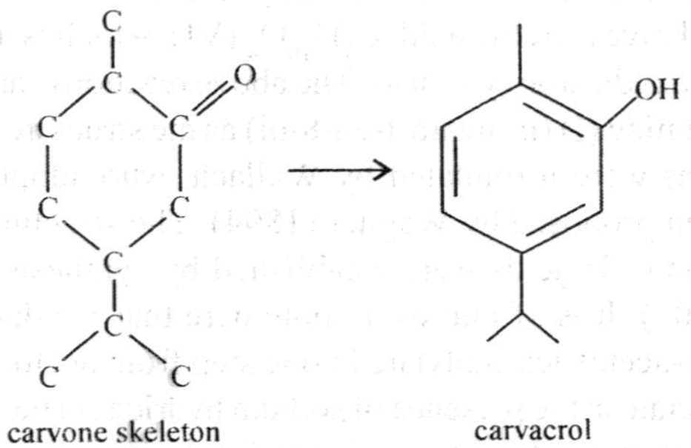

carvone skeleton → carvacrol (OH)

Fig. 9

The addition of nitrosyl chloride to α-terpineol (I) produces α-terpineol nitrosochloride (II), the addition occurring according to Markownikoff's rule (the chlorine is the negative part of the addendum).

This nitrosochloride rearranges spontaneously to the oximino compound (III) (see nitroso-compounds, it might be noted that this rearrangement proves the orientation of the addition of the nitrosyl chloride to the double bond; addition the other way could not give an oxime, since there is no hydrogen atom at position 1 in α-terpineol). Removal of a molecule of hydrogen chloride from (III) by means of sodium ethoxide produces (IV) and this, on warming with dilute sulphuric acid, loses a molecule of water with simultaneous hydrolysis of the oxime to form carvone (V).

Thus, according to this interpretation of the reactions, carvone is p-menth-5,8 dien-2-one. Actually, these reactions show that carvone has the same carbon skeleton as α-terpineol, and also confirm the position of the keto group. They do not prove conclusively the positions of the two double bonds; instead of position 6 (in (IV)), the double bond could have been 1(7), and instead of position 8 (as in (V)), the double bond could have been 4(8).

Thus the above reactions constitute an ambiguous synthesis of carvone (α-terpineol has already been synthesised). The exact positions of these two double bonds have been determined analytically as follows. The double bond in the 8-position. The following reactions were carried out by Tiemann and Semmler (1895).

$$\underset{(V)\,C_{10}}{\text{Carvone}} \xrightarrow[(+4H)]{Na\,/\,C_2H_5OH} \underset{(VI)\,C_{10}}{\text{Dihydrocarveol}} \xrightarrow[KMnO_4]{1\%\ alk} \underset{(VII)\,C_{10}}{\text{Trihydroxy Compund}} \xrightarrow[CH_3CO_2H]{CrO_2}$$

$$\underset{(VIII)\,C_9}{\text{Ketonic alcchol}} \xrightarrow{NaOBr} \underset{(IX)\,C_8}{\text{Hydroxy acid}} \xrightarrow[190°C]{KMnO_4} \text{(X)}$$

(X): benzene ring bearing CH_3, OH and CO_2H

Reduction of carvone (V) with sodium and ethanol gives dihydrocarveol, $C_{10}H_{18}O$ (VI); this is a secondary alcohol and contains one double bond, *i.e.*, the keto group and one of the two double bonds in carvone have been reduced. Hydroxylation of the double bond in dihydrocarveol by means of 1 per cent alkaline permanganate produces the trihydroxy compound $C_{10}H_{20}O_3$ (VII). Oxidation of (VII) with chromic acid causes scission of the glycol bond to produce a compound $C_9H_{16}O_2$ (VIII); this was shown to contain a keto group and a hydroxyl (alcoholic) group.

The action of sodium hypobromite on (VIII) caused the loss of one carbon atom to produce the compound $C_8H_{14}O_3$ (IX); this was shown to be a hydroxymonocarboxylic acid, and since one carbon is lost in its formation, its precursor (VIII) must therefore be a methyl ketone. Finally, dehydrogenation of (IX) by heating with bromine-water at 190°C under pressure produced m-hydroxy-p-toluic acid (X) (a known compound). Tiemann and Semmler explained these reactions on the assumption that one double bond in carvone is in the 8-position. Thus:

Had the double bond been in the 4(8)-position (structure (Va)), then compound (VIII), and consequently (X), could not have been obtained, since three carbon atoms would have been lost during the oxidation.

It might be noted in passing that (V) contains a chiral centre, whereas (Va) is symmetric and so cannot exhibit optical activity. Since carvone is known in optically active forms, structure (Va) must be rejected on these grounds.

The double bond in the 6-position. Carvone adds on one molecule of hydrogen bromide to form carvone hydrobromide, $C_{10}H_{15}OBr$ (XI), and this, on treatment with zinc dust and methanol, is converted into carvotanacetone, $C_{10}H_{16}O$ (XII), by replacement of the bromine atom by hydrogen. Thus the final result of these reactions is to saturate one of the two double bonds in carvone. Carvotanacetone, on oxidation with permanganate, gives isopropylsuccinic acid (XIII) and pyruvic acid (XIV) (Semmler, 1900). These products are obtainable only if the ring contains the double bond in the 6-position. Had the double bond been in the 1(7)-position, formic acid and not pyruvic acid would have been obtained. Further support for the 6-position is provided by the work of Simonsen *et al.* (1922), who obtained 3-isopropylglutaric acid and acetic acid on oxidation of carvotanacetone with permanganate.

The ultraviolet absorption spectrum is in agreement with the structure of an α, β-unsaturated ketone, but does not distinguish (V) from (Va); λ_{max} 235 (ε 19000) nm, and the calculated value (for both (V) and (Va)) is 237 nm (see 3vii). Dihydrocarveol (VI) does not show any maximum in the region 220-250 nm, and therefore the α, β-unsaturated carbonyl system is absent in this compound. On the other hand, carvotanacetone (XII) has λ_{max} 233 (E 9150) nm and is therefore an α, β-unsaturated carbonyl compound.

The NMR spectrum of carvone shows a multiplet signal at t 3.25 for the proton at C-6, a value which is characteristic of a β-proton in α, β-unsaturated carbonyl compounds. On the other hand, the multiplet signal for the C-8(9) methylene group has τ 4.78, which is in the normal range for olefinic protons.

RIBONUCLEIC ACIDS

These are polymers of ribonucleotides, and hydrolysis by alkali or by certain enzymes results in a mixture of ribonucleotides. Hydrogen-ion titrations on purified RNAs showed that secondary phosphate ionisations are absent. This suggests that the individual ribonucleotides are linked together by phosphodiester bonds. As we have seen, the attachment of the phosphate is at the 3′-position in the ribose molecule. Hence possible internucleotide bonds are 2′-3′ and 3′-5′. The answer has been obtained by various means, the most important being the use of enzymes which are known to hydrolyse specific ester bonds in nucleotides. Thus, it has been shown that:

(a) the enzyme spleen phosphodiesterase (specific for the C-5′-OP bond) converts RNAs into a mixture of ribonucleoside 3′-phosphates;

(b) snake venom phosphodiesterase (specific for the C-3′-OP bond) hydrolyses RNAs to a mixture of ribonucleoside 5′-phosphates. Hence, RNAs have a linear structure of units linked by 3′-5′ bonds. There appears to be little, if any, branched chains.

As we have seen, the common bases in RNAs are adenine, guanine, uracil, and cytosine. Early work on the base composition of nucleic acids led to the conclusion that the four bases were present in equimolar proportions. Subsequent work, as a result of accurate methods of analysis (chromatography, etc.), has shown that the molar proportions of the bases vary considerably according to the source of the nucleic acid : ribosomal (r) and transfer (t) RN As and messenger (m) RNAs. The less common bases are widespread in tRNAs. It has also been shown that the keto-bases (guanine and uracil) and the amino-bases (adenine and cytosine) are present in all RNAs in roughly equal amounts.

A great deal of work has been done to elucidate the sequence of the bases in RNAs and methods are, in principle, similar to those used in the determination of the primary structure of proteins. End groups have been determined by enzyme hydrolysis of the RNA with snake venom phosphodiesterase. Among the nucleotides (mainly nucleoside 5′-phosphates) will be some nucleosides. Some nucleoside 3′,5′-diphosphates. These can be identified and estimated by means of chromatographic methods. Hence, the end groups are determined and the length of the polynucleotide chain can be estimated.

The nucleotide sequence has been determined for some of the relatively short RNAs by the use of different enzymes, end-group analysis, and the application of the overlapping method. A point of interest in this connection is that RNAs are synthesised in association with DNAs. Hence it can be expected that there will be some correspondence in the base sequence between the DNA and its complementary RNA.

The abbreviated forms are also given; in these the letters refer to the nucleoside, *e.g.*, G = guanosine; U = uridine; etc. Various methods have been used to determine the molecular weights of purified nucleic acids, *e.g.*, end-group assay (see above), ultracentrifugation, light scattering, etc.. Values obtained for RNAs range from about 2×10^4 to 2×10^6. The secondary structure of RNAs has also been investigated. The results (mainly from X-ray analysis) appear to indicate that RNAs exist as single strands which contain helical segments stabilised by hydrogen bonding. There are, however, some examples of RNAs which exist as double strands (double helical structure).

Limonene, $C_{10}H_{16}$, b.p. 175.5-176.5°C

This is optically active; the (+)-form occurs in lemon and orange oils, the (–)-form in peppermint oil, and the (±)-form in turpentine oil. The racemic modification is also produced by racemisation of the optically active forms at about 250°C. The racemic modification is also known as dipentene; this name was given to the inactive form before its relation to the active form (limonene) was known.

Since limonene adds on four bromine atoms, it therefore contains two double bonds. (+)-Limonene may be prepared by dehydrating (+)-α-terpineol with potassium hydrogen sulphate, and limonene (or dipentene) may be converted into a-terpineol on shaking with dilute sulphuric acid. Thus the carbon skeleton and the position of one double bond in limonene are known.

The position of the other double bond, however, remains uncertain from this preparation; (I) or (II) is possible. Proof of position 8. Structure (I) contains a chiral centre C-4, and hence can exhibit optical activity. (II) is symmetric and so cannot be optically active. Therefore (I) must be limonene.

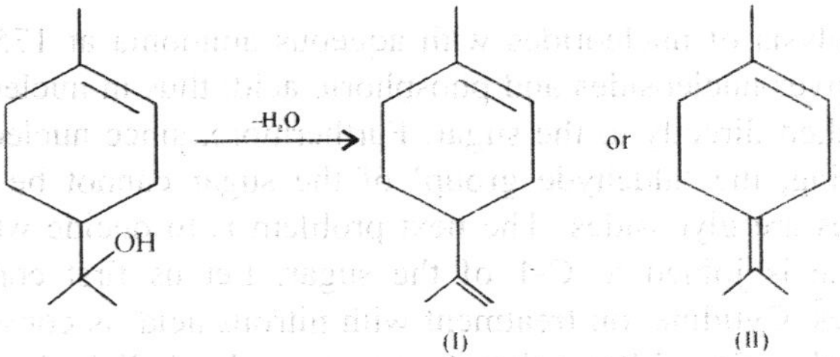

-H₂O

or

(I) (II)

Fig. 10

Chemical proof for position 8 is afforded by the following reactions:

Limonene (I) $\xrightarrow{NOCl}$ Limonene nitrosochloride (III) $\xrightarrow[EtOH]{KOH}$ carvoxime (IV

Since the structure of carvoxime is known, it therefore follows that (I) must have one double bond in position 8; t

Sylvestrene, $C_{10}H_{16}$, b.p. 175-178°C

This compound exists in (+)-, (−)- and (±)- forms; the racemic modification is also known as carvestrene (cf. limonene and dipentene, 13). The (−)-form of sylvestrene was first obtained from Swedish pine needle oil (Attenberg, 1877), and was shown to contain the m-cymene carbon skeleton (Baeyer *et al.*, 1898).

Thus, sylvestrene appeared to be the only monocyclic monoterpenoid which did not have the p-cymene structure and was obtainable from natural sources. Although, the m-cymene structure can be divided into two isoprene units (Wallach's isoprene rule), these two units are not joined head to tail. Subsequent work, however, showed that sylvestrene does not occur in pine oil. In the extraction of sylvestrene, the pine oil is heated with hydrogen chloride to give dipentene dihydrochloride (I) and sylvestrene dihydro-chloride (II).

These two compounds were shown by Simonsen *et al.* (1923, 1925) to be produced by the action of hydrogen chloride on car-3-ene, *i.e.*, these workers showed conclusively that the terpene originally present in Swedish pine oil is car-3-ene.

Sylvestrene may be obtained from its dihydrochloride by heating the latter with aniline; removal of hydrogen chloride from the ring can give rise to two possible positions for the ring double bond.

NUCLEOSIDES STRUCTURE

Hydrolysis of nucleotides with aqueous ammonia at 175°C under pressure gives nucleosides and phosphoric acid; thus in nucleosides the base is linked directly to the sugar. Furthermore, since nucleosides are non-reducing, the 'aldehyde group' of the sugar cannot be free, *i.e.*, nucleosides are glycosides. The next problem is to decide which atom of the base is joined to C-1 of the sugar. Let us first consider the pyrimidines. Cytidine, on treatment with nitrous acid, is converted into uridine; it therefore follows that the sugar residue is linked in the same position in both of these nucleosides.

The point of linkage cannot be 3 or 4, since cytidine has a free amino-group at position 4 and consequently there cannot be a hydrogen atom on N-3. Also, since uridine forms a 5-bromo derivative, C-5 must be free (Levene *et al.*, 1912). When uridine is treated with an excess of bromine, followed by the addition of phenyl hydrazine, a uridine derivative is obtained which contains two phenyl hydrazino groups.

This compound wasven structure (I) since work by Levene (1925) showed that this type of compound can be obtained only if uracil is substituted in position 1 and positions 5 and 6 are free. Thus the sugar is attached to N-1. In a similar way, it has been shown that the other pyrimidine nucleosides (ribosides and deoxyribosides) have the sugar residue linked at N-1.

Todd *et al.* (1947) have synthesised uridine and cytidine, and thereby have confirmed the linkage at N-1. This linkage has also been confirmed by the X-ray analysis of cytidine (Furberg, 1950). Now let us consider nucleosides containing purine bases. Adenosine has a free amino-group at position 6; therefore the sugar cannot be at C-6 or N-1 (cf. cytidine). Similarly, since guanosine has a free amino-group at position 2, the sugar cannot be at C-2 or N-3. Now Levene found that the two purine ribosides are equally readily hydrolysed by dilute acids and by the same enzyme.

He therefore assumed that the sugar residue is linked at the same place in both nucleosides. On this basis, only positions 7, 8 and 9 are possible points of attachment. Position 8 was then excluded since this point would involve a carbon-carbon bond, a linkage which would be very stable, whereas nucleosides are very readily hydrolysed by dilute acids (see also below). Thus positions 7 or 9 are free. This is supported by the following evidence (Levene, 1923). When guanosine is treated with nitrous acid, xanthosine is produced and this, on methylation with diazomethane followed by hydrolysis, gives theophylline (1,3-dimethylxanthine). Thus positions 1 and 3 are free in guanosine, and so the sugar must be attached at positions 7 or 9.

The evidence so far does not permit a decision to be made between these two positions since the system (in the imidazole nucleus) is tautomeric. It should be noted that had the sugar residue been attached to C-8, then a trimethylxanthine would have been obtained instead of theophylline (cf. above).

The ultraviolet absorption spectrum of guanosine is very similar to that of 9-methylguanine and differs from that of 7-methylguanine; hence it appears likely that guanosine is the 9-guanine glycoside (Gulland *et al.*, 1935, 1938). Todd *et al.* (1947, 1948) have synthesised guanosine and adenosine in which the sugar is known to be in the 9-position, and showed that their synthetic compounds are identical with the natural products; *e.g.*, the synthesis of adenosine.

A difficulty with the nitrous acid deamination is that it deaminates both aminopurines and aminopyrimidines in nucleic acids. Shapiro *et al.* (1970), however, have shown that cytosine but not adenine or guanine can be deaminated by means of sodium hydrogen sulphite under suitable conditions of pH. It might be noted, in passing, that glycosides are compounds formed by the linking of a sugar (at C-1) with a COH group. Thus the nucleosides are, strictly speaking, not glycosides; they should be called ribosylpyrimidines and ribosylpurines.

The final problem to be elucidated in connection with the structure of nucleosides is the nature of the ring in the sugar residue and the type of linkage (α or β). Degradative experiments have shown that the sugar is present as the furanose form, *e.g.*, methylation of a pyrimidine riboside, followed by hydrolysis, gives a trimethylribose which, on oxidation, forms dimethylmesotartaric acid. This product shows that the ribose ring is furanose; had the ring been pyranose, then the final product would have been trimethoxyglutaric acid. This is confirmed by the fact that on oxidation with periodic acid one molecule of the reagent was consumed to form a dialdehyde with no loss of carbon.

$$>N_1 - \overbrace{CHCHOHCHOHCH}^{O}CH_2OH \xrightarrow{(CH_2)_2SO_4}$$

$$>N_1 - \overbrace{CHCH(OCH_3)CH(OCH_3)CH}^{O}CH_2OCH_2 \xrightarrow{acid}$$

$$\overline{\text{CHOHCH(OCH}_3\text{)CH(OCH}_3\text{)CHCH}_2\text{OCH}_3} \xrightarrow{[O]} \text{CO}_2\text{CCH(OCH}_3\text{)CH(OCH}_3\text{)CO}_2\text{H}$$

Deoxyribose has also been shown to be of the furanose type, *e.g.*, Lythgoe *et al.* (1950) found that pyrimidine deoxyribosides consume a negligible amount of periodic acid; this agrees with the 2-deoxyribofuranose structure since, in this state, the molecule does not contain two adjacent hydroxyl groups.

$$> N_1 — \overline{\text{CHCH}_2\text{CHOHCH}}\text{CH}_2\text{OH}$$

These results have been confirmed by other work (see below).

The configuration of the furanoside link has been shown to be β-by various means, *e.g.*, Todd *et al.* (1947) oxidised adenosine with periodic acid, and showed that the product is identical with that from the oxidation of 9-β-D-mannopyranosidyladenine (a synthetic compound). Furberg (1950) has shown by means of the X-ray analysis of cytidine that the sugar residue is attached to N-1 and is β-D-ribofuranoside. Since other ribonucleosides exhibit the same general pattern, it is inferred that all are furanosides with the β-configuration. Manson *et al.* (1951), from absorption spectra measurements, have shown that deoxyribonucleosides also exist in the β-configuration.

NUCLEIC ACIDS STRUCTURE

Nucleic acids are colourless solids, all of which contain the following elements: carbon, hydrogen, oxygen, nitrogen and phosphorus. The following chart shows the nature of the products obtained by hydrolysis under different conditions.

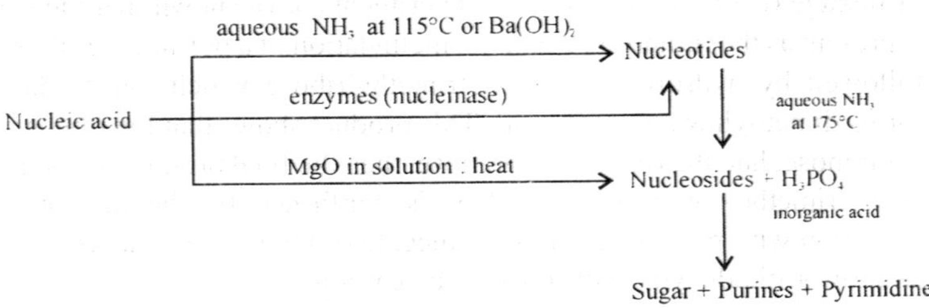

Fig. 11

These earlier methods have been modified by later workers in this field of study. Complete hydrolysis of the purine nucleotides by dilute acid occurs relatively easily, but the pyrimidine nucleotides usually require heating under pressure. On the other hand, complete hydrolysis of nucleic acids may be carried out by heating with 12 N perchloric acid or with formic acid. Alkaline hydrolysis results in the formation of ribonucleoside 2′- and 3′-phosphates. Enzymic hydrolysis produces nucleoside 3′- and 5′-phosphates, the actual product depending on the nature of the enzyme.

Separation and isolation of the various types of hydrolytic products of nucleic acids are now carried out by chromatographic methods and by counter current distribution. The purine and pyrimidine bases are readily separated and isolated by means of ion-exchange chromatography. Paper chromatography is particularly useful when dealing with small amounts of nucleic acids, and paper electrophoresis is very useful for the separation of small amounts of nucleotides. Column chromatography, counter current distribution, etc., have been used to separate and purify polynucleotides.

2

GENERAL NATURE OF THE ANTHOCYANINS

INTRODUCTION

Most of the anthocyanidins are derivatives of 3,5,7-trihydroxy-flavylium chloride. Thus, the hydroxylation patterns in the natural anthocyanidins fall into the three basic groups of pelargonidin, cyanidin and delphinidin. A far less common type is the 3-deoxyanthocyanidin group (the 3-hydroxyl group is absent), *e.g.*, luteolinidin (3-deoxycyanidin).

Various sugars (mono-, di- and trisaccharides) have been found in anthocyanins ; the most common are glucose, galactose and rhamnose, and the most important of these is glucose. Some pigments as well as being glycosides, are also acylated derivatives. The most common acids appear to be derivatives of cinnamic acid. The isolation of anthocyanins depends on the plant source. The earlier methods used solvent extraction (ethanol, ether, acetone and light petroleum), but nowadays chromatography is the main method. With column chromatography (cellulose powder, silica gel, ion-exchange resins, etc.), the solvent used depends on the nature of the adsorbent. Furthermore, since anthocyanins are coloured, a series of bands is produced on the column. The identity of the bands may then be determined by specific colour tests for the anthocyanins. However, a better way of identifying these compounds is to use paper chromatography, together with known compounds for comparison. Moreover, paper chromatography is particularly useful for micro-scale work. Other methods used for separation are paper electrophoresis and counter-current distribution.

The anthocyanins are characterised by two absorption bands: Band I, 475-560 nm (visible region), and Band II, 275-280 run (ultraviolet region). The actual colour (Band I) depends on the number and positions of the hydroxyl and methoxyl groups, and when these are fixed, the colour then depends on pH and solvent.

HAEMOGLOBIN PRODUCTS

For many years, hxmoglobin, ham, etc., were written as shown, but the tendency now is to write them as haemoglobin, haem, etc. There is also a tendency to spell them as hemoglobin, heme, etc. (the latter is common American practice). Two of the most important compounds of the natural porphyrins are haemoglobin and chlorophyll. The bile pigments, which are formed mainly in the liver, are degradation products of haemoglobin. Haemoglobin and chlorophyll act as catalysts (biological) in many biological processes.

It is a chromoprotein, the protein part being globin (94 per cent), and the prosthetic group being haem (6 per cent). The composition of haemoglobin varies slightly, depending on the species from which it is isolated; the variation occurs only in the globin part of the molecule. The way in which the globin part is bound to haem has been the subject of much discussion. Globin consists of four polypeptide chains, and in human haemoglobin the chains are of two types which have different terminal acid groups: a-chain, valyl-leucyl end-group; β-chain, valyl-histidylleucyl end-group.

Normal adult haemoglobin contains two α-chains (141 amino-acids each) and two β-chains (146 amino-acids each) [Pauling *et al.*, 1957]. Braunitzer *et al.* (1961) have now worked out the amino-acid sequence in these chains. Haemoglobin occurs in all vertebrates (with certain exceptions) and in many invertebrates; it has also been found in certain strains of yeasts, moulds, etc. Haem is an iron-protoporphyrin complex. When the iron atom is in the ferrous state, the complex is called ferrous protoporphyrin, ferroprotoporphyrin, protohaem, or haem, and the molecule is electrically neutral.

When the iron atom is in the ferric state, the complex is called ferric protoporphyrin, ferriprotoporphyrin, or haemin, and the molecule carries a unit positive charge (and is consequently associated with an anion). In the anima body, haemoglobin readily combines with oxygen to form ox vhaemoglohin and this, when treated with glacial acid, forms haemutin, $[C_{34}H_{32}N_4O_4Fe(III)]^+ OH^-$. The chloride of haematin is haernin chloride, $[C_{34}H_{32}N_4O_4Fe(III)^+Cl^-$ Haemin may be prepared by warming blood with acetic acid and sodium chloride (Teichmann, 1853). The iron atom can be removed from haernin and from haem, but it is easier to do this with the latter. Hence, in general, haernin is reduced to haem, ($C_{34}H_{32}N_4O_4Fe$), by, *e.g.*, sodium hyposulphite followed by treatment with acid (hydrochloric or sulphuric acid; see also below). In haem the four ligands (the four pyrrole groups) form a square-planar complex (via the nitrogen atoms). The two remaining positions of co-ordination are perpendicular to this plane (*i.e.*, the plane of the porphyrin ring). Now, haemoglobin contains four molecules of haem for each molecule of globin (which consists of two α- and two β-chains; see above). Each iron atom (ferrous) has formed a square-planar complex with the protoporphyrin molecule and a fifth position is occupied by an imidazole ring (of the histidine amino-acid residue). It appears that the iron atom is hound to histidine-87 in the α-chain and to histidine-92 in the β-chain. Furthermore, it has been shown that each haem molecule is embedded in one of the four chains of the globin molecule.

If the sixth valency of the ferrous ion is unoccupied, the arrangement is a square pyramid. This is considered to be the case with haemoglobin, but it is possible that the sixth position is occupied by a water molecule, resulting in an octahedral complex. In either case, when haemoglobin combines with one molecule of oxygen to form oxyhaemoglobin, it is this sixth position which co-ordinates with the oxygen molecule (the iron atom is still in the ferrous state); the water molecule, if present in haemoglobin, is readily displaced.

The mechanism of the incorporation of a metal ion into porphyrin molecules is uncertain. Hambright et a!. (1972) have proposed a mechanism in which the first stage involves deformation of the porphyrin, and in the second stage the metal is introduced by attack on this deformed species.

Since haernin forms a diester with methanol, the molecule therefore contains two carboxyl groups. Also, since haernin absorbs two molecule of hydrogen when catalytically reduced (palladium), two ethylenic double bonds are thus probably present in the molecule. When subjected to vigorous reduction with hydriodic acid and phosphonium iodide or hydriodic acid and acetic acid, haernin is degraded into the four pyrrole derivatives opsopyrrole (I), haemopyrrole (II), cryptopyrrole (III), and phyllopyrrole (IV).

All four compounds have been synthesised by means of the Knorr pyrrole synthesis (1884, 1886); this is the condensation between an α-aminoketone and a β-diketone or β-keto-ester. The general reaction may be written as shown:

Detailed studies of this reaction have shown that the yields depend on the nature of the R group, When R^3 is an alkyl group, the yields are poor (or the reaction fails). When R^2 and R^3 are acyl or carbalkoxyl groups, the yields are usually very good. The α-aminoketone is generally prepared in situ, *e.g.*,

$$\begin{matrix} MeCO \\ | \\ EtO_2CCH_2 \end{matrix} \xrightarrow{HNO_2} \begin{matrix} MeCO \\ | \\ EtO_2CC{=}NOH \end{matrix} \xrightarrow[AcOH]{Zn} \begin{matrix} MeCO \\ | \\ EtO_2CCHNH_2 \end{matrix}$$

Various modifications of the Knorr pyrrole synthesis are used now, *e.g.*, t-butyl or benzyl oximino-acetoacetic esters are used instead of the corresponding ethyl ester; removal of the ester group is facilitated. Also, reduction is carried out with sodium dithionite instead of zinc and acetic acid.

As an example of this synthesis, let us consider the preparation of opsopyrrole (I) and cryptopy-role (III). Opsopyrrole may be synthesised by condensing aminoacetone with ethyl 2,4-diketopentanoate, and then subjecting the product to the Wolfh-Kishner reduction, *i.e.*, first converting the product into the hydrazone and then heating the latter with sodium ethoxide at 160°C. By this means a keto-group is converted into a methylene group. By using an excess of sodium ethoxide, decarboxylation is also effected at the same time.

When reduced with tin and hydrochloric acid, haemin is again degraded into four pyrrole derivatives, but in this case the products are all carboxylic acids in which each of the four pyrroles (I)-(IV) contains a carboxyl group attached to the ethyl group: The propionic acid residue can be introduced into the β-position of pyrrole in several ways, *e.g.*, (note the use of the Gattermann aldehyde synthesis and the Knoevenagel reaction):

When oxidised with chromic acid, haemin gives two molecules of haematinic acid (IX). On the other hand, mesoporphyrin (see below) gives, on oxidation, two molecules of ethylmethylmaleimide (X). The treatment of haemin with iron dust and formic acid results in the removal of the iron atom and the formation of protoporphyrin, $C_{34}H_{34}N_4O_4$. The iron atom is also removed from haemin by the action of hydrobromic acid in acetic acid, but in this case the product is haematoporphyrin, $C_{34}H_{38}N_4O_6$. If, however, haemin is treated with hydriodic acid in acetic acid, the iron atom is again removed and mesoporphyrin, $C_{34}H_{38}N_4O_4$, is obtained. Finally, when porphyrins containing two carboxyl groups are decarboxylated, the products obtained (after reduction, if necessary) are known as aetioporphyrins, *e.g.*, when protoporphyrin is decarboxylated, and the product then reduced, the final product is aetioporphyrin, $C_{32}H_{38}N_4$, which is also a degradation product of chlorophyll. Thus haemin and chlorophyll are closely related chemically. From the foregoing evidence (the molecular formula and the degradation products of haemin), it is reasonable to infer that haemin contains four substituted pyrrole nuclei linked together. The isolation of the pyrroles (I)-(IV) suggests that each of the four pyrrole nuclei contains a methyl group in the β-position. The isolation of the oxidation products (IX) and (X) [oxidation at the α-position], and of the reduction products (I)-(VIII) [appearance of a methyl group at the α-position], suggests that the pyrrole nuclei are linked at the α-positions ria one carbon atom.

The isolation of two molecules of (IX) suggests the presence of two propionic acid residues each in the α-position of two pyrrole nuclei (this would also account for the two carboxyl groups present in haemin). The appearance of ethyl groups in (I)-(IV) on the reduction of haemin could be explained by the presence of two vinyl groups in the β-position of two pyrrole nuclei (haemin contains two ethylenic double bonds). A possible structure for haemin is thus a ring structure containing four pyrrole nuclei linked at the α-positions via one carbon atom, with four β-positions occupied by methyl groups, two β'-positions by vinyl, groups and the remaining two β'-positions by propionic acid residues. Kuster (1912) was the first to propose that the four pyrrole nuclei formed a cyclic structure, and this has been proved correct by synthetic work; the porphyrins so obtained had the same absorption spectra as the natural porphyrins. At the same time, this synthetic work established the nature and the positions of the substituent groups.

These methods, reductive and oxidative degradations, were used by the earlier workers, but the latter is now the preferred method. The difficulty with reductive degradation was that it gave rise to many products, the separation of which was difficult. More recently, however, the pyrroles have been isolated and identified by means of GLC. A difficulty with oxidative degradation is that only a few imides are obtained because others, which would have been formed from pyrrole nuclei containing β-substituents such as formyl, vinyl, etc., are further oxidised. This problem has been overcome by first converting these sensitive β-substituents into more stable groups, *e.g.*, vinyl into ethyl. After oxidation, the imides are separated by GLC. A semi-micro oxidative degradation has also been developed. Oxidation is carried out with potassium permanganate-potassium hydrogen carbonate solution and the resulting pyrrolecarboxylic acids are separated and identified by paper chromatography (the R_F values have been obtained from synthetic specimens).

Vitamin

Vitamin *A (Retinol, Axerophthol)* is also usually referred to as vitamin A_1 since a second compound, known as vitamin A_2, has been isolated. These vitamins are diterpenoids; they are usually classified as belonging to the apocarotenoid group.

Vitamin A_1 influences growth in animals, and also apparently increases resistance to disease. Night blindness is due to vitamin A_1 deficiency in the human diet, and a prolonged deficiency leads to xerophthalmia (hardening of the cornea, etc.). Vitamin A_1 occurs free and as esters in fats, in fish livers and in blood. It was originally isolated as a viscous yellow oil, but later it was obtained as a crystalline solid, m.p. 63-64°C (Baxter *et al.*, 1940). Vitamin A_1 is estimated by the blue colour reaction it gives with a solution of antimony trichloride in chloroform (the Carr-Price reaction); it is also estimated by light absorption (λ_{max} 325 (ε 51000) nm).

The IUPAC name of vitamin A (A_1) is retinol; that of the corresponding aldehyde is retinal (retinene, retinene$_1$); and that of the corresponding acid is retinoic acid. The traditional names are still widely used.

Carotenoids are converted into vitamin A, in the intestinal mucosa, and feeding experiments showed that the potency of α- and γ-carotenes is half that of β-carotene. This provitamin nature of β-carotene led to the suggestion that vitamin A_1 is half the molecule of β-carotene.

The biological conversion of β-carotene into vitamin A_1 is still not certain. There is evidence that one molecule of β-carotene undergoes central fission to give two molecules of vitamin A_1. This, however, does not appear to be the only path of breakdown. There is also evidence that indicates a stepwise oxidative degradation which starts at one end of the molecule and results in the formation of one molecule of vitamin A_1.

On catalytic hydrogenation, vitamin A_1 is converted into perhydrovitamin A_1, $C_{20}H_{40}O$; thus vitamin A_1 contains five double bonds. Since vitamin A_1 forms an ester with p-nitrobenzoic acid (this ester is not crystallisable), it follows that vitamin A_1 contains a hydroxyl group. Hence the parent hydrocarbon of vitamin A_1 is $C_{20}H_{40}$, and consequently the molecule contains one ring. Ozonolysis of vitamin A_1 produces one molecule of geronic acid per molecule of vitamin A_1 and so there must be one β-ionone nucleus present (Karrer, 1931, 1932). Oxidation of vitamin A_1 with permanganate produces acetic acid; this suggests that there are some $-C(CH_3)=$ groups in the chain. All of the foregoing facts are in keeping with the suggestion that vitamin A_1 is half the β-carotene structure.

When heated with an ethanolic solution of hydrogen chloride, vitamin A_1 is converted into some compound (II) which, on dehydrogenation with selenium forms 1,6-dimethylnaphthalene (III) (Heilbron *et al.*, 1932). Heilbron assumed (I) as the structure of vitamin A_1, and explained the course of the reaction. Perhydrovitamin A_1 has been synthesised from β-ionone (Karrer, 1933), and was shown to be identical with the compound obtained by reducing vitamin A_1; thus there is evidence to support the structure assigned to vitamin A_1. Final proof of structure must rest with a synthesis of vitamin A_1 itself, and this has now been accomplished by several groups of workers. The following synthesis is that of Isler *et al.* (1947). This starts with methyl vinyl ketone to produce compound (IV), one stage of the reactions involving an allylic rearrangement. Also note the formation of the cisproduct. The preparation of (V) has been described in 3 (structure XV). In the hydrogenation of (VI) to (VII), barium sulphate is used to act as a poison to the catalyst to prevent hydrogenation of the double bonds. Partial acetylation of (VII) (primary alcoholic groups are more readily acetylated than secondary) protects the terminal alcoholic group from an allylic rearrangement in the conversion of (VIII) to (IX). It should also be noted that dehydration with iodine is accompanied by isomerisation to the all-trans configuration.

The crude vitamin A, obtained in the above synthesis was purified via its ester with anthraquinone-2-carboxylic acid, and was thereby obtained in a crystalline form which was shown to be identical with natural vitamin A_1.

Lindlar (1952) has shown that triple bonds may be partially hydrogenated in the presence of a Pd–$CaCO_3$ catalyst that has been partially inactivated by treatment with lead acetate; better results are obtained by the addition of quinoline. Thus the hydrogenation of (VI) gives (VII) in 86 per cent yield when the Lindlar catalyst is used. Another method of synthesising vitamin A_1 is due to van Dorp et al. (1949), who prepared retinal, (X), which was then reduced by means of lithium aluminium hydride to vitamin A_1; β-ionone and ethyl γ-bromocrotonate were the starting materials.

Vision

Two types of light receptor exist in the retina of the eye: rods (for vision in dim light) and cones (for vision in bright light and for colour vision). A chromoprotein rhodopsin (or visual purple), which is a highly photosensitive protein complex with retinal$_1$, accumulates in the rods in the dark, and when the retina is illuminated the rhodopsin is bleached.

Bleaching occurs by rhodopsin undergoing a series of changes, one part of the sequence involving the isomerisation of retinal$_1$ into all-trans-retinal$_1$. It is the latter form which is produced together with the protein opsin in the dissociation of rhodopsin.

Experimental work has shown that the 11-cis form of retinal$_1$ (see neovitamin b, above) is the geometrical isomer of rhodopsin and is isomerised to the all-trans form when bleaching occurs. Since the 11-cis isomer is the only one which can combine with opsin, once the all-trans form is produced, recombination is not possible. Complex-formation to regenerate rhodopsin, however, occurs after a time in the dark, due to the action of an enzyme which catalyses the isomerisation of the trans- into the cis-retinal$_1$.

Vitamin A_1 is the source of all-trans-retinal$_1$ and hence its deficiency results in night blindness. The rhodopsin cycle may be represented as shown (no enzymes are given).

An interesting point about this problem of vision is that 11-cis-retinal$_1$ is sterically hindered (see 1). There would also appear to be some connection with the fact that this isomer, a 'bent' molecule, can form a complex with opsin, whereas the all-trans isomer, a 'straight' molecule, cannot.

Xanthophylls

These are naturally occurring carotenoids which contain an oxygen function. Many have been synthesised.

Cryptoxanthin, $C_{40}H_{56}O$, m.p. 169°C, is 3-hydroxy-β-carotene; it has pro-vitamin-A activity. Rubixanthin, $C_{40}H_{56}O$, m.p. 160°C, is 3-hydroxy-γ-carotene, and lycoxanthin, $C_{40}H_{56}O$, m.p. 168°C, is 3-hydroxylycopene.

Rhodoxanthin, $C_{40}H_{52}O_2$, m.p. 219°C, is 3,3'-diketo-retro-β-carotene. The prefix retro indicates that the positions of the double bonds in the central chain of 'normal' carotenoids have been reversed, *i.e.*, the linkage to a terminal ring is a double bond.

Lutein, $C_{40}H_{56}O_2$, m.p. 193°C, was formerly known as xanthophyll; it is 3,3'.dihydroxy-a-carotene. Zeaxanthin, m.p. 205°C, and lycopbyll, m.p. 179°C, are the corresponding 3,3'-dihydroxy derivatives of β-carotene and lycopene, respectively.

SPIRILLOXANTHIN (RHODOVIOLASCIN), M.P. 218°C

This has been isolated from various sources. Karrer *et al.* (1935-1940) showed that its molecular formula was $C_{42}H_6O_{92}$ and that it contained two methoxyl groups (Zeisel method) and thirteen conjugated double bonds (hydrogenation; u.v. spectrum). Oxidation with permanganate gave b.xindial and a higher dialdehyde. These workers then proposed a structure for spirilloxanthin, but were not certain of the positions of the two methoxyl groups. Weedon *et al.* (1959,1966) re-investigated this problem and assigned structure (I). They examined the NMR spectrum in the s 9-5-6-5 region. The absence of signals τ 8.35 and 8.41 indicated the absence of isopropylidene end-groups. The signal at τ 6.78 indicated two methyl groups (integration) and the signal at τ 8.02 indicated six in-chain methyls. A signal at τ 7.70 (d, J 6.8 Hz) was assigned to two methylene groups, each being adjacent to one vicinal proton, *i.e.*, $-CH_2-CH=$. The usual τ-value for such a methylene group is 8.3-8.7; the shift from this position (to 7.70) was attributed largely to the presence of an adjacent carbon atom carrying an oxygen atom (OMe; deshielding). Finally, the signal at τ 8.83, because it was exceptionally sharp, was assigned to methyl groups (four; integration) attached to saturated carbon atoms. However, since a methyl group of this type normally absorbs at τ 9.10-9.15, the downfield shift (deshielding) was attributed to the proximity of the oxygen atom (in OMe). These data led to the proposal of the structure given for spirilloxanthin (I). This structure has also been inferred from chemical evidence (Jensen, 1959), and has been confirmed by synthesis (Surmatis *et al.*, 1963).

Capsorubin, m.p. 218°C

This occurs in red peppers. The early work showed that capsorubin contained four oxygen atoms, two of which were present in hydroxyl groups. The results of microhydrogenation and the similarity of the visible spectrum of capsorubin with those of β-carotene and bixindial suggested that the two remaining oxygen atoms were ketonic. This led to the proposal that capsorubin had the partial structure (I). This was supported by the fact that on heating with aqueous alkali, capsorubin gave crocetindial (II) (this is the reversal of the aldol condensation). Further support was afforded by the observation that the infrared spectrum of capsorubin showed a band which was also present in the spectra of authentic compounds of the type (I).

Also, the reduction of capsorubin with potassium borohydride produced shifts in the u.v. spectrum consistent with selective reduction of two terminal carbonyl groups to give a nonaene chromophore (Weedon *et al.*, 1958). Compound (III) was synthesised and examination of its NMR spectrum showed that the τ values of the methyls were upfield (0.12-0.16) with respect to those in capsorubone. This is due to deshielding by the presence of the carbonyl group in capsorubone. Hence, the hydroxyl group in capsorubin was tentatively placed at position 3 or 4. Synthetic work then established that the hydroxyl group was at position 4. Complete synthesis of (±)-capsorubin confirmed this structure and also established that the hydroxyl group at position 4 and the keto group at position 6 were trans (as previously suggested from infrared and NMR data). The structure (and absolute stereochemistry) of capsorubin is (IV) [Weedon *et al.*, 1962].

The relative configurations of the 4-OH and 1-Me were established independently as follows. The (±)-form of (V) was synthesised and, on reduction, gave (VI)-(±)-trans and (VII)-(±)-cis. These were separated and differentiated by the fact that only (VII) formed a lactone.

Each of these ((VI) and (VII); $CO_2H \rightarrow COMe$) was condensed with crocetindial, and it was shown that it was (VII) that gave a product identical with natural capsorubin. Also, the absolute configuration of C-1 has been established by correlation with that of C-1 of (+)-camphor.

CAPSANTHIN, M.P. 175-176°C

This occurs in red peppers together with capsorubin but in far greater amount. The structure of capsanthin was actually elucidated alongside that of capsorubin (Weedon *et al.*, 1958, 1962). The early work showed that when heated with aqueous alkali capsanthin gave β-citraurin (I), a known compound. It was also found, from ultraviolet spectroscopic studies, that capsanthin contained a decaenone chromophore, and Cholnoky *et al.* (1957) showed that the molecular formula of capsanthin was

$C_{40}H_{56}O_3$. Thus, capsanthin contains two rings (cf. cap sorubin, above). The Oppenauer oxidation of capsanthin gave capsanthone, a hydroxydiketone, the infrared spectrum of which indicated the presence of a cyclopentanone ring and a hydroxyl group (Weedon *et al.*, 1960). The NMR spectrum of capsanthone was examined and some of the signals observed were the same as those for capsorubin, but only half the intensity of the latter. Thus, both compounds have a common end-group.

This, as we have seen, is the 4-hydroxy-l,2,2-trimethylcyclopentyl residue. The NMR spectrum of capsanthin showed methyl signals at τ 9.02, 8.95, 8.77, 8.65, 8.3, and 8.02 in the approximate ratios 1:2:1:1:1:4 (the italicised values are the same as those shown by capsorubin). However, the signals at τ 8.95 and 8.3 are characteristic of methyls in cyclohexene rings in, *e.g.*, zeaxanthin). This is in agreement with the formation of (I) from capsanthin. Hence, from the data given, it follows that the structure of capsanthin is (II). A point to note is the failure of the Oppenauer oxidation to convert the secondary hydroxyl group in the cyclohexene end-group into the ketonic group.

Fucoxanthin, m.p. 166-167°C

The characteristic carotenoid of brown algae and diatoms; it appears to be the most abundant xanthophyll in nature. Weedon *et al.* (1964), from mass spectrometry, established the molecular formula as $C_{42}H_{58}O_6$, and the presence of two hydroxyl groups and one acetoxy-group (these groups had also been shown to be present by Jensen, 1961). Since fucoxanthin formed only a monoacetate, this suggested that one hydroxyl group was tertiary. On reduction with lithium aluminium hydride, fucoxanthin gave a mixture of 'fucoxanthols', the u.v. spectrum of which indicated the presence of a conjugated octaene. Potassium permanganate oxidation of fucoxanthin gave dimethylmalonic acid and 2,2-dimethylsuccinic acid. Thus, fucoxanthin contains the grouping (I) and this was confirmed by examination of the i.r. spectrum, which also indicated the presence of an allenic grouping.

Partial oxidation of fucoxanthin with zinc permanganate gave four aldehydes. Structure (II) was assigned to one of these (based on molecular formula and u.v. spectrum); (III) was a known compound. The other two aldehydes had molecular formulae $C_{25}H_{34}O_4$ (τ 0.55) and $C_{27}H_{36}O_4$ (τ 0.45; d, J 7.5 Hz). These formulae were obtained by mass spectrometry. The C_{25}- and C_{27}- aldehydes were shown to be a pentaenealdehyde and a hexaenealdehyde, respectively, by examination of their u.v. spectra before and after lithium aluminium hydride reduction (cf. capsorubin).

In addition to these aldehydes, there was also a mixture of 'allenic aldehydes', which were classified as such because their i.r. spectrum exhibited bands in the region normally associated with allenes (see also above). Further oxidation of the C_{25}- and C_{27}-aldehydes gave (II) and (III), respectively. On the other hand, complete oxidation of the C_{27}-aldehyde and the mixture of 'allenic aldehydes' gave 2,2-dimethylsuccinic acid; the former also gave dimethylmalonic acid.

The NMR spectra of fucoxanthin and the C_{27}-aldehyde showed a number of common signals, *e.g.*, τ 6.32 and 7.40; J 18 Hz, which were attributed to a methylene group adjacent to an oxo group. It was also found that the C_{27}- aldehyde formed a monoacetate, and condensed with the geranyl Wittig reagent to give a conjugated octaenone (as shown by the u.v. spectra before and after hydride reduction).

On the basis of this work, Weedon *et al.* proposed (IV) as the structure of the C_{27}- aldehyde (and the C_{25}- aldehyde as its lower vinylogue). They also tentatively formulated fucoxanthin as (V). This was confirmed by later work of Weedon *et al.* (1966). The mixture of 'allenic aldehydes' was separated (chromatography) into (VI), (VII), and (VIII).

All of these exhibited spectral properties (u.v., visible, i.r., NMR) in keeping with the structures assigned.

Carotenoid Acids (Apocarotenoids)

These are compounds which do not contain 40 carbon atoms.

Bixin, $C_{25}H_{30}O_4$

Natural bixin is a brown solid, m.p. 198°C, and is the cis-trans-form; it is readily converted into the more stable all-trans-form, m.p. 216-217°C, by iodine in benzene solution.

When boiled with potassium hydroxide solution, bixin produces one molecule of methanol and a dipotassium salt which, on acidification, gives the dibasic acid norbixin, $C_{24}H_{28}O_4$. Thus bixin is a monomethyl ester, and can be esterified to give methylbixin.

On catalytic hydrogenation, bixin is converted into perhydrobixin, $C_{25}H_{48}O_4$; thus there are 9 double bonds present in the molecule (Liebermann *et al.*, 1915). Perhydrobixin, on hydrolysis, forms perhydronorbixin. Oxidation of bixin with permanganate produces four molecules of acetic acid (Kuhn *et al.*, 1929); hence there are four $-C(CH_3)=$ groups in the chain. Furthermore, since the parent hydrocarbon of perhydronorbixin, $C_{24}H_{46}O_4$, is $C_{22}H_{46}$ (the two carboxyl groups are regarded as substituents), the molecule is acyclic.

The thermal decomposition of bixin produces toluene, m-xylene, m-toluic acid and the methyl ester of this acid (Kuhn *et al.*, 1932). Hence the following assumptions may be made regarding the nature of the chain (cf. β-carotene). It has been known for some time, based on infrared, ultraviolet, and visible absorption spectroscopy, that bixin had a cis-configuration about one of its double bonds. Also, the ultraviolet data excluded the possibility that the cis-isomer was one of the hindered type (see structure (II)).

Hence, the cis-isomerism cannot be at the 6,7 or the 14,15 double bond (see structure of bixin, above). Furthermore, the central double bond (10,11) was eliminated by synthesis. One suggestion was that the 2,3-double bond was involved, and the other suggestion was that it was the 4,5. Weedon *et al.* (1960) have settled the problem by the examination of the NMR spectra of natural and all-trans methylbixin. Some of the relevant data are given here. In the all-trans isomer the ends are identical and so only one signal is given by H-2 and H-2′, and one signal by H-3 and H-3′ (note that the a-proton absorbs upfield with respect to the β-proton in α, β-unsaturated oxo compounds).

Since four signals were observed for the end protons in natural methylbixin, these ends must be different. However, the J values $J_{2,3}$ and $J_{2',3'}$ were the same (15.8 Hz). This indicates that the configurations about the end double bonds (2,3 and 2′,3′) are unaffected (trans). As pointed out above, 6,7- (14,15-) and 10,11-double bonds have been excluded.

This leaves the 4,5- and the 8,9-double bond as possibilities. If cis-isomerism occurred at the 8,9-double bond, it is then difficult to explain the large change in τ value for H-3. It therefore follows that the difference between the natural and all-trans methylbixin is at the 4,5-double bond. However, since bixin is unsymmetrical at its ends, the isomerism could therefore be at the 16,17-double bond (instead of 4,5). This uncertainty was settled by the examination of the NMR spectrum of cis-apo-l-norbixinal methyl ester, obtained by oxidation of natural bixin. The spectrum showed doublets at τ 4.08 and τ 2.04. Hence, the isomerism occurs at the 4,5-double bond at the ester end in bixin, *i.e.*, bixin is

Crocetin, $C_{20}H_{24}O_4$

This occurs in saffron as the digentiobioside, crocin. The structure of crocetin was elucidated by Karrer *et al.* (1928) and Kuhn *et al.* (1931). Crocetin behaves as a dicarboxylic acid and has seven double bonds (as shown by catalytic hydrogenation to perhydrocrocetin, $C_{20}H_{38}O_4$). On oxidation with chromic acid, crocetin gives 3-4 molecules of acetic acid per molecule of crocetin; thus there are 3-4 methyl side-chains. The structure of crocetin was finally shown by the degradation of perhydrcnorbixin, $C_{24}H_{46}O_4$, by means of the following method

$$RCH_2CO_2H \xrightarrow{Br_2-P} RCHBrCO_2H \xrightarrow{\text{hydrolysis}} RCHOHCO_2H \xrightarrow{CH_2N_2}$$

$$RCHOHCO_2CH_2 \xrightarrow{CH_3MgI} RCHOHC(OH)(CH_3)_2 \xrightarrow{(CH_3CO_2)_4Pb}$$

$$RCHO \xrightarrow{[O]} RCO_2H$$

This set of reactions was performed twice on perhydronorbixin, thereby resulting in the loss of four carbon atoms (two from each end); the product so obtained was perhydrocrocetin, $C_{20}H_{38}O_4$. On these results, crocetin is therefore.

This structure is supported by the fact that the removal of two carbon atoms frcm perhydrocrocetin by the above technique (one carbon atom is lost from each end) resulted in the formation of a diketone. The formation of this compound shows the presence of an a-methyl group at each end of the molecule. The structure of crocetin is further supported by the synthesis of perhydrocrocetin, and by the synthesis of crocetin diesters by Isler *et al.* (1957). It appears, however, that part of the crocetin in crocin is present in the form of the 2,3-cis isomer. The evidence for this has been obtained from a study of the NMR spectrum of the dimethyl ester (Weedon *et al.*, 1960). This cis-form is readily converted into the all-trans form by iodine. The trans-crocetin dimethyl ester has also been synthesised by the Wittig reaction between the dialdehyde (I) and two molecules of the phosphorane (Buchta *et al.*, 1959, 1960) [structure (II), for the preparation of (I)].

Anthocyanins

Introduction : The anthocyanins, which are water-soluble pigments, generally occur in the aqueous cell-sap, and are responsible for the large variety of colours in flowers; red violet blue. Willstatter *et al.* (1913) showed that the various shades of colour exhibited by all flowers are due to a very small number of different compounds. Furthermore, these different compounds were shown to contain the same carbon skeleton, and differed only in the nature of the substituent groups. The anthocyanin pigments are amphoteric; their acid salts are usually red, their metallic salts usually blue and in neutral solution the anthocyanins are violet. Anthocyanins are natural plant pigments; they are glycosides and their aglycons, *i.e.*, the sugar-free pigments, are known as the anthocyanidins.

In addition to anthocyanins, the colour of flowers depends on the presence of co-pigments such as flavones, flavonols, etc., and to metal chelation, particularly with iron and aluminium. The colour (due to chelation) of the anthocyanins is affected to a large extent only when the molecule contains two hydroxyl groups in the o-position. Geissman *et al.* (1952) have applied the term flavonoids to embrace all compounds whose structure is based on flavone. Thus the anthocyanins are one group of flavonoid compounds.

PORPHYRINS

The various compounds described in are all derivatives of the parent substance known as porphin (I), and this may be written as (II) (Fischer). (I) is now usually written as (1a) and the alternative method of numbering (IUPAC rules, 1960) is also given. The carbon atoms forming the methine bridges are labelled α, β, γ, and 6 in (I) or numbered 5, 10, 15, and 20 in (Ia). The hydrogen atoms attached to these methine carbon atoms are referred to as mess hydrogens. The pyrrole rings have been labelled A, B, C, and D in (Ia), but in the earlier literature they have also been labelled (I), (II), (III), and (IV), respectively (as in (1)).

Examination of structure (I) or (Ia) shows the presence of an inner eighteen-membered ring containing a complete arrangement of conjugated double bonds. Thus many resonating structures contribute to this molecule, and consequently its stability will be great; this is observed in practice, the molecule has a very large heat of combustion. Also, the resonance gives rise to the colour in porphin derivatives: porphin itself does not occur naturally.

The geometry of porphin has been investigated by various workers, and the more recent results are in conflict with the earlier ones. The resonance energy of porphin would be a maximum if the molecule had structure (I) or (Ia) and was flat. Earlier work, based on analogy with the X-ray data of the phthalocyanines, led to the conclusion that the porphin molecule is planar, and this was also in agreement with magnetic measurements. Crute (1959) examined nickel aetioporphyrin II by X-ray analysis and the results indicated a flat molecule. On the other hand, Fleischer (1963) has shown, from X-ray analysis, that nickel aetioporphyrin I is not planar: the alternate pyrrole rings are bent up and down from the mean plane.

Fleischer *et al.* (1965) have also examined porphin by X-ray analysis, and found that the molecule is nearly planar, the observed small deviations from planarity not being large enough to be significant. These authors have also proposed an alternative electronic structure (Ib) for porphin. All 3,4-bonds in the pyrrole nuclei are represented as double bonds. The inner ring it-cloud has 16 atoms, each atom in the ring contributing one electron to the π-system of this ring.

Since porphin with no protons or metal ions bonded to the nitrogens is a dianion, two more electrons are therefore added to the n-system of the 16-membered ring. Thus the total number of electrons is 18, and this is consistent with the (4n + 2) electrons required by Huckel's rule for a stable aromatic system. The inner ring is now similar to the cyclo-octatetraene dianion (which has 10 electrons). Addition of two protons or a di-positive metal ion to the centre of the dianion ring produces a neutral species.

The authors suggest that a better description of the molecule may be a combination of the two electronic structure, (Ia) and (Ib). The NMR spectrum ($CDCl_3$) of coproporphyrin I methyl ester (all P groups are methyl esters: see Table) has been examined (Becker *et al.*, 1959). Two singlets were observed at τ 0.04 and 1389. The low-field singlet was assigned to the four 'outer' meso hydrogens, the large shift to lower field of olefinic protons being due to the presence of a large ring current. On the other hand, the very high-field signal was attributed to the imino hydrogens (NH) which are inside the ring (see the ring current effect). The existence of a ring current is evidence for aromaticity.

Substituted porphins are known as porphyrins. Let us first consider the aetioporphyrins, $C_{32}H_{38}N_4$. These are tetra-ethyltetramethylporphins, and on the assumption that two identical groups do not occur in the same pyrrole nucleus there will be four possible isomers (H. Fischer). All four have been synthesised by Fischer and they are known as aetioporphyrins I, II, II and IV, respectively.

The degradation of haemin gives aetioporphyrin III. Some common porphyrins are listed in Table : all the aetioporphyrins have been given, since these illustrate the distributions of the substituents. When there are three different substituent groups and each pyrrole nucleus has two different types, then fifteen isomers are possible. Some porphyrins occur free in nature, *e.g.*. protoporphyrin IX and coproporphyrins I and III occur in blood. Inspection of the Table shows that protoporphyrin IX, coproporphyrin III and uroporphyrin III have the same 'pattern' as aetioporphyrin III. All the porphyrins which possess some biological function belong to this pattern.

SPECTRAL PROPERTIES OF PORPHYRINS

Infrared studies of porphyrins have shown that most of the functional groups absorb in the expected regions and so may be detected. *e.g.*, OH, CO, H, CO_2R. etc. The band for the NH group (str.) occurs around 3300 cm^{-1}, and since it is almost unaffected in a dilute solution in carbon tetrachloride, this is strong evidence that there is intramolecular hydrogen bonding. Also, this band is consistent with there being hydrogen atoms attached to opposite nitrogen atoms and each being bonded to an adjacent nitrogen atom (cf. formula (I)). The visible spectra of porphyrins have been studied in great detail. The neutral metal-free porphyrins usually have four absorption bands between 500 and 650 nm. These hands have been numbered I, II, III and IV, the wavelengths of the maxima decreasing from I to IV, (diagrammatically) these spectra and the intensity of any given band depends on the porphyrin under investigation. In practice four types have been observed. As pointed out above, the aetio 'pattern' is the most important one, three examples are listed in Table 1.

Table 1

	Band λ_{max} ***nm*** (ε)			
Porphgrin	***I***	***II***	***III***	***IV***
Deuteroporphyrin	618(4330)	565 (6800)	525 (8 750)	495 (15 950)
Mesoporphyrin	620(5410)	567 (6590)	528 (9 820)	496 (14 240)
Protoporphyrin	630(5580)	575 (6 780)	537 (11 580)	503 (14640)

In addition to the four bands (I-IV), there is a hand in the region of 400 nm. This is known as the Soret hand and is characteristic of all conjugated tetrapvrroles. It is absent when this conjugation is broken; *e.g.*, bile pigments do not show this band. The Soret band has about twenty times the intensity of the strongest visible band and can be used to characterise porphyrins. The spectra of porphyrin mono-cations usually consist of a Soret band and three visible bands (I-III) and those of porphyrin di-cations usually consist of a Soret band and two visible bands (I and II). Metalloporphyrins generally exhibit two maxima in the visible region. These are known as the α- (nearer the red) and β-bands, and there is also the Soret band (γ-band) near 400 nm.

Substituent groups in which a 'saturated' carbon atom is directly joined to the pyrrole nucleus have very little effect on the spectrum, *e.g.*, groups such as Me, Et, CH_2CO_2H, $CH_2CH_2CO_2H$. In cases such as these the 'resonance path' of the porphin molecule is very little changed. When, however, the substituent group can give rise to different resonance paths (by extended conjugation), then both the wavelengths and intensities of the bands are changed. Such groups are $-CH{=}CH_2$, CO_2H, CHO.

Examination of the visible spectra of a large number of porphyrins of known structure has resulted in correlations being made between the nature of the side-chain and the position and intensity of its visible bands. These correlations are very important as a means of identifying porphyrins, and are also of great help in elucidating structures of unknown porphyrins. The ultraviolet and visible absorption maxima of haemin itself are: 390, 505, 540, 578 and 659 nm.

Abraham *et al.* (1961, 1966) have studied the NMR spectra of porphyrins in trifluoroacetic acid and showed that the τ-values of the protons in various groups lie between ranges which clearly distinguish them from one another: meso-hydrogen, τ-1.22 to -0.98; methyl hydrogens, τ 6.24-6.16; ethyl hydrogens: CH_2, τ 5.71-5.69; CH_3, 8.18-8.15; etc.

Furthermore, it was shown that the τ-values of meso-hydrogens were affected by the nature of the adjacent β-substituents, *e.g.*, in porphin itself, τ of meso-hydrogen = −1.22; for a meso-hydrogen flanked on both sides by β-propionate residues, τ = − 1.19. Further work by Abraham *et al.* (1966) showed that NMR spectra measured in chloroform solutions could be used to distinguish various porphyrin isomers from one another.

The mass spectra of many porphyrins have been examined and it is possible to deduce structural information of these compounds (Jackson *et al.*, 1965; Whitten *et al.*, 1966). The molecular ion is usually the base peak (in the absence of labile side-chains).

BILE PIGMENTS

Several pigments occur in bile, *e.g.*, bilirubin, mesobilirubin, etc.; the most important one is bdirubin, $C_{33}H_{36}N_4O_6$. On vigorous oxidation, bilirubin gives haematinic acid; and on vigorous reduction, it gives cryptopyrrole and cryptopyrrolecarboxylic acid. When catalytically reduced, bilirubin gives mesobilirubin, $C_{33}H_{40}N_4O_6$, which, on reduction with hydriodic acid in acetic acid, forms, among other products, bilirubic acid, $C_{17}H_{24}N_2O_3$, and neobilirubic acid, $C_{16}H_{22}N_2O_3$.

Finally, the reduction of bilirubic acid gives crypto-pyrrolecarboxylic acid as the main product, and the reduction of neobilirubic acid gives haemopyrrolecarboxylic acid. From this evidence it is reasonable to conclude that bilirubin contains the four pyrrole nuclei that occur in haemoglobin.

Bilirubin is an orange solid, and its ultraviolet spectrum shows one maximum at 450-455 nm. This spectrum is totally different from those of the porphyrins, and so it may be inferred that bilirubin has an openchain structure. On the basis of its relation to haem (the same four pyrrole nuclei) and further degradative and synthetic work, bilirubin has been assigned the structure shown (M, P and V have the meanings given in Table).

Until recently, the two end α-positions were considered to carry hydroxyl groups. It has now been shown that α-hydroxypyrroles are unstable and rapidly tautomerise to the stable keto form (Plieninger et al., 1956). Plieninger *et al.* (1962) have also studied the NMR spectra of 3,4-dialkyl α-hydroxypyrroles and showed these compounds exist as pyrrolin-2-ones. These results are supported by the ultraviolet spectroscopic examination of the titration curves of a number of bile pigments (Gray *et al.*, 1961).

SYNTHESIS OF THE PORPHYRINS

The first step in the synthesis, of porphyrins is usually the synthesis of dipyrrylmethenes. This was the method adopted by the earlier workers, but pure porphyrins were obtained in poor yield. Later syntheses start with dipyrrylmethanes; the yield of porphyrin is high.

(i) Dipyrrylmethenes may be prepared by the bromination of a 2-methvlpvrrole in which position 5 is vacant (H. Fischer, 1915); *e.g.*,

(ii) When pyrroles, in which the 5-position is vacant, are coupled by means of formic acid in the presence of hydrobromic acid, dipyrrylmethenes are produced (H. Fischer *et al.*, 1922); *e.g.*,

(iii) Unsymmetrical dipyrrylmethenes, *i.e.*, those containing two different substituted pyrroles, may be prepared as follows (Piloty *et al.*, 1912, 1914; H. Fischer *et al.*, 1926), *e.g.*:

(iv) Another method of preparing unsymmetrical dipyrrylmethenes is (H. Fischer *et al.*, 1928):

(v) Dipyrrylmethanes have been prepared as shown in method (i).

Many methods were introduced by H. Fischer to convert dipyrrylmethenes into porphyrins, but the most useful were: (i) Condensation between the hydrobromides of two 5-bromo-5′-methyldipyrrylmethenes; (ii) condensation between the hydrobromides of a 5,5′-dibromo- and a 5,5′dimethyldipyrrylmethene. The condensations were carried out by heating with succinic acid at 220°C.

The particular value of these two methods was that only one product was obtained. Many of the other methods resulted in a mixture of isomers. These were difficult to separate, but Marks *et al.* (1960) have separated such mixtures by counter-current distribution between dilute hydrochloric acid and ether.

Synthesis of Haemin (H. Fischer *et al.*, 1929)

The approach to this synthesis has already been discussed, viz., synthesis of pyrroles containing methyl groups and propionic acid residues in the correct positions, followed by their comhination to dipyrrylmethenes, and then formation of the porphyrin. However, at the time, there was no existing method for synthesising porphyrins with unsaturated side-chains.

Fischer solved this problem by introducing acetyl groups into the required positions by means of a Friedel-Crafts reaction. An interesting point about this reaction was that Fischer found that it gave better yields when carried on the iron complex than on the free porphyrin. The reduction of the acetyl group proved to be very difficult, and Fischer finally succeeded by the use of ethanol and potassium hydroxide, a reducing reagent that had been discovered by Dumas *et al.* (1840).

BIOSYNTHESIS OF PORPHYRIN

The progress made in this field is one of the outstanding examples of the use of isotopes. Tracer syntheses in vivo and in vitro and degradation methods have established the origin of all the carbon and nitrogen atoms in protoporphyrin (of haem), and have also established the nature of the pyrrole precursors. The results are the outcome of a large volume of work, but in the following account only a few experiments have been mentioned. These indicate, to some extent, the lines of research pursued.

Bloch *et al.* (1945), using acetic acid labelled with deuterium atoms, showed that deuteriohaemin was produced. Thus at least the methyl carbon of acetic acid is involved in the biosynthesis of haem. Then Shemin *et al.* (1950) and Neuberger *et al.* (1950) carried out experiments with $^{14}CH_3CO_2H$ and $CH_3{}^{14}CO_2H$, and showed that both carbon atoms of acetate participate in the synthesis of haem.

The latter authors also showed that with $^{14}CH_3CO_2H$, about half of the radioactive tracer atom appeared in the two pyrrole nuclei carrying the vinyl groups, and the other half in the two pyrrole nuclei carrying the propionic acid residues. When, however, $CH_3{}^{14}CO_2H$ was used as the precursor, then about 20 per cent of the tracer atom appeared in the vinyl pyrrole nuclei and 80 per cent in the propionic acid pyrrole nuclei. In neither case of the labelled acetates was there any significant radioactivity in the methine carbon of the haem. Thus the carbons of the methine bridges do not originate from acetate.

Shemin *et al.* (1945, 1946) carried out experiments with [^{15}N] glycine, and showed that all the and showed that the carboxyl group of glycine is not incorporated into protoporphyrin. On the other hand, Altman et al. (1948), using $^{14}CH_2NH_2CO_2H$, showed that the α-carbon atom of glycine is used in the protoporphyrin synthesis. This was confirmed by Shemin *et al.* (1950) who used $^{14}CH_2{}^{15}NH_2CO_2H$ and showed that for each nitrogen used for haem synthesis two α-carbon atoms of glycine were also incorporated into the molecule. Similar results were obtained by Neuberger *et al.* (1950) who also showed that the α-carbon atom of glycine is used in the formation of the methine bridge. Thus all the carbon atoms of protoporphyrin, except eight derived from the α-carbon of glycine, originate from acetate.

We may now summarise the foregoing results for the biosynthesis of protoporphyrin IX (and haems in general).

(i) All four nitrogen atoms are derived from the nitrogen of glycine.

(ii) The four methine carbon atoms are derived from the a-carbon atoms of glycine (see also below).

(iii) Four carbon atoms, one in each pyrrole ring (a-position), are derived from the α-carbon atoms of glycine (see also below).

(iv) The carbon atom of the carboxyl group in glycine is not incorporated into protoporphyrin IX.

(v) The remaining 26 carbon atoms in protoporphyrin IX are derived from either the methyl group or the carboxyl group of acetic acid.

A detailed study of the degradation products of the labelled protoporphyrins showed that it was very probable that the two sides of the pyrrole nuclei were synthesised from identical intermediates. It also seemed very reasonable that a common pyrrole of the type (I) was formed first. Also, consideration of the distribution of the radioactivity of the carbon atoms of the propionic acid residue and the (pyrrole) nuclear carbon to which it was attached led to the suggestion that succinic acid was a precursor.

The tracer distribution of the labelled succinic acid could arise by acetate entering the Krebs cycle. Shemin *et al.* (1952) tested this succinic acid hypothesis by using $HO_2{}^{14}CCH_2CH_2{}^{14}CO_2H$ and $HO_2C^{14}CH_2{}^{14}CH_2CO_2H$, and showed that haem contained the labelled carbon.

In 1952, Westall isolated porphobilinogen from the urine of humans suffering from acute porphyria. Based on this, Shemin *et al.* (1953) now proposed that 6-aminolaevulic acid (II) can replace 'active' succinate and glycine in porphyrin synthesis. It should be noted that (i) the α-carbon atom of the pyrrole nucleus not attached to the aminomethyl side-chain is the only one in the ring which is derived from the α-carbon atom of glycine; (ii) the methine carbon atom is also derived from the α-carbon atom of glycine.

This pyrrole synthesis is supported by various experiments, *e.g.*, Shemin *et al.* (1954) used [$\delta^{14}C$]δ–aminolaevulic acid as precursor, and showed that half of the radioactivity is equally distributed among the four pyrrole nuclei and the other half is in the methine-bridge carbons. This distribution is in agreement with the equation given. Furthermore, Falk *et al.* (1953) have shown that porphobilinogen is the common precursor in porphyrin synthesis.

The enzyme ALA synthetase, which catalyses the reaction between succinyl-coenzyme A and glycine to give δ-aminolaevulic acid (ALA), was first isolated by Neuberger *et al.* (1958) and later by Kikuchi *et al.* (1958) and by Burnham (1962) in a purer form. Hence, the conversion of succinic: acid (as succinyl-coenzyme A) and glycine into porphobilinogen may be formulated as shown.

Battersby *et al.* (1972) have synthesised porphobilinogen labelled with ^{13}C at the carbon atom of the aminomethyl side-chain ($-^{13}CH_2NH_2$) and used this as a precursor in the biosynthesis of protoporphyrin IX. Examination of the signals in the ^{13}C-NMR spectrum of the porphyrin (as methyl ester) showed that all four meso-carbon atoms (α, β, γ, and δ) were equally labelled with ^{13}C.

The final steps in the pathway are not certain, but there is a large amount of evidence to show that the sequence is probably as follows.

Porphobilinogen is converted mainly into uroporphyrinogen III (this is uroporphyrin with methylene bridges instead of methine and to a much smaller extent into uroporphyrinogen I. Uroporphyrinogen III then undergoes stepwise decarboxylation of the acetic acid residues to methyl groups to give coproporphyrinogen III which, on oxidative decarboxylation of the propionic acid residues in rings A and B (see formula (Ia)) to vinyl groups, thereby forms protoporphyrinogen IX.

This is now oxidised to protoporphyrin IX (*i.e.*. methine bridges are produced), and a ferrous ion is then incorporated to form haem which finally links up with globin to form haemoglobin. In the chart the symbols A, M, P, and V.

Chlorophyll

Introduction : Chlorophyll is the green colouring matter of leaves and green stems, and its presence is essential for photosynthesis. Photosynthesis is the process in which light energy is used by plants to synthesise carbohydrates, proteins and fats. In green plants it is the chlorophyll which absorbs the light energy. The name chlorophyll was given to the green pigment in leaves by Pelletier and Caventou (1818). There the matter rested until 1864, when Stokes showed, from spectroscopic evidence, that chlorophyll was a mixture. This paper apparently did not attract much attention, and it was not until Willstatter entered the field that any progress in the chemistry of chlorophyll was made.

When an ethereal solution of chlorophyll is shaken with methanolic potassium hydroxide solution, various colour changes occur. With chlorophyll-a the green colour immediately changes to yellow; with chlorophyll-h, from green to carmine red; and with a mixture of the two chlorophylls, from green to yellowish brown. Then, after a few moments, the green colour reappears in the lower layer; the ether is colourless. This set of colour reactions is known as the phase test, and chlorophyll preparations that fail to give it are said to be allomerised. It is now known that the phase test involves the cyclopentanone ring in chlorophyll (see formula (XI)); the alkali forms traces of some intermediate which undergoes slow oxidation by oxygen (atmospheric).

When dried leaves are powdered and then digested with ethanol, a 'crystalline' chlorophyll is obtained after concentration of the solvent. If, however, ether or aqueous acetone is used instead of ethanol, then the product is 'amorphous' chlorophyll (Willstatter *et al.*, 1908). The extraction of chlorophyll is also accompanied by the extraction of two other pigments, carotene and xanthophyll. Willstatter *et al.* (1920) then showed that 'crystalline' chlorophyll was produced during the extraction of chlorophyll by means of ethanol, a molecule of phytyl alcohol being replaced by ethanol under the influence of an enzyme, chlorophyllase (which is present in leaves).

Willstatter *et al.* (1911) originally gave chlorophyll the molecular formula $C_{55}H_{72}M_4N_4O_6$, but in 1912 Willstatter *et al.* showed that chlorophyll, obtained from a wide variety of sources, was a mixture of two compounds, chlorophyll-a and chlorophyll-b. The separation was effected by shaking a light petrol solution of chlorophyll with aqueous methanol; chlorophyll-a remains in the light petrol, and chlorophyll-b passes into the aqueous methanol. Chlorophyll-a is a bluish-black solid, giving a green solution in organic solvents; chlorophyll-b is a dark green solid, also giving a green solution in organic solvents. The two components occur in proportions of approximately 3 of a to 1 of b in natural chlorophyll. The chlorophylls (and chlorophyll degradation products) are now separated by chromatography. Column chromatography (alumina, sugar, starch, etc.) may be used on a large scale, and partition, paper and thin-layer chromatography are used for small-scale work, but the partition technique has also been used on a preparative scale.

The molecular formulae that have been assigned to chlorophyll-a and chlorophyll-b are $C_{55}H_{72}N_4O_4Mg$ and $C_{55}H_{70}N_4O_6Mg$, respectively (Willstatter, 1913); the two compounds have different absorption spectra: chlorophyll-a, 380, 418, 428, 510, 580 and 700 nm; chlorophyll-b, 428, 464 and 675 nm.

These characteristic absorption maxima have been used to estimate the amounts of each chlorophyll in a mixture. Infrared spectroscopy has been used to detect functional groups, and Oster *et al.* (1964), from their examination of the infrared spectra of the chlorophylls, have shown that these spectra provide a means of detecting trace amounts of chlorophyll-b in samples of chlorophyll-a and, at the same time, offer a means of estimating the proportions of the two components.

The hydrolysis of both chlorophylls with cold dilute potassium hydroxide solution gives one molecule of phytol, $C_{20}H_{40}O$, one molecule of methanol, and one molecule of chlorophyllide-a (chlorophyllin-a) [I] or chlorophyllide-b (chlorophyllin-b) [II]. Thus the chlorophylls are diesters. When either chlorophyll is heated with an ethanolic solution of hydrated oxalic acid, the magnesium atom is replaced by two hydrogen atoms to produce phytyl phaeophorbide-a (III) or b ((IV); these phytyl phaeophorbides are also known as phaeophytins a and b, and `crystalline' chlorophyll is ethyl chlorophyllide).

NOMENCLATURE OF THE CHLOROPHYLL DEGRADATION PRODUCTS

Porphyrins are substituted porphitls. Phyllins, phyllides and chlorophylls contain magnesium, whereas phorbins, phorbides and phytins are magnesium-free compounds, the magnesium atom having been removed and replaced by two hydrogen atoms. 7,8-Dihydroporphin is the nucleus of the chlorin series of compounds (tricarboxylic derivatives) which are derived from chlorophyll-a; rhodins are the corresponding compounds derived from chlorophyll-b.

The introduction of the extra ring-two methylene groups across the 6,γ-positions–gives rise to the phorbins. The prefix phaeo designates those compounds which have the same substituents that occur in chlorophyll. Chlorin itself is dihydroporphin, and the natural red porphyrin pigments are derivatives of porphin, whereas the green chlorophylls and their derivatives are derivatives of chlorin.

In some cases a subscript is used to indicate the number of oxygen atoms in the molecule, *e.g.*, phaeoporphyrin-a5 contains five oxygen atoms.

Structure of Chlorophyll-a

When phytyl phaeophorbide-a is hydrolysed with boiling methanolic potassium hydroxide (30 seconds), the product is chlorin-e. This is a tricarboxylic acid (*e.g.*, it forms a trimethyl ester), and its molecular formula may thus be written as $C_{31}H_{33}N_4(CO_2H)_3$. Chlorin-e, on oxidation with chromic acid or with Caro's acid, gives haematinic acid (I) and ethylmethylmaleimide (II) [Willstatter *et al.*, 1910]. When chlorin-e is reduced with hydriodic acid in acetic acid, haemopyrrole (III) and phyllopyrrole (IV) are produced (Willstatter *et al.*, 1911).

When phylloporphyrin is reduced under the same conditions, the products are now (III), (IV) and cryptopyrrole (V). From these results it is reasonable to infer that chlorophyll-a contains four pyrrole nuclei, probably having a methyl group in the β-position (see (II)-(V)). It is also reasonable to suppose that at least one pyrrole nucleus contains a propionic acid residue in the β'-position (see (I)). It also appears likely that a vinyl group is present in the molecule (this would account for the presence of an ethyl group on reduction; at the same time, the presence of an ethyl group, as such, is not excluded). Furthermore, the isolation of (I) and (II) on oxidation (giving oxidation at the a-position), and of (III) and (IV) on reduction (the appearance of a methyl group at the α-position), can be interpreted as meaning that the four pyrrole nuclei are joined to each other at their α-positions via one carbon atom.

Thus a possible skeleton structure for chlorin-e could be a cyclic one, (VI); the positions of the various substituent groups cannot be assigned on the evidence obtained so far, *e.g.*, a methyl group at I and a propionic acid residue at 2 would produce the same oxidation product (I) had the positions of the two groups been interchanged in (VI). It is also necessary to fit a second carboxyl group into this structure (VI), since chlorophyll-a forms chlorophyllide-a on hydrolysis (the latter compound contains two carboxyl groups). Furthermore, since chlorophyllide-a, on further hydrolysis, forms chlorin-e, a tricarboxylic acid, some group must be present which can give rise to this third carboxyl group. Such a group could be a lactone; it must be cyclic since no carbon atoms are lost after the hydrolysis. By the further degradation of chlorin-e, *e.g.*, heating in a sealed tube with ethanolic potassium hydroxide, various porphyrins are obtained. Three of these are pyrroporphyrin, rhodoporphyrin and phylloporphyrin.

Pyrroporphyrin $C_{30}H_{33}N_4(CO_2H)$ has an absorption spectrum closely resembling that of mesoporphyrin; this agrees with the tentative skeleton structure (VI) proposed for chlorin-e. Pyrroporphyrin, on bromination followed by oxidation with chromic acid, gives bromocitraconimide (VII) as one of the products (Treibs *et al.*, 1928). It therefore follows that at least one of the pyrrole nuclei in pyrroporphyrin has a free β-position available for bromination. Synthetic work then showed that pyrroporphyrin has structure (VIII) [H. Fischer *et al*., 1929, 1930, 1933]; thus the positions of the four methyl groups and the position of the propionic acid group are now established.

Rhodoporphyrin $C_{30}H_{32}N_4(CO_2H)_2$, on heating with sodium ethoxide, readily loses one carboxyl group to form pyrroporphyrin (VIII). From a detailed study of the haemin series, it was obscrved that a carboxyl group in a side-chain of a pyrrole nucleus was difficult to remove. Hence it is probable that the carboxyl group lost from rhodoporphyrin is attached directly to a pyrrole nucleus.

The only position for this carboxyl group is at 6 (see structure (VIII)); elimination of the carboxyl group from rhodoporphyrin would then give one pyrrole nucleus with a free β-position (6), *i.e.*, pyrroporphyrin.

Furthermore, comparison of the absorption spectra of rhodoporphyrin with compounds of known structure showed that the two carboxyl groups are in positions 6 and 7 (the latter is the propionic acid residue), and this was confirmed by the synthesis of rhodoporphyrin.

Phylloporphyrin $C_{31}H_{35}N_4(CO_2H)$ contains one CH, group more than pyrroporphyrrin, and may be converted into the latter by heating with sodium ethoxide. It therefore follows that the alkyl groups in both compounds occupy similar positions. Synthetic work then showed that phylloporphyrin contains a methyl group attached to the y-methine carbon atom (H. Fischer *et al*., 1930, 1933).

Consideration of the information obtained from the structures of the porphyrins described above shows that the skeleton structure (IX) is present in chlorin-e. Now chlorin-e contains three carboxyl groups and one more carbon atom than the structure shown in (IX).

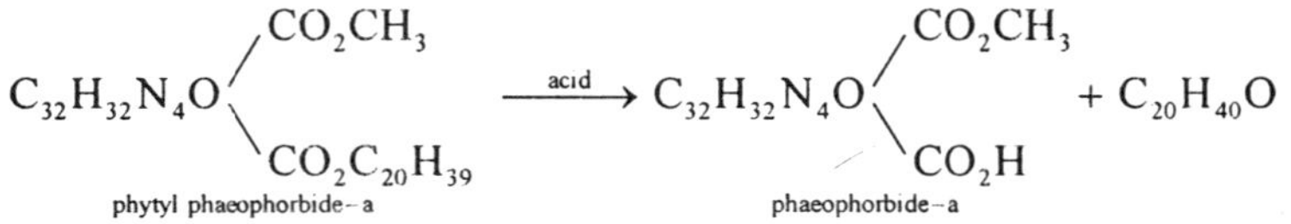

$$C_{32}H_{32}N_4O\begin{cases}CO_2CH_3\\CO_2C_{20}H_{39}\end{cases}\xrightarrow{\text{acid}}C_{32}H_{32}N_4O\begin{cases}CO_2CH_3\\CO_2H\end{cases}+C_{20}H_{40}O$$

phytyl phaeophorbide-a　　　　phaeophorbide-a

Fig. 1

The formation of a methyl group (at the y carbon atom) could be explained by assuming a carboxyl group is attached as shown in structure (X). When phytyl phaeophorbide-a (III) is hydrolysed with acid, the phytyl group is removed to form phaeophorbide-a. When phaeophorbide-a is treated with hydriodic acid in acetic acid and followed by atmospheric oxidation, the product is phaeophorphyrin-a_5.

This, on further treatment with hydriodic acid in acetic acid, forms phylloerythrin, $C_{33}H_{34}N_4O_3$, by loss of the carbomethoxyl group; phylloerythrin has the same absorption spectrum as that of the porphyrins, and so the porphin structure is still present. Now both phaeophorbide-a and phylloerythrin contain a keto group (as is shown by the formation of an oxime, etc.), and so when the carbomethoxyl group is hydrolysed, the elimination of carbon dioxide can be expected if the keto group is in the β-position with respect to the carboxyl group (produced on hydrolysis).

Furthermore, the hydrolysis of phaeophorbide-a with methanolic potassium hydroxide gives chlorin-e. In this reaction, apart from the hydrolysis of the carbomethoxyl group, the keto group is lost and a carboxyl group is introduced without the loss of any carbon atoms.

This may be explained by assuming that this carboxyl group (the third one in chlorin-e) is produced by the fission of a cyclic ketone, and not from a lactone as suggested previously (see above). Thus a possible skeleton structure for phaeophorbide-a is (XI); if the ketone ring is opened, then the formation of (X) can be expected.

Also, the hydrolysis of (XI) would produce a β-keto-acid, which can be expected to lose carbon dioxide readily to form phylloerythrin. Phaeophorbide-a can be reduced catalytically to its dihydro-derivative in which the keto group remains intact. This suggests the presence of a readily reducible double bond. Oxidation experiments on phaeophorbide-a and dihydrophaeophorbide-a showed the presence of one vinyl group in the former.

Furthermore, the existence of a vinyl group in the ester of chlorin-e was shown by the reaction with diazoacetic ester to give a cyclopropane derivative, which was isolated by the oxidation of the addition product (H. Fischer *et al.*, 1935). Thus one of the ethyl groups (see pyrroporphyrin (VIII)) must have been a vinyl group before reduction. Further degradative and synthetic work by H.

Fischer *et al.* (1934-1936) showed that phaeophorbide-a is (XII) and that phytyl phaeophorbide-a is (XIII). The replacement of the two imino hydrogen atoms in (XIII) by a magnesium atom would therefore give chlorophyll-a; this is (XIV). Chlorophyll-b has been assigned structure (XV).

SYNTHESIS OF CHLOROPHYLL

The following is a brief account of the total synthesis of chlorophyll-a by Woodward *et al.* (1960). Their approach to the synthesis was guided by some of the known reactions of chlorophyll, *e.g.*, an unusual property of chlorophyll is that, although a chlorin (7,8dihydroporphyrin), it is difficult to oxidise to the porphyrin.

Since the two groups (methyl and the propionic acid residue) are in the same positions in haemin, it was argued that the substituent on the γ-carbon atom (structure (VI)) must create large steric effects when all three groups lie in one plane. In the chlorin, the groups at 7 and 8 are now trans and so the steric effects are minimised. On the basis of this argument and also for other reasons, porphyrin (I) was chosen as the key intermediate of the synthesis and its conversion to the chlorin left to a later stage.

This chlorin, chlorin-e trimethyl ester (XVII), had previously been converted into chlorophyll-a; hence the Woodward synthesis is a total synthesis. The malondinitrile fragment in ring B was introduced to protect the aldehyde group in acidic media. The condensation of (VII) with (V) gave the required porphyrin (I), but condensation also occurred in the opposite way.

Woodward therefore devised a method to give only (I). The formyl group in (V) was converted into the 5-thioformyl group (VIII). This reacted rapidly with the amino-group in (VII) to form a Schiff base which, on treatment with methanolic hydrogen chloride, gave (I).

The mechanism of the conversion of (X) into ((XI); a purple compound of the group known as purpurins) is uncertain. (X) underwent tautomeric change to give an equilibrium mixture with the cyclic form (XI) predominating. This predominance can be partly explained by the relief of steric strain in the formation of (XI) [see the earlier discussion of the steric effects in (I)].

Biosynthesis of Chlorophyll: The steps outlined in for the biosynthesis of protoporphyrin IX are also believed to constitute the pathway for chlorophyll biosynthesis.

This porphyrin then incorporates magnesium to form magnesium protoporphyrin IX which, by a series of steps (some completely hypothetical so far), is converted into chlorophylls a and b.

RECENT SYNTHESES OF PORPHYRINS

In more recent methods, one approach to porphyrin synthesis has been via tetrapyrrolic intermediates. These compounds are usually split by acids, but were stabilised in this work by the presence of an internuclear oxo group (Kenner *et al.*, 1965); *e.g.* (MeP = $-CH_2CH_2CO_2Me$; Bzl = $PhCH_2-$):

The pyridinium derivative of (I) was condensed with the lithium salt of (II) to give (III). Reduction of (III) gave (IV) which, on cyclisation, gave mesoporphyrin IX dimethyl ester (V). (VI) was ring-closed to the corresponding porphyrin by heating in o-dichlorobenzene.

A number of new chlorophylls have been isolated recently, *e.g.*, chlorophyll-d (vinyl group at position 2 in chlorophyll-a is replaced by CHO); bacteriochlorophyll-a (this is chlorophyll-a with an acetyl group instead of the vinyl group (at 2) and with the 3,4-double bond in ring B reduced in trans configuration). There are also the Chlorobium chlorophylls. These have been classified into two groups, the '650' and '660' series, these numbers being the respective 'red' absorption maxima in ether solution. Both groups differ from chlorophyll-a in a number of ways, *e.g.*, the carbomethoxy group is absent at position 10 (see formula (XI)) and they contain farnesol instead of phytol. Each series consists of six members, each compound differing from its related compound by a CH_2 group.

PREPARATION OF THE PHTHALOCYANINES

Phthalocyanines are a very important class of organic dyes and pigments; they are coloured blue to green. They were discovered by accident at the works of Scottish Dyes Ltd. in 1928. It was there observed that some lots of phthalimide, manufactured by the action of ammonia on molten phthalic anhydride in an iron vessel, were contaminated with a blue pigment. The structure and method of formation of this compound were established by Linstead and his co-workers (1934).

The phthalocyanines form metallic complexes with many metals, and the colour depends on the nature of the metal (copper, magnesium, lead, etc.); greener shades are obtained by direct chlorination or bromination. The metal phthalocyanines are insoluble in water, and are used as pigments. They are made water-soluble by sulphonation, and these soluble salts are used as dyes. Metal phthalocyanines have great thermal stability and sublime (usually without melting) at about 550°C. Decomposition occurs at higher temperatures.

Metal phthalocyanines may be prepared as follows:

(i) By passing ammonia into molten phthalic anhydride or phthalimide in the presence of a metal salt.

(ii) By heating o-cyanobenzamides or phthalonitriles with metals or metallic salts.

(iii) By heating phthalic anhydride or phthalimide with urea and a metallic salt, preferably in the presence of a catalyst such as boric acid.

Phthalocyanine (I) the parent substance of this group, may be prepared by heating phthalonitrile with a little triethanolamine. It can be seen from formula (I) that phthalocyanine contains four isoindole nuclei joined in a ring by means of nitrogen atoms. If we ignore the benzene nuclei, then we have four pyrrole nuclei linked by nitrogen atoms, a structure similar to the porphyrins, in which the pyrrole nuclei are linked by methine groups ((II) is porphin).

Both types of compounds are coloured, and both contain two imino hydrogen atoms which can be replaced to form metal complexes. Because of these similarities the phthalocyanines are often known as the tetra-azaporphyrins. The first commercial phthalocyanine pigment was Monastral Fast Blue BS; this is copper phthalocyanine (III).

Structure of the Phthalocyanines

Analysis showed that the phthalocyanines had an empirical formula $C_{32}H_{16}N_8M$, where M is a bivalent metal, *e.g.*, copper, magnesium, etc. The molecular weight determination of magnesium phthalocyanine by the ebullioscopic method with naphthalene as solvent showed that the empirical formula was also the molecular formula (Linstead *et al.*, 1934). This has been confirmed by means of X-ray measurements (Robertson, Linstead *et al.*, 1935).

Linstead showed that the phthalocyanines can be obtained by reaction between a metal and phthalonitrile (I), o-cyanobenzamide (II), phthalamide (III), but not with, for example, terephthalonitrile (IV), homophthalonitrile (V), or o-xylylene dicyanide (VI). It is therefore reasonable to infer that in the formation of phthalocyanines, the two nitrile groups involved must be in the ortho-position.

Thus there are probably four $C_8H_4N_2$ units, each having an isoindole structure (VII) or a phthalazine structure (VIII). (VIII) was shown to be untenable since no phthalocyanine could be prepared from compounds containing this skeleton.

The oxidation of phthalocyanines with hot nitric acid, cold acid permanganate or ceric sulphate produces phthalimide and ammonium salts, the amount of phthalimide being that which would correspond to the presence of four isoindole units. The problem then is: How are these units joined together? The treatment of magnesium phthalocyanine with sulphuric acid replaces the magnesium atom by two hydrogen atoms.

$$(C_8H_4N_4)_4Mg \xrightarrow{H_2SO_4} (C_8H_4N_2)_4H_2$$

This suggests that in metal phthalocyanines, the metal has replaced two imino hydrogen atoms. A reasonable structure for phthalocyanine is one in which the four isoindole units are joined through nitrogen atoms to form a cyclic structure (IX). On the other hand, an open-chain structure could also be produced by joining four isoindole units through nitrogen atoms (X); in this case the molecular formula would be $(C_8H_4N_2)_4H_4$. It seems unlikely that (X) could be rejected on these grounds alone, since in a large molecule of this type it is difficult to estimate the hydrogen with certainty ((IX) contains approximately 3.5 per cent hydrogen, and (X) 3.9 per cent). (X), however, is unlikely, since phthalocyanine is a very stable substance; the presence of an imino group at the end of the molecule could be expected to render the compound unstable to, *e.g.*, acid reagents. Furthermore, the oxidation of phthalocyanine with ceric sulphate in dilute sulphuric acid proceeds according to the following equation (over 90 per cent of the phthalimide has been isolated).

$$(C_8H_4N_2)_4H_2 + 7H_2O + [O] \rightarrow 4C_8H_5NO_2 + 4NH_3$$

This agrees with (IX), but had the structure been (X) then the molecule would have required two atoms of oxygen.

$$(C_8H_4N_2)_4H_4 + 6H_2O + 2[O] \rightarrow 4C_8H_5NO_2 + 4NH_3$$

Thus (IX) represents best the known properties of phthalocyanine. The two imino hydrogen atoms are replaceable by a bivalent metal, and the remaining two nitrogen atoms form co-ordinate links (see formula (III)).

The most common metals in phthalocyanines are in the bivalent state, *e.g.*, Cu^{2+}, Fe^{2+}, Ni^{2+}, etc. It is also possible for the central metal atom to have a valency of three or four, and in this case one or two anions are also attached to the metal, *e.g.*, bromoaluminium and dichlorostannic phthalocyanines (cf. haem and haemin). Phthalocyanines containing two alkali metal ions, *e.g.*, Na, K (each metal atom joined to an adjacent pair of pyrrole N atoms) have been prepared, but these are readily hydrolysed by dilute acids to the metal-free phthalocyanine. Finally, the metal may have a zero valency. Thus, copper phthalocyanines are known in which the copper is copper (0), copper (I), or copper (II).

In metal phthalocyanines resonance is possible, and so all four nitrogen atoms linked to the metal atom would be equivalent (18n-electrons; 4n + 2; see (I) and (II)). Phthalocyanines (with and without a central metal atom) have been examined by means of X-ray analysis (Robertson, 1936), and the results shows that these compounds are large flat molecules with a centre of symmetry. The bond lengths of the C–N bonds indicate resonance, ds do those of the benzene ring (all the lengths are equal). Robertson also showed that for nickel phthalocyanine, if the radius of the nickel atom be assumed, then the positions of the other atoms in the molecule are exactly those obtained by chemical evidence.

Phthalocyanines are known to exist in at least three polymorphic forms : α, β and γ. The β-form is the most stable one, and is the one produced by sublimation of the phthalocyanine. Kendall (1953) has distinguished between the three forms by means of their infrared spectra, and Kahn *et al.* (1965) have shown the difference in the polymorphic structures between α- and β-cobalt phthalocyanines by means of their ESR spectra. Phthalocyanines can act as catalysts, *e.g.*, they catalyse the combination of hydrogen and oxygen to form water, the decomposition of hydrogen peroxide, the isomerisation of dimethyl maleate to dimethyl fumarate, etc.

Carbohydrates Classification

Sugars, cellulose and starches are collectively called carbohydrates. They form. one of the most important classes of natural products. These are widely distributed m plant kingdom and are of wry great use to both plants and animals. These substances silpport the plant tissue and can be concerted into a number of useftil products such as artificial silk, papery explosives, lacquers and phstics. Moreover the sugars and starches are the sources of energy to animals. The ptants can syntfaesise carbohydrates from CO_2 by the process of photosynthesis.

As the term carbohydrate indicates, these are the hydrates of carbon i.e. in these hydrogen and oxygen are present in the same proportion as in water, so the general formula for carbohydrate may be written as $C_x(H_2O)_y$ e.g. glucose $C_6(H_2O)_6$, cane-sugar $C_{12}(H_2O)_{11}$, etc. But recently some compounds are found to behave like carbohydrate yet they do not correspond to the general formula $C_x(H_2O)_y$ e.g. rhamnose ($C_6H_{12}O_5$), rhamnohaxose ($C_7H_{14}O_6$). Moreover there are certain compounds which conform to the general formula and are not carbohydrate e.g. formaldehyde CH_2O, acetic acid ($C_2H_2O_2$) and lactic acid ($C_2H_6O_3$) etc. Hence this definition is discarded. Therefore the carbohydrates may now be defined as poly-hydroxy aldehydes or ketones or compounds which yield these on hydrolysis and contain at least eae asymmetric C-atom.

Carbohydrate may be classified as :

1. *Monosaccharides*—These are the simplest carbohydrates which cannot be hydrolysed into simple molecules. They are further divided into aldose and ketoses depending upon whether they have an aldehydic or a ketonic group, e.g. glucose $CH_2OH(CHOH)_4CHO$, fructose $CH_2OH(CHOH)_2COCH_2OH$ respectively.

2. *Oligosaccharides* can be sub-divided into following classes.

   (a) *Disaccharides*—These give two molecules of monosaccharides on acid hydrolysis e.g. sucrose, maltose ($C_{12}H_{22}O_{11}$) etc.

   (b) *Trisaccharides*— These yield, on hydrolysis, three molecules of monosaccharides e.g. raffinose, $C_{18}H_{22}O_{16}$.

   (c) *Tetrasaccharides*—They yield four molecules of monosaccharides on hydrolysis e.g. stachyose, $C_{24}H_{42}O_{21}$.

3 *Polysaccharides*—These are the carbohydrates which on hydrolysis give a number of molecules of monosaccharides e.g. cellulose, and starch, $(C_6H_{10}O_5)_n$

DISACCHARIDES

The methanol forms methyl glycosides with the monosaccharides, so can other hydroxy-compounds form similar unions with the monosaccharides. Since the latter are themselves hydroxy-compounds, it is possible that they can link up with themselves to form acetals, *i.e.*, glycosides in which the aglycon is another sugar molecule. A number of such compounds occur in nature, *e.g.*, sucrose, maltose and lactose. All the disaccharides are crystalline solids, soluble in water, and fall into two classes, the *reducing* sugars and the *non-reducing* sugars.

Sucrose is *not* a reducing sugar, *e.g.*, it will not reduce Fehling's solution; it does not form an oxime or an osazone, and does not undergo mutarotation. This indicates that neither the aldehyde group of glucose nor the ketonic group of fructose is free in sucrose. Thus a tentative structure of sucrose is one in which two molecules, glucose and fructose, are linked by the aldehyde group of the former and ketonic group of the latter. This has been amply confirmed by further work, and sucrose has been shown to be α-D-glucopyranosyl-β-D-fructofuranoside (*i.e.*, α-glucose is linked to β-fructose).

It should be noted that the fructose molecule in sucrose exists as the furanose form, and that when sucrose is hydrolysed, it is the pyrapose form of fructose which is isolated.

Maltose (*malt sugar*), C_{12},$H_{22}O_{11}$, is produced by the action of malt (which contains the enzyme diastase) on starch:

$$(C_6H_{10}O_6)_8 \xrightarrow{\text{diastase}} \frac{n}{2}C_{12}H_{22}O_{11}$$

Maltose is a white crystalline solid, m.p. 160-165°C, soluble in water, and is dextrorotatory.

When it is hydrolysed with dilute acids or by the enzyme *maltose*, maltose yields two molecules of D(+)-glucose. Maltose is a reducing sugar, *e.g.*, it reduces Fehling's solution; it forms an oxime and an osazone, and undergoes mutarotation. This indicates that at least one aldehyde group (of the two glucose molecules) is free in maltose. Further work has shown that maltose is 4-O-α-D-glucopyranosyl-D-glucopyranose (*i.e.*, the reducing half is linked to the non-reducing half by an to-link). It should also be noted that the enzyme maltase splits only α-glycosidic links, and hence is used to ascertain the presence of this bond in disaccharides and polysaccharides.

Cellobiose, $C_{12}H_{22}O_{11}$, may be obtained from cellulose by acetylating good filter paper (which is almost pure cellulose) with acetic aphydride in the presence of concentrated sulphuric acid. The octa-acetate of cellobiose so obtained is hydrolysed with potassium hydroxide or with sodium ethoxide, whereupon cellobiose is produced. Cellobiose is a white crystalline solid, m.p. 225°C, soluble in water, and dextrorotatory.

When hydrolysed with dilute acids or by the enzyme *emulsin*, it yields two molecules of D(+)-glucose. It is a reducing sugar, forms an oxime and osazone, and undergoes mutarotation. Cellobiose is 4-O-β-D-glucopyranosyl-D-glucopyranose. Thus, the only difference between cellobiose and maltose is that in the former the glycosidic link is β, whereas in the latter it is α. Furthermore, the enzyme emulsin splits only β-glycosidic links (cf. maltase).

Lactose (*milk-sugar*), $C_{12}H_{2},O$, occurs in the milk of all animals, and is prepared commercially from whey by evaporation to crystallisation; whey is obtained as a by-product in the manufacture of cheese.

Lactose is a white crystalline solid, m.p. 203°C (with decomposition), soluble in water, and is dextrorotatory. It is hydrolysed by dilute acids or by the enzyme lactose, to an equimolecular mixture of D(+)-glucose and D(+)-galactose. Lactose is a reducing sugar, forms an oxime and osazone and undergoes mutarotation. It is 4-O-β-D-galactopyranosyl-D-glucopyranos It should also be noted that lactase is a P-glycosidase. *i.e.*, splits 8-glycosides (it has been shown to be identical with emulsin).

Sucrose maltose and lactose are three disaccharides which occur naturally. Cellobiose may be prepared from cellulose. Two other disaccharides which have been prepared are **melibiose** (from the trisaccharide *raffinose*) and **gentiobiose** (from the trisaccharide *gentianose*). These differ from the other disaccharides in that the two monosaccharide molecules are linked by the sixth carbon atom (the aldehyde carbon atom being number one) of the reducing mono-saccharide.

Sucrose, cane-sugar, $C_{12}H_{22}O_{11}$, is one of the most important compounds commercially, and Is obtained from the sugar-cane and sugar-beet. In addition to crystalline sucrose, a syrup is always obtained which will not crystallise. This group, known as *molasses*, is also a commercial product.

Sucrose is a white crystalline solid, m.p. 180°C, soluble in water. When heated above its melting point, it forms a brown substance known as caramel. Concentrated sulphuric acid chars sucrose, the product being almost pure carbon. Sucrose is dextrorotatory, its specific rotation being +66-5°. On hydrolysis with dilute acids sucrose yields an equimolecular mixture of D(+)-glucose and D(–)-fructose:

$$C_{12}H_{22}O_{11} + H_2O \xrightarrow{HCl} \underset{\text{glucose}}{C_6H_{12}O_6} + \underset{\text{fructose}}{C_6H_{12}O_6}$$

Since D(–)-fructose has a greater specific rotation than D(+)-glucose, the resulting mixture is laevorotatory. Because of this, the hydrolysis of cane-sugar is known as *the inversion of cane-sugar* (this is not to be confused with the *Walden inversion*), and the mixture is known as *invert sugar*. The inversion (*i.e.*, hydrolysis) of cane-sugar may also be effected by the enzyme *invertase* which is found in yeast. Controlled oxidation of sucrose in alkaline solution with air gives D-arabonic acid. Oxidation of sucrose with nitric acid under different conditions gives either oxalic acid (80%), tartaric acid (40%), or glucaric acid (30%). Hydrogenation of sucrose under controlled conditions gives a mixture of mannitol and sorbitol (these may be separated by fractional crystallisation).

POLYSACCHARIDES

The molecular weight of cellulose varies between 20000 and 500000, and the compound consists of an unbranched chain. The polysaccharides are high polymers of the monomeric sugars and have molecular weights that may range from a lew thousand to several millions.

Alginic acids occur in marine algae, and are polymannuronic acids (linked 1,4).

Cellulose, $(C_6H_{10}O_5)_n$, is the main constituent of the cell-wall of plants, and also occurs in certain animal tissues. It is the most widely distributed organic compound, and its main source is cotton (almost pure cellulose) and wood (which also contains *lignin*, which is not α-polysaccharide).

Cellulose is a white solid, insoluble in water but soluble in ammoniacal copper hydroxide solution (*Schweitzer's reagent*). Careful hydrolysis of cellulose gives cellobiose; it is also possible to isolate *cellotriose* (trisaccharide) and *cellotetrose* (tetrasaccharide). All of these saccharides, on further hydrolysis, yield only D(+)-glucose which exists in the β-form in cellulose (c. starch).

Artificial silk. The term rayon is used collectively to cover all synthetic or manufactured fibres from cellulose, but it is usually most often applied to viscose yarns.

Cellulose nitrates (*nitrocellulose*). 'The highest nitrate of cellulose theoretically possible is the trinitrate, $(C_6H_7N_3O_{11})_n$, which contains 14.4 percent nitrogen (each glucose unit contains three hydroxyl groups). The nitrates are prepared by the reaction of cellulose with a mixture of nitric and sulphuric acids, and the degree of 'nitration' depends on the concentrations of the acids and the time of the reaction. Cellulose trinitrate (12.2-13.2% N) is known as *gun-cotton* and is used in the manufacture of blasting explosives and smokeless powders.

The lower nitrates, in the solid state, are known as *pyroxylin*, and their use depends on the nitrogen content, *e.g.*, (i) pyroxylin (10.8-11.0%N) mixed with camphor as plasticiser is converted into *celluloid*; (ii) *nitrocellulose lacquers* are made by dissolving pyroxylin (11.5-12.2%N), resins, pigments and plasticisers in a mixture of solvents.

Cellulose acetate (*celanese silk*). When acetylated with acetic anhydride in the presence of sulphuric acid, cellulose is converted into cellulose triacetate.

When the reaction is complete, water is added to decompose the triacetate into the diacetate (approximately). The diacetate is washed, dried and dissolved in a mixture of organic solvents (of which acetone is generally the main constituent).

The solution is then forced through a spinneret into a warm chamber, whereupon the solvent evaporates leaving behind fine threads of cellulose acetate.

Cellulose acetate is also used for making non-inflammable photographic and motion-picture films non-shatterable glass, lacquers and varnishes.

Esters other than the acetate are also used, *e.g.*, cellulose acetate butyrate (prepared by the esterification of cellulose with a mixture of acetic anhydride, acetic acid, and butyric anhydride) mixed with plasticisers forms a thermoplastic which is used for many purposes, *e.g.*, film and sheeting.

Viscose rayon. In the viscose process, cellulose is digested with sodium hydroxide solution, and there carbon disulphide is passed into the solution. A mixture of sodium cellulose xanthates, soluble if sodium hydroxide, is formed: R = cellulose:

$$R\text{—}OH + CS_2 + NaOH \longrightarrow S{=}C\begin{matrix} \diagup OR \\ \diagdown SNa \end{matrix} + H_2O$$

This alkaline solution has a high viscosity, and hence the silk obtained by this process was name viscose rayon. The viscose solution is forced through a spinneret into a sulphuric acid bath, whereupon cellulose is precipitated as fine threads.

Cellophane is made by extruding a viscose solution through a long narrow slit into an acid bath where upon cellulose is precipitated as very thin sheets. These sheets are made moisture-proof be coating with a transparent nitrocellulose lacquer. Some cellulose ethers, particularly the ethyl ether, are used in the manufacture of plastics.

When agitated with about 20 per cent aqueous sodium hydroxide, cellulose swells; apparent) cellulose combines with the sodium hydroxide to form a sodio-cellulose. This is unstable, and is readily decomposed into cellulose by the addition of water, but the cellulose has a number of it physical properties changed, *e.g.*, it absorbs dyes more readily than untreated cellulose. This 'reactive' form of cellulose is known as *mercerised cellulose.*

Cellulose may be oxidised by nitrogen dioxide to give a product which retains its original form a much of the original tensile strength. It appears that about half of the CH_2OH groups (alternately), oxidised to carboxyl, thereby giving a compound readily soluble in sodium hydroxide. This oxidic cellulose is useful in medicine *e.g.*, it possesses haemostatic properties, and is useful as sterile gauze

Inulin occurs in many plants, *e.g.*. in the roots of the dandelion, in the tubers of the dahlia and in certain lichens. It is a white powder, insoluble in cold water, and in hot water forms a colloidal solution which does not form a gel on cooling. Inulin solutions do not give any colour with iodine. Inulin is hydrolysed by dilute acids to D(--)-fructose. Thus insulin is $(C_6H_{10},O_5)_n$. and; has a value of about 30 (obtained by chemical methods).

Starch, $(C_6H_{10}C_5)_8$, occurs in all green plants; the commercial sources of starch are maize wheat, barley, rice, potatoes and sorghum.

Starch consists of two fractions, one being known as α-**amylose** (the 'A' fraction), and the other as β-**amylose** or **amylopectin** (the 'B' fraction); the former comprises 10-20 per cent of starch, and the latter 80-90 per cent. a-Amylose is insoluble in water, and the solution gives a blue colour with iodine.

This blue colour is believed to be due to the formation of an inclusion complex. An aqueous solution of α-amylose slowly forms a precipitate, since σ-amylose has a strong tendency to 'revert' to the insoluble state in solution. Amylopectin is soluble in water, is stable in contact with water, and gives a violet colour with iodine. α-Amylose and amylopectin are both hydrolysed to maltose by the enzyme diastase, and to D(+)-glucose by dilute acids (amylopectin gives about 50 per cent of maltose).

Thus, both contain D-glucopyranose units joined by the α-glucosidic linkage. α-Amylose consists of an unbranched chain, with a molecular weight varying between 10000 (n = ~60) and 1000000 (n = ~ 6000). The value of n depends-on the source and treatment of α-amylose. Amylopectin differs from α-amylose in that it contains *branched chains*, the branching occurring through 1,6-linkages (and other linkages) and the length of the unbranched sections being about 24-30 glucose units. The molecular weights recorded for amylopectin vary between 50000 and 10000000.

The dextrins, $(C_6H_{10}O_5)_n$, are produced by the partial hydrolysis of starch by boiling with water under pressure at about 250°. They are white powders, and are used for making adhesives and confectionery, for sizing paper, etc.

Glycogen, $(C_6H_{10}O_5)_n$, is found in nearly all animal cells, occurring mainly in the liver; it is the reserve carbohydrate of animals, and so is often known as '*animal starch*'. It has also been isolated from plant sources. Glycogen is a white powder, soluble in water, the solution giving a purplish-red colour with iodine. On hydrolysis with dilute acid, glycogen gives D(+)-glucose. The molecular weight of glycogen has been given as 1000000 to 5000000, and glycogen contains highly branched chains.

Pectins are found in plant and fruit juices. Their characteristic property is the ability of their solutions to gelate, *i.e.*, form jellies. They have a high molecular weight, and are polygalacturonic acids (linked 1,4) with the carboxyl groups partially esterified with methanol.

KINDS OF CARBOHYDRATES

Carbohydrates are divided into two main classes, *sugars* and *polysaccharides* (*polysaccharoses*). Sugars are crystalline substances with a sweet taste and soluble in water. Polysaccharides are more complex than the sugars, their molecular weights being far greater. Most of them are non-crystalline substances which are not sweet, and are insoluble or less soluble in water, than the sugars.

The names of the simpler carbohydrates end in *-ose*, carbohydrates with an aldehydic structure are known as *aldoses*, and those with ketonic, *ketoses*. The number of carbon atoms in the molecule is indicated by a Greek prefix, *e.g.*, a *tetrose* contains four carbon atoms, a *pentose* five, a *hexose* six, etc.

Sugars are subdivided into two major groups:

1. *Monosaccharide* (*monofacchuroses*). These are sugars which cannot be hydrolysed into smaller molecules. Their general formula is $C_nH_{2n}O_n$, (there are exceptions), where n is 2—10 (but see later), and the most important are the pentoses and hexoses.

2. *Oligosaccharides.* These yield 2-10 monosaccharide molecules on hydrolysis, *e.g.*, *disaccharides* (two monosaccharide molecules which may, or may not, be the same); *trisaccharides* (three monosaccharide molecules): etc.

Polysaccharides (*polysaccharoses*) are carbohydrates which yield a large number of mono-saccharide molecules on hydrolysis. The most widely spread polysaccharides have the general formula $(C_6H_{10}O_5)_n$, *e.g.*, starch, cellulose, etc., a group of polysaccharides which are not so widely spread in nature is the *pentosans*, $(C_5H_8O_4)_n$.

MONOSACCHARIDES

The simplest monosaccharide is glycolaldehyde, $HOCH_2CHO$. This does not contain an asymmetric carbon atom and is therefore not optically active. Since all naturally occurring sugars are optically active, it is more satisfactory to exclude glycolaldehyde from the group of sugars, and to define sugars as optically active polyhydroxy-aldehydes or ketones.

Triose, $C_3H_6O_3$, *Glyceraldehyde,* $HOCH_2CHOHCHO$, is an aldotriose. It contains one asymmetric carbon atom, and can therefore exist in two optically active forms, D- and L-; both are known. Glyceraldehyde has been chosen as the standard configuration in sugar chemistry.

Dihydroxyacetone, $HOCH_2COCH_2OH$, is not optically active, and hence, by the definition given above, is not a sugar. If the definition be rejected, *i.e.*, the proviso that a sugar is always optically active is rejected, then dihydroxyacetone will be a ketotriose.

Tetroses, $C_4H_8O_4$. The structure of an aldotetrose is $HOCH_2CHOHCHOHCHO$. There is only one structure possible for a ketotetrose, viz., $HOCH_2COCHOHCH_2OH$. This contains one asymmetric carbon atom, and corresponds to D- and L-erythrulose.

Pentoses, $C_5H_{10}O_5$. The *aldopentoses* are an important group of sugars (monosaccharides), and their structure has been elucidated by methods similar to those used for glucose and fructose. Their structure is:

$$OCHCHOHCHOHCHOHCH_2OH.$$

This contains three structurally different asymmetric carbon atoms, and can therefore exist in eight optically active forms. All are known, and correspond to the D- and L-forms of *arabinose*, xylose, ribose and lyxose

L(+)-*Arabinose*, m.p. 158°C, occurs naturally as pentosans (arabans) in various gums, *e.g.*, gum arabic. It is usually obtained by the hydrolysis of cherry-gum with dilute sulphuric acid.

D(+)-*Arabinose* occurs in certain glucosides.

O(+)-*Xylose* (wood-sugar), m.p. 145°C, occurs as entosans (xylans) in wood gums, bran and straw, from which it may be obtained by hydrolysis with dilute sulphuric acid.

D(+)-*Ribose*, m.p. 95 C, occurs in plant nucleic acids, and in liver and pancreas nucleic acids.

All the other aldopentoses are synthetic compounds.

The chemical properties of the aldopentoses are similar to those of the aldohexoses (see glucose, below); pentoses do not undergo fermentation. All the aldopentoses are converted quantitatively into *furfural* when warmed with dilute acid: A very important aldopentose is 2-deoxyribose, $HOCH_2$ $CHOHCHOHCH_2CHO$ (the prefix 'deoxy' indicates the replacement of a hydroxyl group by hydrogen;). It occurs in nucleic acids.

Ketopentoses with the structure $HOCH_2COCHOHCHOHCH_2OH$ have been prepared. The molecule contains two structurally different asymmetric carbon atoms, and can therefore exist in four optically active forms.

All are known, and correspond to D- and L-forms of ribulose and xylulose.

Hexoses, $C_6H_{12}O_6$. The *aldohexoses* are another important group of monosaccharides, and their structure has been elucidated as follows:

(i) Analysis and molecular-weight determinations show that the molecular formula of the aldohexoses is $C_6H_{12}O_6$.

(ii) When treated with acetic anhydride, aldohexoses form the penta-acetate. This indicates the presence of five hydroxyl groups.

(iii) Aldohexoses form an oxime when treated with hydroxylamine, and therefore contain a carbonyl group.

(iv) When an aldohexose is oxidised with bromine-water, a pentahydroxy-acid of formula $C_6H_{12}O_7$ is obtained. This indicates that the carbonyl group is present in an *aldehydic* group.

(v) When reduced with concentrated hydriodic acid and red phosphorus at 100°C, aldohexoses give a mixture of 2-iodohexane and n-hexane. This indicates that the six carbon atoms in an aldohexose are in a straight chain.

This conclusion is supported by ascending the series by the Kiliani reaction and reducing the product, a polyhydroxy acid, with hydriodic acid; n-heptanoic acid is produced.

The foregoing reactions show that the structure of the aldohexoses is:

$$OCHCHOHCHOHCHOHCHOHCH_2OH.$$

This contains four structurally different asymmetric carbon atoms, and can therefore exist in sixteen optically active forms. All are known, and correspond to the D- and L-forms of glucose, *mannose*, *galactose*, *allose*, *altrose*, *gulose*, *idose* and *talose*.

Although the foregoing evidence indicates an open-chain structure, there is further evidence which shows that the sugars actually exist as *six-membered rings* in the *solid* state.

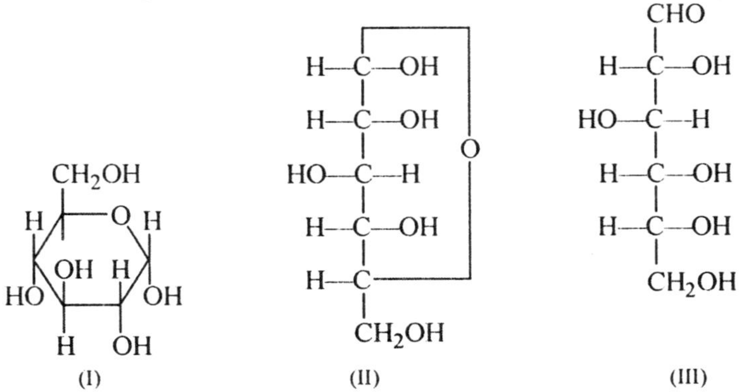

Fig. 2

The ring structure is hexagonal (I), but it is convenient to use the planar formula (II), *e.g.*, α-D-glucose, (II), corresponds to the open-chain formula (III), and it should be noted that the way of drawing the conventional planar formula of an aldose is with the carbon chain vertical and the aldehyde group at the top.

In the D-series the hydroxyl group is always to the right on the bottom asymmetric carbon atom. The L-series is then the enantiomers of the D-series.

D (+)-*Glucose, dextrose* (*grape-sugar*), is found in ripe grapes, honey and most sweet fruits; it is also a normal constituent of blood, and occurs in the urine of diabetics. Commercially, pure D(+)-glucose is manufactured by heating starch with dilute hydrochloric acid under pressure:

$$(C_6H_{10}O_5)_n + nH_2O \xrightarrow{HCl} nC_6H_{12}O_6$$

D(+)-Glucose, m.p. 146°C, has a sweet taste, but is not as sweet is cane-sugar. Naturally-occurring glucose is dextrorotatory (hence name *dextrose*). It is a strong reducing agent, reducing both Fehling's solution and ammoniacal silver nitrate. When heated with sodium hydroxide, an aqueous solution of glucose turns brown. Glucose forms a cyanohydrin with hydrogen cyanide (also the Kiliani reaction), and an oxime with hydroxylamine. All of these reactions are typical of aldehydes, but heating with phenyl-hydrazine in acetic acid does not give the phenylhydrazone; an *osazone* is formed.

OXIDATION OF SUGARS

The nature of the product formed by the oxidation of aldoses depends on the oxidising agent used. Mild oxidising reagents such as bromine-water or an alkaline solution of iodine oxidise only the aldehyde group to give *aldonic acids*. Thus, glucose give: *gluconic acid*, m.p. 131°C.

$$HOCH_2(CHOH)_4CHO + [O] \xrightarrow[(50\%)]{Br_2/H_2O} HOCH_2(CHOH)_4CO_2H$$

This type of oxidation may also be carried out with mercuric oxide in the presence of calcium carbonate (calcium salt is obtained). Aldonic acids (also as calcium salt), however, are best prepared by the electrolysis of a solution of the aldose in the presence of calcium bromide and calcium carbonate. Gluconic acid is manufactured by the oxidation of glucose by the enzyme glucose oxidase, which is present in *Aspergillus niger*, *Penicillium chrysogenum*, etc.

The method of naming aldonic acids is to add prefixes to ionic to indicate the parent aldose. *e.g.*, galactonic and arabonic acids. Oxidation of aldoses with strong oxidising reagents, *e.g.*, nitric acid (and under suitable conditions), produces α, ω-dicarboxylic acids known as saccharic or aldaric acids, *e.g.*, glucose gives glucosaccharic (still often called saccharic acid) or glucaric acid (some oxalic acid is also obtained):

$$HOCH_2(CHOH)_4CHO \xrightarrow[HNO_3]{[O]} HO_2C(CHOH)_4CO_2H \ (40\text{–}46\%)$$

The general method of naming these acids is to indicate the parent aldose by a prefix, *e.g.*, mannosaccharic or mannaric acid, ribosaccharic or ribaric acid. Aldaric acids readily form monolactones, but in some cases the dilactone may be produced

Uronic acids are produced by oxidation of only the terminal —CH_2OH group in an aldose, *e.g.*, glucose forms glucuronic acid:

$$HOCH_2(CHOH)_4CHO \xrightarrow{[O]} HO_2C(CHOH)_4CHO$$

```
CHOH                      CHO\
 |                         |   CMe2
CHOH                      CHO/
 |                         |
CHOH      Me2CO           CHOH        AcOH
 |       -------->         |        -------->
CHOH      HCl             CH
 |                         |
CH                        CHO\
 |                         |   CMe2
CH2OH                     CH2O/

CHO\                                   CHO
 |   CMe2                               |
CHO/                                   CHOH
 |                                      |
CHOH          (i) O2/Pt-C              CHOH
 |         ------------------->         |
CH            (ii) HCl                 CHOH
 |                                      |
CHOH                                   CHOH
 |                                      |
CH2OH                                  CH2H
```

Fig. 3

This oxidation may be carried out in several ways. A general method is to protect the aldehyde group, usually as the isopropylidene derivative, which is then oxidised with oxygen in the presence of Pt-C as catalyst, *e.g.*, glucuronic acid from glucose (cyclic structures are used). (III) is 1,2:5,6-di-isopropylideneglucofuranose and (IV) is 1,2-isopropylidene-glucofuranose.

Acetone usually condenses with *cis*-5-hydroxyl groups on adjacent carbon atoms and hence, to form the di-isopropylidene derivative, the ring changes size (the OH groups on C-3 and C-4 are *trans*, see also mutarotation. Aldoses (and ketoses) are oxidised by periodic acid with complete breakdown of the carbon chain.

REDUCTION OF SUGARS

As with oxidation, the product of reduction depends on the nature of the reagent.

Sodium amalgam, sodium borohydride, high-pressure catalytic hydrogenation (Ni—H_2), or electrolytic reduction in acid solution all convert aldoses into the corresponding alcohols, *e.g.*, glucose is converted into *sorbitol*.

$$HOCH_2(CHOH)_4CHO \xrightarrow{e,H^+} HOCH_2(CHOH)_4CH_2OH$$
(50%)

Reduction of glucose with concentrated hydriodic acid and red phosphorus at 100°C produces 2-iodohexane; prolonged heating finally gives n-hexane.

Sugar derivatives. The sugars form a variety of derivatives, *e.g.*, ethers, esters, isopropylidene compounds (see above), etc. Because of the easy removal of the iso-propylidene group (dilute acid), these derivatives are very useful for protecting alcoholic groups (they are stable to alkali).

Aldohexoses, when heated with concentrated hydrochloric acid, first give hydroxymethyl-furfural, which then decomposes into laevulic acid and other products.

Glucose is fermented by yeast to ethanol and forms glucosates with various metallic hydroxides, *e.g.*, with calcium hydroxide it forms calcium glucosate, $C_6H_{12}O_6.CaO$; the structure of these glucosates is uncertain. Aldohexoses (and aldopentoses) have a hydroxyl group at C-1 in the ring structure. This does not behave as an alcoholic group, *e.g.*, glucose combines with monohydric alcohols, in the presence of hydrochloric acid, to form glucosides.

D(+)-*Mannose*, m.p. 132°C, may be obtained by carefully oxidising with nitric acid:

$$HOCH_2(CHOH)_4CH_2OH + [O] \quad HOCH_2(CHOH)_4CHO + H_2O$$

Commercially, mannose is prepared by hydrolysing the polysaccharide seminine with boiling dilute sulphuric acid. Seminine (a *mannan*) occurs in many plants, particularly in the shell of the ivory nut which is used as the starting material for D(+)-mannose.

D(+)-*Galactose* (m.p. of monohydrate, 118°C) is found in several polysaccharides (galactans), and is combined with glucose in the disaccharide *lactose*. Galactose may be prepared by hydrolysing lactose and separating it from glucose by fractional crystallisation from water, in which glucose is more soluble than galactose.

Rhanmoase, $CH_3(CHOH)_4CHO$, is a 6-deoxyhexose; it occurs in several glycosides.

Ketohexoses. The only important ketohexose is D(–)-fructose, m.p. 102°C, the structure of which has been elucidated as follows:

(i) Analysis and molecular-weight determinations show that the molecular formula of fructose is $C_6H_2O_6$...

(ii) When treated with acetic anhydride, fructose forms the penta-acetate. This indicate the presence of five hydroxyl groups.

(iii) Fructose forms an oxime when treated with hydroxylamine, and therefore contains a carbonyl group.

(iv) When oxidised with nitric acid, fructose is converted into a mixture of trihydroxyglutaric, tartaric and glycollic acids. Since a mixture of acids each containing fewer carbon atoms than fructose is obtained, the carbonyl group in fructose must be present in a *ketonic* group.

(v) Fructose may be reduced to a hexahydric alcohol, *sorbitol*, which, on reduction with hydriodic acid and red phosphorus at 100°C, gives a mixture of 2-iodohexane and n-hexane. The formation of the latter two compounds indicates that the six carbon atoms in fructose are in a straight chain.

(vi) On ascending the series by the *Kiliani reaction*, and reducing the product with hydriodic acid, 2-methylhexanoic acid is obtained. This shows that the ketonic group in fructose is adjacent to one of the terminal carbon atoms.

The foregoing reactions show that the structure of fructose is:

$$HOCH_2CHOHCHOHCHOHCOCH_2OH$$

This contains three structurally different asymmetric carbon atoms, and can therefore exist in eight optically active forms. Of these the following six are known: D(–)- and L(+)-*fructose*, D(+)- and L(–)-*sorbose*, D(+)-*tagatose* and L(–)-*psicose*.

Naturally-occurring fructose (*fruit-sugar*) is laevorotatory, and is therefore also known as laevulose (cf. dextrose). D(–)-Fructose is found in fruits and honey, and occurs combined with glucose in the disaccharide *cane-sugar*, from which it may be obtained by hydrolysis and fractional crystallisation. Fructose is prepared commercially by hydrolysis of *inulin*, a polysaccharide which occurs in dahlia tubers and Jerusalem artichokes:

$$(C_6H_{10}O_5)_n + nH_2O \xrightarrow{HCl} nC_6H_{12}O_6$$

The chemical reactions of fructose are similar to those of glucose, any differences being due to the fact that the former is a ketone and the latter an aldehyde. Thus, the most noticeable difference is the behaviour towards oxidising agents; fructose is not oxidised by bromine-water or an alkaline solution of iodine, and oxidation with nitric acid gives a mixture of acids, each containing less than six carbon atoms.

Cyclic Structures of Monosaccharides

When a monosaccharide is dissolved in water, the optical rotatory power of the solution gradually changes until it reaches a constant value, *e.g.*, a freshly prepared solution of glucose has a specific rotation of +111°; when this solution is allowed to stand, the specific rotation falls to + 52.5°, and remains constant at this value. The final stage can breached more rapidly either by heating the solution or by adding some catalyst which may be an acid or a base. This change in value of the specific rotation is known as *mutarotation*. All *reducing* sugars (except some ketoses) undergo mutarotation. To account for mutarotation, Tollens (1883) suggested an oxide ring structure for D(4-)-glucose, whereby *two* forms would be produced since, in the formation of the ring, another asymmetric carbon atom (which can exist in *two* configurations) is produced (*cf.* the Kiliani reaction). Tollens assumed that a five-membered ring (the furanose form) was produced, (I) and (II).

The difficulty of this suggestion was that there was no experimental evidence for the existence of these two forms. Tanret (1895), however, isolated two isomeric forms of D(+)-glucose, thus apparently verifying Tollens' supposition (see later). The two forms are called α- and β-D (-l-)-glucofuranose; (I) is the α-form, and (II) the β-.

(I)				(II)
H—C—OH ┐		CHO		HO—C—H ┐
H—C—OH O		H—C—OH		H—C—OH O
HO—C—H │	⇌	HO—C—H	⇌	HO—C—H │
H—C ———┘		H—C—OH		H—C ———┘
H—C—OH		H—C—OH		H—C—OH
CH_2OH		CH_2OH		CH_2OH

Ring formation of a sugar is really hemiacetal formation, one alcohol group of the sugar forming a hemiacetal with the aldehyde group of the *same* molecule, thus producing a ring structure. This equilibrium between the open and ring forms is an example of ring-chair tautomerism. The carbon atoms in the chain tend to form a loop, and this results in the OH on C_4, or C_5 being very close to the CHO group.

α-D(+)-glucopyranose				α-D(+)-glucopyranose
H—C—OH ┐		CHO		HO—C—H ┐
H—C—OH O		H—C—OH		H—C—OH O
HO—C—H │	⇌	HO—C—H	⇌	HO—C—H │
H—C—OH │		H—C—OH		H—C │
H—C ———┘		H—C—OH		H—C ———┘
CH_2OH		CH_2OH		CH_2OH

Fig. 4

With ketoses, it is the OH group on C_5 or C_6 that is close to the CO group. In both aldoses and ketoses, therefore, the geometry of the molecule is very favourable for hemi-acetal formation.

Later work by Haworth, Hirst and their co-workers (1926 onwards) has shown that glucose (and other sugars) exists, not as a five-membered ring but as a six-membered ring, the two forms being α- and β-D(+)-glucopyranose. Such pairs ofstereoisomeric ringlforms of any sugar are known as *anomers*.

The widely accepted view is that monosaccharides in solution exist mainly as an equilibrium mixture of the α- and β-anomeric pyranoses, a small amount of the open-chain form together with small amounts of the α- and β-anomeric furanoses. The presence of the furanose forms has been inferred from the fact that monosaccharides undergo some reactions which lead to the formation of furanose derivatives (isopropylidene derivatives).

The mechanism of mutarotation has been the subject of a great deal of research, but it is still uncertain. A widely accepted mechanism involves a concerted attack by base and acid (water is an amphiprotic solvent) to produce the *open-chain structure*, and then this recloses to the anomeric form:

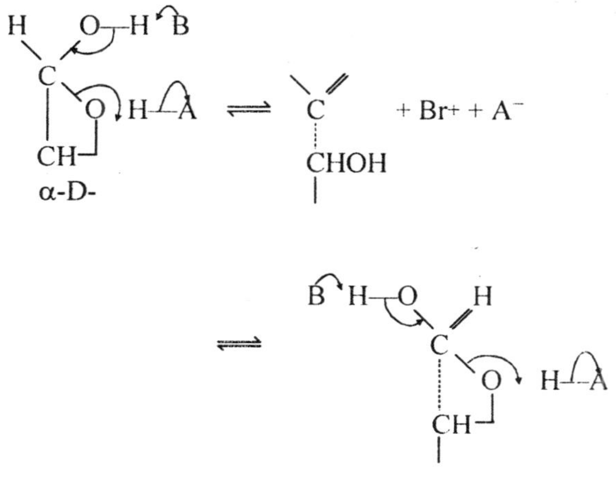

H O–H B
C O H–A ⇌ C + Br+ + A^-
CH– CHOH
α-D-

⇌ B H–O H
C
O H–A
CH–

Fig. 5

The ordinary form of D(+)-glucose is the α-anomer. m.p. 146°C, specific rotation +111°, and may be prepared by crystallisation of glucose from cold aqueous solution. The β-anomer, m.p. 148-150°C, specific rotation + 19.2°, may be obtained by crystallising glucose from hot saturated aqueous solution. Both anomers show mutarotation, the final value of the specific rotation being 52.5°. This corresponds to about 36 per cent of the α-anomer and 64 per cent of the β. It is therefore assumed that whenever a sugar is formed in solution, it immediately changes into a mixture of the two pyranose anomers. the open chain isomer (and furanoses) being present in extremely mall amount. The cyclic structure of the sugars accounts for the following facts: (i) the existence of two isomers, *e.g.*, α- and β-glucose; (ii) mutarotation; (iii) glucose and other aldoses do not give certain characteristic reactions of aldehydes, *e.g.*, Schiff's reaction, do not form a bisulphite or an aldehyde-ammonia compound.

Haworth (1926) proposed a six-membered ring formula (hexagonal formula) based on the pyran ring which is almost planar. The *pyranose* structure is applicable to nearly all the sugars, and is supported by X-ray crystal analysis (of the sugars). Thus α-D(+)-*glucopyranose* is (a). Reeves (1950) has shown that the conformation of α-D(+)-glucopyranose is (b) and that of the β-anomer is (c). 'Both have the chair form, but in the former the glycosidic hydroxyl is axial and in the latter equatorial. As is explained on, it is a general rule that the conformation which has the greatest number of large groups in equatorial orientation is the most stable form. Thus the β-anomer will be more stable than the α-, and so will predominate in the equilibrium mixture. It has also been shown (Irvine, Haworth *et al.*) that glucose, fructose, etc. can exist as five membered rings, which may be regarded as derivatives of *furan*. So far, the *furanose* sugar (corresponding to Tollens' suggestion) have not been isolated in the free state, but some of their derivatives have been prepared. Thus α-D(+)-glucofuranose would (if it existed) have perspective (*i.e.*, pentagonal) formula (III); methyl α-D(+)-glucofuranoside (a known compound) has formula (IV) Pentoses also normally exist in the pyranose form, and derivatives of the furanose form have been prepared.

Conversion of the plane-diagrams into the perspective formulae may be done as follows. (V) is α-D-glucopyranose, and if the H on C_5 is interchanged with the group CH_2OH, then a Walden inversion has been effected; and if the H is now interchanged with the point of attachment of the oxygen ring, another Walden inversion is effected, and so the original configuration of (V) is retained; thus we now have (VI) (with no change in configuration). Since all horizontal bonds indicate groups lying above the plane of the paper, and vertical bonds groups lying behind this plane, then by twisting (VI) so that the oxide ring is perpendicular to the plane of the paper, and placing the oxygen atom as shown, (VII) is obtained.

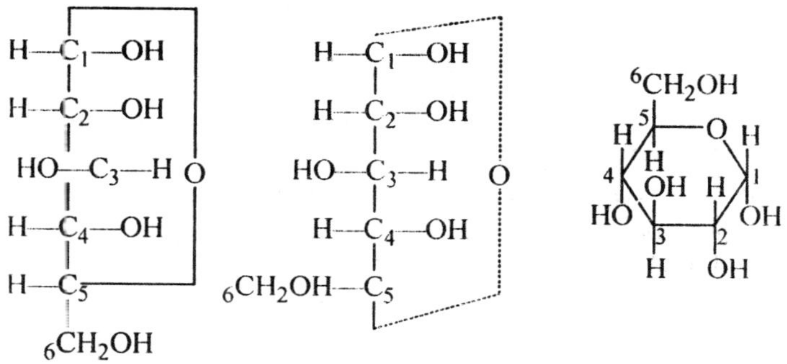

(V) (VI) (VII)

Fig. 6

Thus, to change from (VI) to (VII), first draw the hexagon (as shown in (VII}). and then place all the groups on the left-hand side in (VI) above the plane of the ring in (VII), and all those on the right-hand side in (VI) below, the plane of the ring in (VII). In a similar way, perspective formulae may be obtained -for the furanose sugars, *e.g.*, methyl β-D(+)-fructofuranoside (in this case, *i.e.*, for ketoses, treat the CH_2OH group adjacent to the keto group as H in the aldoses).

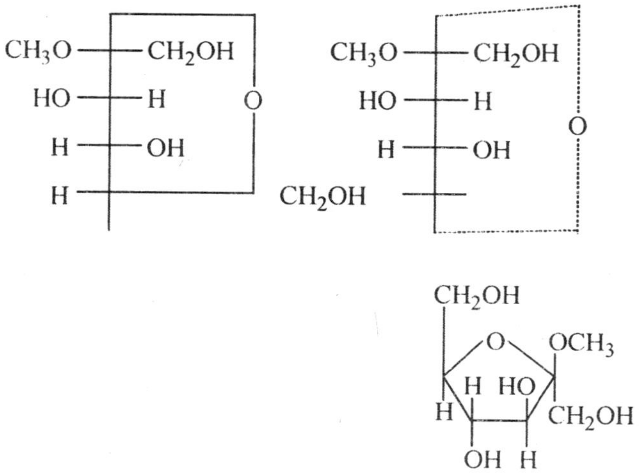

Fig. 7

NMR spectroscopic studies of monosaccharides and their derivatives have led to several generalisations (for the meanings of *axial* and *equatorial*).

(i) Anomeric protons (*i.e.*, H on C-l) occur at lower field than any others (C-1 is linked to *two* oxygen atoms).

(ii) Axial hydrogens appear upfield with respect to equatorial hydrogens.

(iii) For *acetylated pyranosides*, the methyl group on axial hydroxy'i groups occurs downfield with respect to those on equatorial hydroxyl groups.

(iv) For *methylated* pyranosides, the methyl group on axial hydroxyl groups occurs upfield with respect to those on equatorial hydroxyl groups.

These generalisations are useful, but may lead to wrong conclusions in rings which have been deformed by, *e.g.*, steric effects.

Just as simple hemiacetals react with another molecule of an alcohol to form acetals, so can the hemiacetal form of a sugar react with a molecule of an alcohol to form the acetal derivative, which is known under the generic name of *glycoside*; those of glucose are known as *glucosides*, of *fructose*, fructosides, etc.

For example, ethyl α-D(+)-glucopyranoside is prepared by refluxing glucose in excess of ethanol in the presence of a small amount of hydrochloric acid. These glycosides are stable compounds, and do not undergo many of the reactions of the sugars, *e.g.*, they show no reducing properties, they do not mutarotate, etc.

The non-sugar part of a glycoside is known as the *aglycon*, and in most of the glycosides which occur naturally the aglycon is a phenolic compound, *e.g.*, the aglycon in *salicin* is *salicyl alcohol*, in *indican*, *indoxyl*.

REACTION OF GLUCOSE AND FRUCTOSE WITH PHENYLHYDRAZINE

Fischer (1884, 1887) showed that both aldoses and ketones react with excess of phenylhydrazine to form *osazones* (these contain two phenylhydrazino groups), aniline and ammonia. The overall equation may be written:

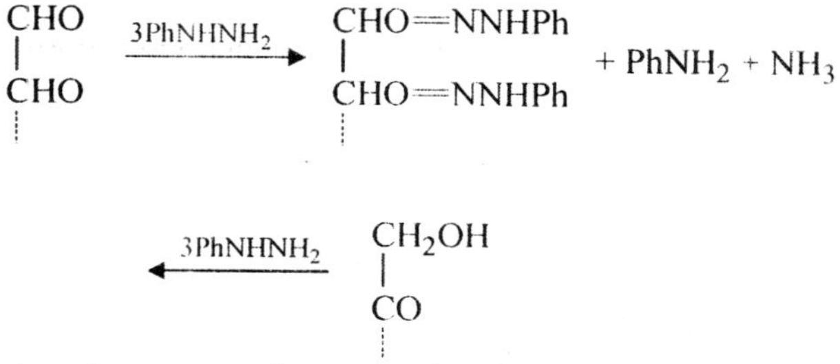

$$\begin{array}{l} CHO \\ | \\ CHO \\ \vdots \end{array} \xrightarrow{3PhNHNH_2} \begin{array}{l} CHO{=}NNHPh \\ | \\ CHO{=}NNHPh \\ \vdots \end{array} + PhNH_2 + NH_3$$

$$\xleftarrow{3PhNHNH_2} \begin{array}{l} CH_2OH \\ | \\ CO \\ \vdots \end{array}$$

Fischer also suggested a mechanism for this reaction, but it has been rejected in favour of that proposed by Weygand (1940).

Glucose

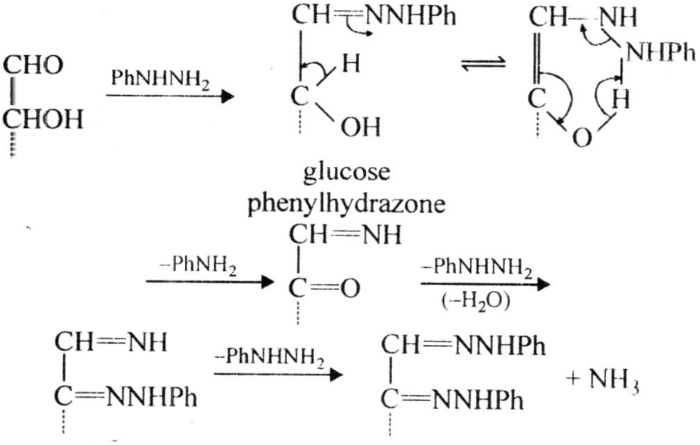

$$\begin{array}{l} CHO \\ | \\ CHOH \\ \vdots \end{array} \xrightarrow{PhNHNH_2} \begin{array}{l} CH{=}NNHPh \\ | \quad H \\ C \\ \vdots \quad OH \end{array} \rightleftharpoons \begin{array}{l} CH{-}NH \\ \| \quad\quad NHPh \\ C \quad\quad H \\ \vdots \quad O \end{array}$$

glucose phenylhydrazone

$$\xrightarrow{-PhNH_2} \begin{array}{l} CH{=}NH \\ | \\ C{=}O \\ \vdots \end{array} \xrightarrow[(-H_2O)]{-PhNHNH_2}$$

$$\begin{array}{l} CH{=}NH \\ | \\ C{=}NNHPh \\ \vdots \end{array} \xrightarrow{-PhNHNH_2} \begin{array}{l} CH{=}NNHPh \\ | \\ C{=}NNHPh \\ \vdots \end{array} + NH_3$$

glucosazone

Fructose

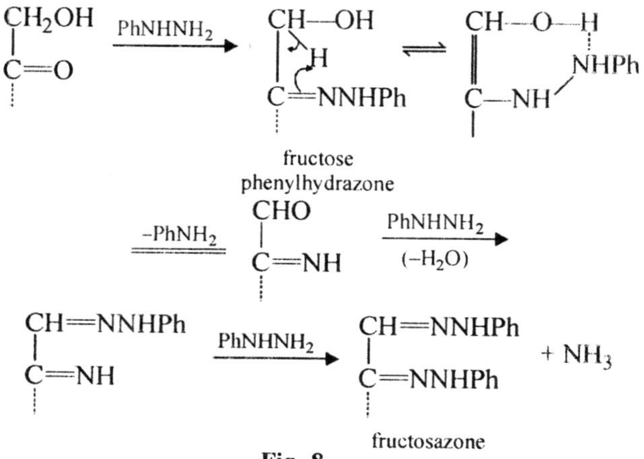

$$\begin{array}{l} CH_2OH \\ | \\ C{=}O \\ \vdots \end{array} \xrightarrow{PhNHNH_2} \begin{array}{l} CH{-}OH \\ | \quad H \\ C{=}NNHPh \\ \vdots \end{array} \rightleftharpoons \begin{array}{l} CH{-}O{-}H \\ \| \quad\quad NHPh \\ C{-}NH \\ | \end{array}$$

fructose phenylhydrazone

$$\xrightarrow{-PhNH_2} \begin{array}{l} CHO \\ | \\ C{=}NH \\ \vdots \end{array} \xrightarrow[(-H_2O)]{PhNHNH_2}$$

$$\begin{array}{l} CH{=}NNHPh \\ | \\ C{=}NH \\ \vdots \end{array} \xrightarrow{PhNHNH_2} \begin{array}{l} CH{=}NNHPh \\ | \\ C{=}NNHPh \\ \vdots \end{array} + NH_3$$

fructosazone

Fig. 8

According to this mechanism, the first step is the formation of a phenylhydrazone, which then undergoes the *Amadori rearrangement* (1926). Weygand had originally proposed two mechanisms, but later (1958-1963) concluded that both were operating. According to Shemyakin *et al.* (1965), however, only one of these mechanisms is operating, viz.,

Weygand's mechanism does not explain why only the first two carbon atoms should be involved. At first sight, it might have been expected that the Amadori rearrangement could carry on down the chain to produce a hexaphenylhydrazone. The failure to undergo further reaction has been explained by stabilisation of the osazone by chelation, *i.e.*, phenylosazones have *ring* structures (Fieser and Fieser, 1944). Two alternatives (1) and (II) were proposed.

Mester (1955, 1965) believed that (1) was the correct structure, but Blair *et al.* (1969), from their chemical and i.r. and u.v. spectra studies, believe that neither is correct, and have proposed (III), the phenylazo structure (at C-2) for the compound in the *solid state*, and that *in solution*, (III) is in equilibrium with the open-chain form.

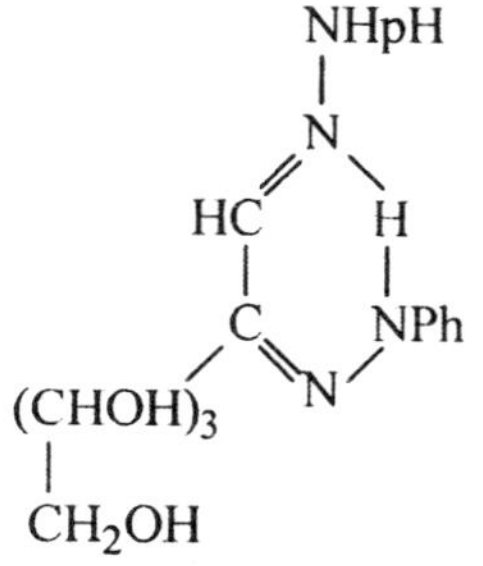

NHpH
N
HC H
C NPh
N
$(CHOH)_3$
CH_2OH

(I)

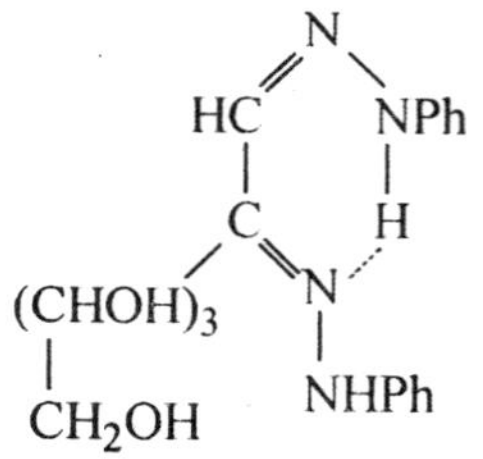

N
HC NPh
C H
N
$(CHOH)_3$ NHPh
CH_2OH

(II)

NHPh
N
CH H
O
CH NPh
N
CH—$(CHOH)_2$
CH_2OH

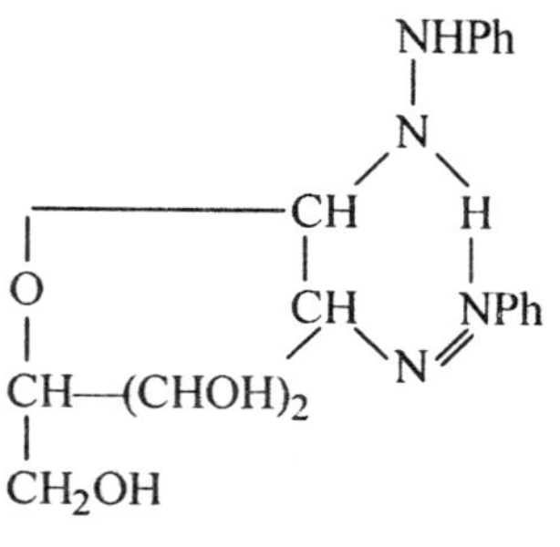

(III)

Fig. 9

Osazones are yellow crystalline solids and are used to characterise the sugars.

They may be hydrolysed with hydrochloric acid, both phenylhydrazino groups being eliminated; the dicarbonyl compound formed is known as an *osone, e.g., glucosazone* forms *glucosone*;

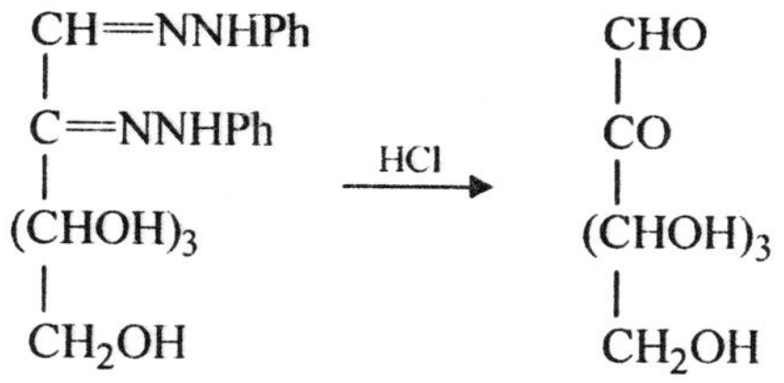

$$\begin{array}{c} CH{=}NNHPh \\ | \\ C{=}NNHPh \\ | \\ (CHOH)_3 \\ | \\ CH_2OH \end{array} \xrightarrow{HCl} \begin{array}{c} CHO \\ | \\ CO \\ | \\ (CHOH)_3 \\ | \\ CH_2OH \end{array}$$

A more convenient method of obtaining the osone is to add benzaldehyde to a solution of the osazone; benzaldehyde phenylhydrazone is precipitated, leaving the osone in solution. Osones react with phenylhydrazine in the cold to form osazones.

EPUNERISATION

Aldoses which produce the same osazones must have identical configurations on all their asymmetric carbon atoms except the *alpha* (since only the aldehyde group and α-carbon atom are involved in osazone formation).

Such sugars are known as *epimers*. Fischer (1890) changed an aldose into its epimervia the aldonic acid. Thealdonic acid was heated with pyridine (or quinoline), whereupon it was converted into an equilibrium mixture of the original acid and its epimer. These were separated, and the epimeric acid lactone reduced to an aldose (the Kiliani reaction below).

The mechanism of epimerisation is not certain, but it is probably similar to racemisation, *e.g.*, epimerisation of glucose into toannose:

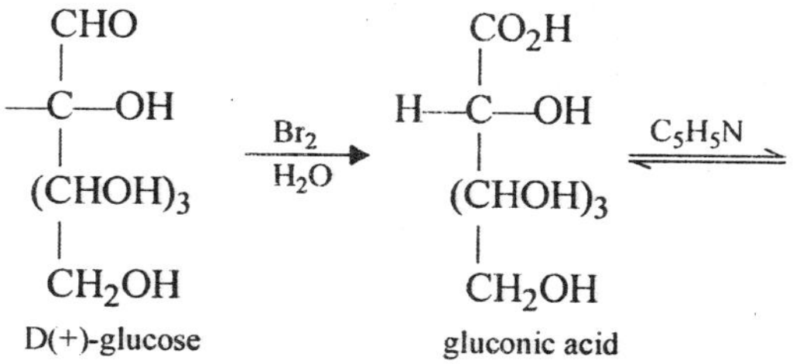

$$\underset{\text{D(+)-glucose}}{\begin{array}{c} CHO \\ | \\ -C-OH \\ | \\ (CHOH)_3 \\ | \\ CH_2OH \end{array}} \xrightarrow[H_2O]{Br_2} \underset{\text{gluconic acid}}{\begin{array}{c} CO_2H \\ | \\ H-C-OH \\ | \\ (CHOH)_3 \\ | \\ CH_2OH \end{array}} \xrightleftharpoons{C_5H_5N}$$

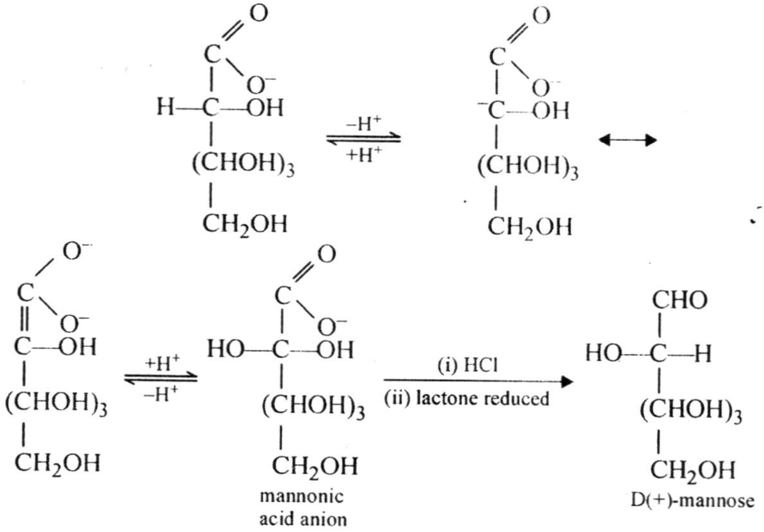

$$\begin{array}{c} O \\ /\!/ \\ C \\ | \quad \backslash O^- \\ H—C—OH \\ | \\ (CHOH)_3 \\ | \\ CH_2OH \end{array} \underset{+H^+}{\overset{-H^+}{\rightleftharpoons}} \begin{array}{c} O \\ /\!/ \\ C \\ | \quad \backslash O^- \\ {}^-C—OH \\ | \\ (CHOH)_3 \\ | \\ CH_2OH \end{array} \longleftrightarrow$$

$$\begin{array}{c} O^- \\ / \\ C \\ \| \quad \backslash O^- \\ C—OH \\ | \\ (CHOH)_3 \\ | \\ CH_2OH \end{array} \underset{-H^+}{\overset{+H^+}{\rightleftharpoons}} \begin{array}{c} O \\ /\!/ \\ C \\ | \quad \backslash O^- \\ HO—C—OH \\ | \\ (CHOH)_3 \\ | \\ CH_2OH \\ \text{mannonic} \\ \text{acid anion} \end{array} \xrightarrow[\text{(ii) lactone reduced}]{\text{(i) HCl}} \begin{array}{c} CHO \\ | \\ HO—C—H \\ | \\ (CHOH)_3 \\ | \\ CH_2OH \\ \text{D(+)-mannose} \end{array}$$

Fig. 10

This change of configuration of one asymmetric carbon atom in a compound containing two - or more asymmetric carbon atoms is known as *epimerisation.*

METHOD OF ASCENDING THE SUGAR SERIES

An aldose may be converted into its next higher aldose, *e.g.*, an aldopentose into an aldohexose, by means of the *Kiliani reaction* (1886). Theoretically, two lactones are possible, since two cyanohydrins may be formed when hydrogen cyanide adds on to the aldopentoge (a new asymmetric carbon is produced), viz.,

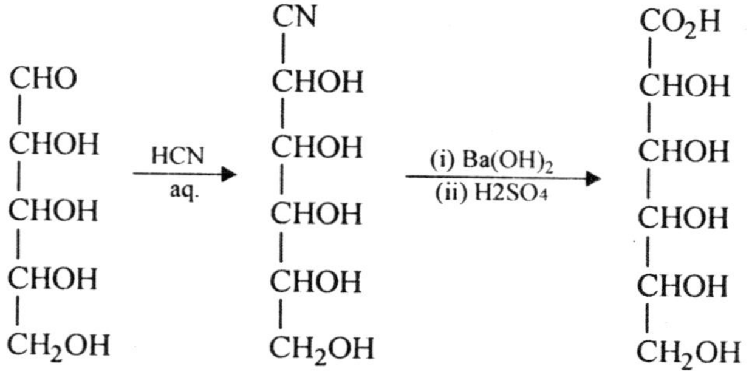

$$\begin{array}{c} CHO \\ | \\ CHOH \\ | \\ CHOH \\ | \\ CHOH \\ | \\ CH_2OH \end{array} \xrightarrow[\text{aq.}]{\text{HCN}} \begin{array}{c} CN \\ | \\ CHOH \\ | \\ CHOH \\ | \\ CHOH \\ | \\ CHOH \\ | \\ CH_2OH \end{array} \xrightarrow[\text{(ii) H2SO4}]{\text{(i) }Ba(OH)_2} \begin{array}{c} CO_2H \\ | \\ CHOH \\ | \\ CHOH \\ | \\ CHOH \\ | \\ CHOH \\ | \\ CH_2OH \end{array}$$

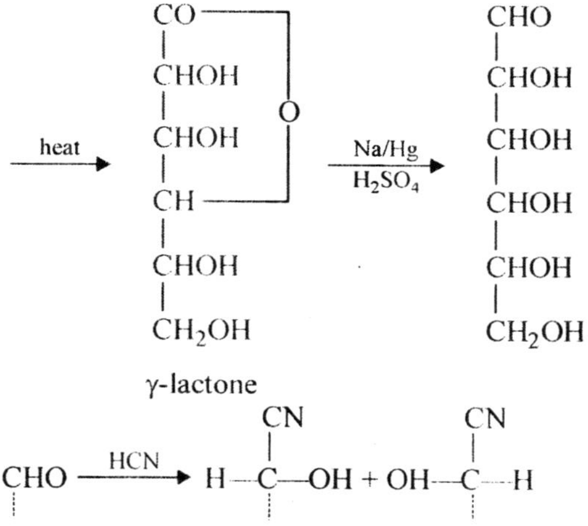

$$\xrightarrow{\text{heat}} \begin{array}{l} CO \\ | \\ CHOH \\ | \\ CHOH \\ | \\ CH \\ | \\ CHOH \\ | \\ CH_2OH \end{array} \text{(ring through O from CO to CH)} \xrightarrow[H_2SO_4]{Na/Hg} \begin{array}{l} CHO \\ | \\ CHOH \\ | \\ CHOH \\ | \\ CHOH \\ | \\ CHOH \\ | \\ CH_2OH \end{array}$$

γ-lactone

$$\underset{\vdots}{CHO} \xrightarrow{HCN} H-\overset{CN}{\underset{\vdots}{\overset{|}{C}}}-OH + OH-\overset{CN}{\underset{\vdots}{\overset{|}{C}}}-H$$

Fig. 11

Thus, two epimeric aldohexoses should be obtained. In practice, one cyanohydrin predominates because the asymmetry present in the molecule produces diastereotopic faces at the double bond. Hence the final product will be mainly one aldohexose, and very little of its epimer.

1,4-Aldonolactones (γ-lactones) may also be reduced to the aldoses by sodium borohydride or lithium aluminium hydride in a mixture of tetrahydrofuran and pyridine.

There are now many other methods available for stepping *et al* the sugars. Sowden *et al.* (1947) used nitromethane as follows, and the proportions of the epimers formed by this method are different from those produced by the Kiliani reaction.

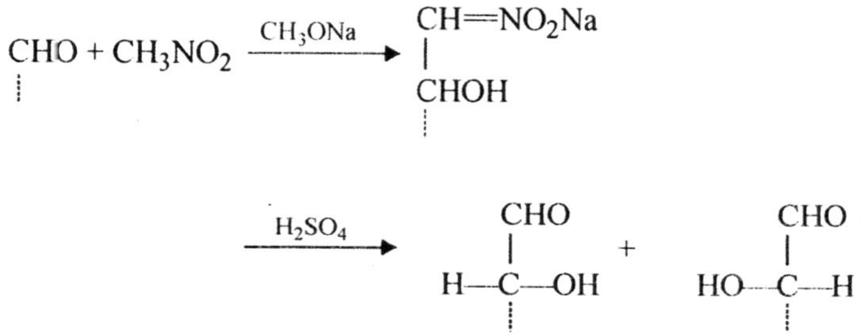

$$\underset{\vdots}{CHO} + CH_3NO_2 \xrightarrow{CH_3ONa} \begin{array}{l} CH{=}NO_2Na \\ | \\ CHOH \\ \vdots \end{array}$$

$$\xrightarrow{H_2SO_4} H-\overset{CHO}{\underset{\vdots}{\overset{|}{C}}}-OH \quad + \quad HO-\overset{CHO}{\underset{\vdots}{\overset{|}{C}}}-H$$

Sowden (1950) has also stepped up an aldose to a *ketose* containing *two* additional carbon atoms using the above method except that 2-nitroethanol is used instead of nitromethane.

$$\begin{array}{c} CHO \\ \vdots \end{array} + \begin{array}{c} CH_2NO_2 \\ | \\ CH_2OH \end{array} \xrightarrow{CH_3ONa} \begin{array}{c} CH_2OH \\ | \\ C{=}NO_2Na \\ | \\ CHOH \\ \vdots \end{array}$$

$$\xrightarrow{H_2SO_4} \begin{array}{c} CH_2OH \\ | \\ CO \\ | \\ H{-}C{-}OH \\ \vdots \end{array} + \begin{array}{c} CH_2OH \\ | \\ CO \\ | \\ HO{-}C{-}OH \\ \vdots \end{array}$$

On the other hand, Wolfrom et at. (1946) have stepped up an aldose to a ketose with one more carbon atom by a modified Arndt-Eistert reaction.

$$\begin{array}{c} CHO \\ | \\ (CHOH)_3 \\ | \\ CH_2OH \end{array} \xrightarrow[H_2O]{Br_2} \begin{array}{c} CO_2H \\ | \\ (CHOH)_3 \\ | \\ CH_2OH \end{array} \xrightarrow[(ii)\ SOCl_2]{(i)\ Ac_2O} \begin{array}{c} COCl \\ | \\ (CHOAc)_3 \\ | \\ CH_2OAc \end{array}$$

$$\xrightarrow{CH_2N_2} \begin{array}{c} CHN_2 \\ | \\ COCl \\ | \\ (CHOAc)_3 \\ | \\ CH_2OAc \end{array} \xrightarrow[(ii)\ Ba(OH)_2]{(i)\ AcOH} \begin{array}{c} CH_2OH \\ | \\ COCl \\ | \\ (CHOAc)_3 \\ | \\ CH_2OAc \end{array}$$

More recently, Kochetkov et at. (1965) have stepped up an aldose to an aldose containing two additional carbon atoms, *e.g.*, the aldose is condensed with carbethoxymethylenephosphorane.:

$$RCHOHCHO + Ph_3P = CHCO_2Et$$

$$\longrightarrow Ph_3PH + RCHOHCH = CHCO_2Et \xrightarrow[HClO_4]{OsO_4}$$

$$[RCHOHCHOHCHOHCO_2H] \longrightarrow RCH(CHOH)_2CO \text{ (O bridge between RCH and CO)}$$

$$\xrightarrow[\text{or } NaBH_4]{Na-Hg} RCHOHCHOHCHO$$

METHOD OF DESCENDING THE SUGAR SERIES

There are various methods of converting a sugar into its next lower sugar. All of these methods start with the aldose, and hence, in order to convert a ketose into the lower aldose, it is first necessary to transform it into an aldose (see later).

Wohls method (1893) starts with the aldose oxime, which is then treated as shown (for an aldohexose).

$$\begin{array}{l} CHO \\ | \\ CHO \\ | \\ (CHOH)_3 \\ | \\ CHO \end{array} \xrightarrow{NH_2OH} \begin{array}{l} CH{=}NOH \\ | \\ CHOH \\ | \\ (CHOH)_3 \\ | \\ CH_2OH \end{array} \xrightarrow{(CH_3CO)_2O} \begin{array}{l} CN \\ | \\ CHOCOCH_3 \\ | \\ (CHOCOCH_3) \\ | \\ CH_2OCOCH_3 \end{array}$$

$$\xrightarrow{(CH_3CO)_2O} \begin{array}{l} CHO \\ | \\ (CHOH)_3 \\ | \\ CH_2OH \end{array} + AgCN\ (55\%)$$

Zemplen (1917) modified Wohl's method by using a solution of sodium methoxide in chloroform instead of an aqueous solution ofammoniacal silver nitrate, to remove hydrogen cyanide and the acetyl groups, and thereby increased the yield of pentose to 60-70 per cent. Weygand *et al.* (1950) have treated the oxime with l-fluoro-2,4-dinitrobenzene in aqueous sodium hydrogen carbonate; the products are the lower aldose (yield: 50-60%), hydrogen cyanide and 2,4-dinitrophenol.

Ruff's method (1898). The aldose is oxidised (by bromine-water) to the corresponding aldonic acid; when the calcium salt of this acid is treated with Fenton's reagent it is converted into the lower aldose (*cf.* oxidation of α-hydroxyacids:

$$\begin{array}{l} CHO \\ | \\ CHO \\ | \\ (CHOH)_3 \\ | \\ CH_2OH \end{array} \xrightarrow{Br_2/H_2O} \begin{array}{l} CO_2H \\ | \\ CHOH \\ | \\ (CHOH)_3 \\ | \\ CH_2OH \end{array} \xrightarrow[H_2O_2/Fe^{2+}]{\text{Ca salt}} \begin{array}{l} CHOH \\ | \\ (CHOH)_3 \\ | \\ CH_2OH \end{array} + CO_2\ (25\%)$$

Berezovski *et al.* (1949) have shown that calcium gluconate can be oxidised with hydrogen peroxide in the presence of ferric sulphate and barium acetate to give D-arabinose (44%).

Weennan's reaction (1913). This is the reaction whereby an α-hydroxy- or α-methoxy-amide is degraded by means of a cold solution of sodium hypochlorite (cf. Hofmann amine preparation)

α-Hydroxyamides

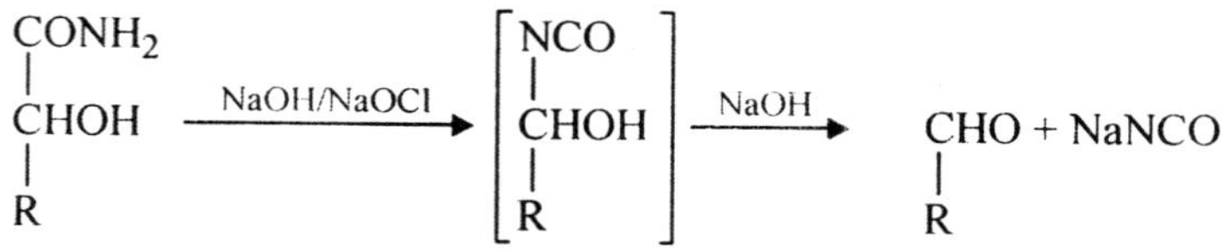

$$\begin{matrix} CONH_2 \\ | \\ CHOH \\ | \\ R \end{matrix} \xrightarrow{NaOH/NaOCl} \left[\begin{matrix} NCO \\ | \\ CHOH \\ | \\ R \end{matrix}\right] \xrightarrow{NaOH} \begin{matrix} CHO + NaNCO \\ | \quad\quad\quad\quad \\ R \quad\quad\quad\quad \end{matrix}$$

α-Methoxyamides

$$\begin{matrix} CONH_2 \\ | \\ CHOCH_3 \\ | \\ R \end{matrix} \xrightarrow{NaOH/NaOCl} \left[\begin{matrix} NCO \\ | \\ CHOCH_3 \\ | \\ R \end{matrix}\right]$$

$$\xrightarrow{NaOH} \begin{matrix} CHO + CO_2 + NH_3 + CH_3OH \\ | \quad\quad\quad\quad\quad\quad\quad\quad\quad\quad \\ R \quad\quad\quad\quad\quad\quad\quad\quad\quad\quad \end{matrix}$$

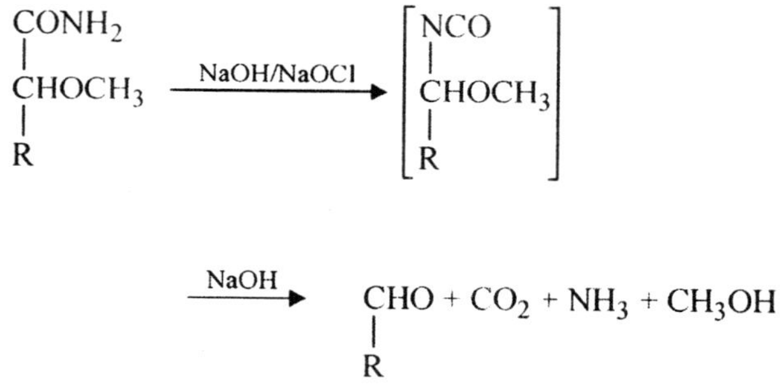

Fig. 12

More recent methods of descending the sugars are also available. Thus, Macdonald *et al.* (1953) have stepped down an aldose as follows:

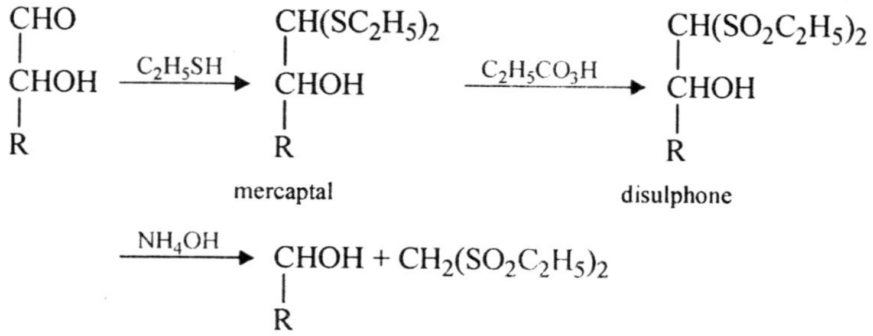

$$\begin{matrix} CHO \\ | \\ CHOH \\ | \\ R \end{matrix} \xrightarrow{C_2H_5SH} \underset{\text{mercaptal}}{\begin{matrix} CH(SC_2H_5)_2 \\ | \\ CHOH \\ | \\ R \end{matrix}} \xrightarrow{C_2H_5CO_3H} \underset{\text{disulphone}}{\begin{matrix} CH(SO_2C_2H_5)_2 \\ | \\ CHOH \\ | \\ R \end{matrix}}$$

$$\xrightarrow{NH_4OH} \begin{matrix} CHOH + CH_2(SO_2C_2H_5)_2 \\ | \quad\quad\quad\quad\quad\quad\quad\quad \\ R \quad\quad\quad\quad\quad\quad\quad\quad \end{matrix}$$

Fig. 13

On the other hand, Perlin (1954) has shown-that aldoses may be stepped down by direct oxidation with lead tetra-acetate or sodium bismuthate, *e.g.*, D-mannose gives D-arabinose (35%).

Conversion of a ketose into an aldose. The ketose is reduced (preferably by catalytic reduction, to the corresponding polyhydric alcohol, which is then oxidised to a monocarboxylic acid (only one of the terminal CH_2OH groups being oxidised). On warming, the acid is converted into the γ-lactone which, on reduction with sodium amalgam in faintly acid solution, is converted into the aldose:

Aldopentoses, $HOCH_2CHOHCHOHCHOHCHO$.

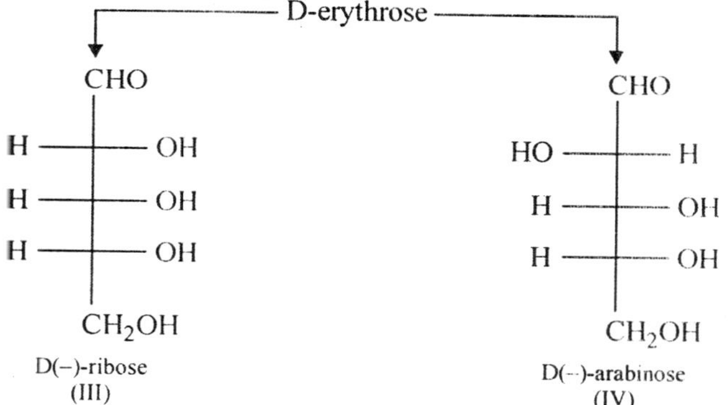

D(–)-ribose (III) D(–)-arabinose (IV)

This structure contains three asymmetric carbon atoms, and so there are eight optical isomers (four pairs of enantiomers).

All are known. D-Erythrose, when stepped up by the Kiliani reaction, gives D(–)-ribose and D(–)-arabinose. Similarly, D-threose gives D(+)-xylose and D(–)-lyxose.

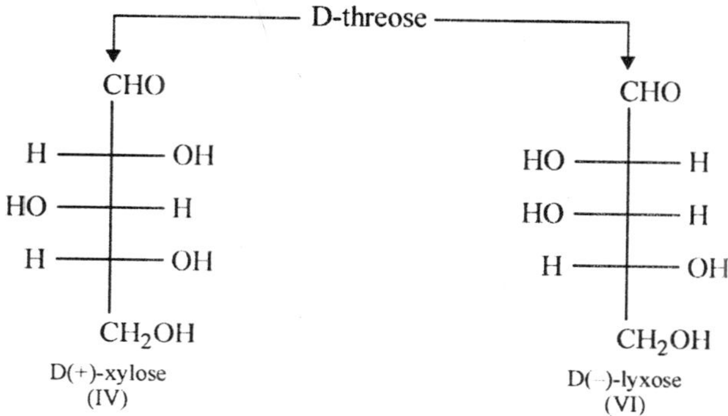

D(+)-xylose (IV) D(–)-lyxose (VI)

(III) and (IV) must be ribose and arabinose, but which is which? On oxidation with nitric acid, arabinose gives an optically active dicarboxylic acid (a trihydroxyglutaric acid), whereas ribose gives an optically inactive dicarboxylic acid.

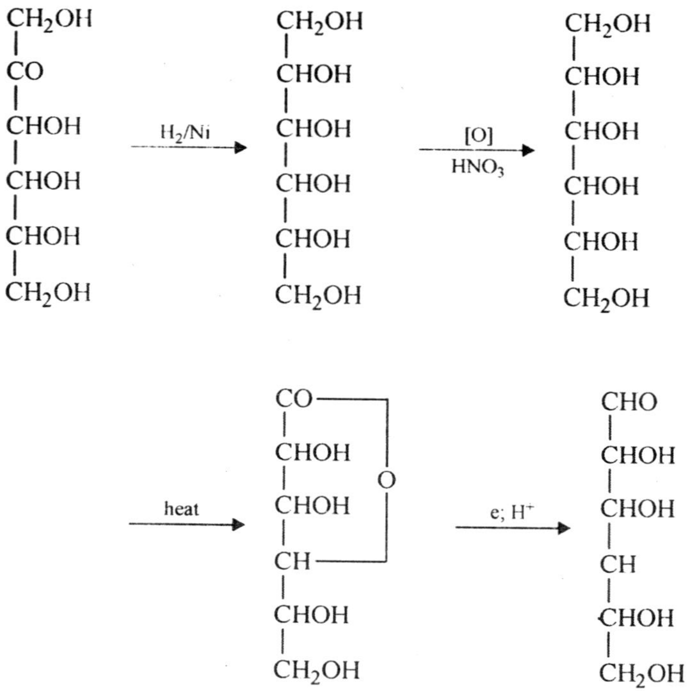

$$\begin{array}{c} CH_2OH \\ | \\ CO \\ | \\ CHOH \\ | \\ CHOH \\ | \\ CHOH \\ | \\ CH_2OH \end{array} \xrightarrow{H_2/Ni} \begin{array}{c} CH_2OH \\ | \\ CHOH \\ | \\ CHOH \\ | \\ CHOH \\ | \\ CHOH \\ | \\ CH_2OH \end{array} \xrightarrow[HNO_3]{[O]} \begin{array}{c} CH_2OH \\ | \\ CHOH \\ | \\ CHOH \\ | \\ CHOH \\ | \\ CHOH \\ | \\ CH_2OH \end{array}$$

$$\xrightarrow{heat} \begin{array}{c} CO\text{—} \\ | \\ CHOH \\ | \\ CHOH \\ | \\ CH\text{—} \\ | \\ CHOH \\ | \\ CH_2OH \end{array} \begin{array}{c} \rceil \\ \\ O \\ \\ \rfloor \end{array} \xrightarrow{e;\ H^+} \begin{array}{c} CHO \\ | \\ CHOH \\ | \\ CHOH \\ | \\ CH \\ | \\ CHOH \\ | \\ CH_2OH \end{array}$$

Fig. 14

The dicarboxylic acid produced from (IV), however, has no plane (or any other element) of symmetry, and so is optically active. Thus (III) is D-ribose and (IV) is D-arabinose. (V) and (VI) must be xylose and lyxose, but which? The former, on oxidation, gives an optically inactive dicarboxylic acid, whereas the latter gives an optically active dicarboxylic acid. Therefore (V) is D-xylose and (VI) is D-lyxose. Theoretically, two polyhydric alcohols may be formed on reduction of the ketose, due to the formation of a new asymmetric carbon atom: In practice, however, one predominates (*cf.* the Kiliani reaction). Furthermore, when these two alcohols are oxidised, oxidation may take place at *either* end of the chain, and hence the final product will be a mixture *of four* aldoses, but these will not be present to the same extent.

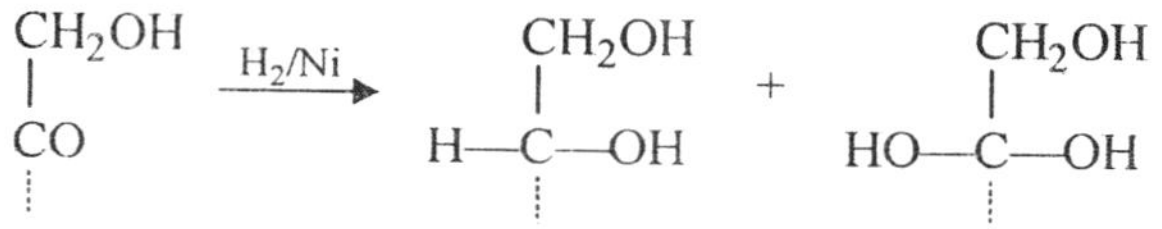

$$\begin{array}{c} CH_2OH \\ | \\ CO \\ \vdots \end{array} \xrightarrow{H_2/Ni} \begin{array}{c} CH_2OH \\ | \\ H\text{—}C\text{—}OH \\ \vdots \end{array} + \begin{array}{c} CH_2OH \\ | \\ HO\text{—}C\text{—}OH \\ \vdots \end{array}$$

Conversion of an aldose into a ketose. The aldose is converted into its osazone, which is then hydrolysed with hydrochloric acid to the osone. On reduction with zinc and acetic acid, the osone is converted into the ketose (an aldehyde group is reduced more readily than a ketonic group):

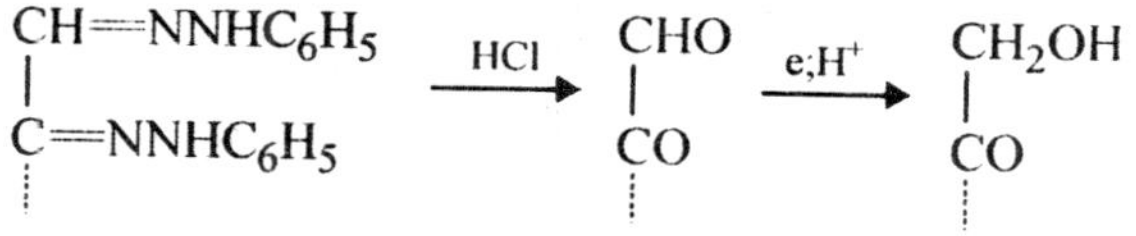

$$\begin{array}{l} CH{=}NNHC_6H_5 \\ | \\ C{=}NNHC_6H_5 \\ \vdots \end{array} \xrightarrow{HCl} \begin{array}{l} CHO \\ | \\ CO \\ \vdots \end{array} \xrightarrow{e;H^+} \begin{array}{l} CH_2OH \\ | \\ CO \\ \vdots \end{array}$$

Lobry de Bruyn-van Ekenstein rearrangement (1890). When warmed with *concentrated* alkali, sugars first turn yellow, then brown and finally resinify *(cf.* aldehydes).

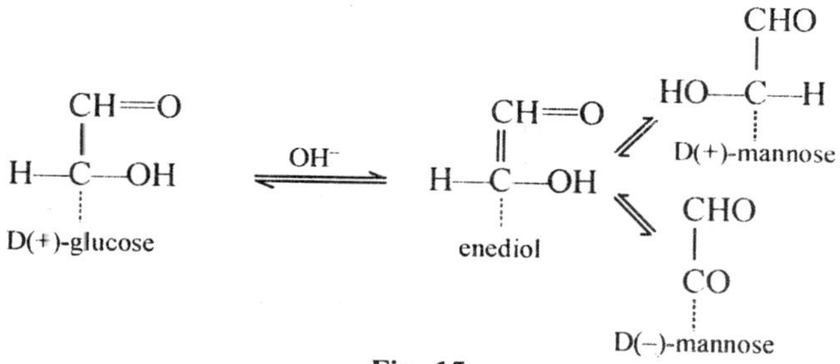

$$\underset{\text{D(+)-glucose}}{\begin{array}{c} CH{=}O \\ | \\ H{-}C{-}OH \\ \vdots \end{array}} \overset{OH^-}{\rightleftharpoons} \underset{\text{enediol}}{\begin{array}{c} CH{=}O \\ || \\ H{-}C{-}OH \\ \vdots \end{array}} \begin{array}{l} \rightleftharpoons \underset{\text{D(+)-mannose}}{\begin{array}{c} CHO \\ | \\ HO{-}C{-}H \\ \vdots \end{array}} \\ \rightleftharpoons \underset{\text{D(−)-mannose}}{\begin{array}{c} CHO \\ | \\ CO \\ \vdots \end{array}} \end{array}$$

Fig. 15

In the presence *of dilute* alkali or amines, sugars undergo rearrangement; *e.g.*, a dilute solution of glucose, in the presence of sodium hydroxide, is converted into an almost optically inactive solution from which have been isolated D(+)-glucose, D(+)-mannose, D(−)-fructose. The same mixture is obtained if the starting material is D(—)-fructose or D(+)-mannose. The details of the mechanism are uncertain, but it is accepted that the rearrangement occurs through an enediol:

Since the Lobry de Bruyn-van Ekenstein rearrangement takes place in alkaline media, it is hest to carry out reactions with the sugars in neutral or acid media.

SYNTHESIS OF THE MONOSACCHARIDES

Photosynthesis is the most important example of biosynthesis and represents the processes whereby plants containing the pigment chlorophyll absorb light energy and utilise it to convert atmospheric carbon dioxide, in the presence of water, to carbohydrates. Photosynthesis has been shown to involving two separate types of reactions. The first involves a photochemical process in which light energy absorbed by chlorophyll, is utilised to form activated compounds. The second involves the reduction of carbon dioxide by the active molecules produced in the first process; the products are oxygen and carbohydrates (and some other compounds). Since this process can proceed in the presence or absence of light, it is referred to as the dark reactions, the first process is distinguished from this as the *light* reactions. Both the dark and light reactions require the presence of various enzymes. The overal equation for photosynthesis may be written as:

$$6O_2 + 6H_2O \longrightarrow 6(CH_2O) + 6O_2$$

Calvin *et al.* using $^{14}CO_2$ as tracer, have worked out the pathway of carbon dioxide in photosynthesis.

In the laboratory, the sugars have been synthesised in various ways:

(i) By the aldol condensation of formaldehyde in the presence of calcium hydroxide; the product is a mixture of compounds among which is a number of hexoses. This hexose mixture, know as *formose*, has been shown to contain DL-fructose (Butlerow, 1861; Loew, 1886).

(ii) By the aldol condensation of glycolaldehyde in the presence of sodium hydroxide; the product formose (E. Fischer, 1887):

$$3HOCH_2CHO \xrightarrow{NaOH} C_6H_{12}O_6$$

(iii) When oxidised with nitric acid or bromine-water, *glycerol yields* a product known as *glycerose*, which contains, among other compounds, glyceraldehyde and dihydroxyacetone. These, in the presence of barium hydroxide, are converted into a mixture of α- and β-*acrose* (E. Fischer, 1887); α-acrose is DL-fructose and /?-acrose is DL-sorbose:

$$HOCH_2CHOHCHO + HOCH_2COCH_2OH$$

$$\longrightarrow HOCH_2CO(CHOH)_3CH_2OH$$

Starting with α-acrose, Fischer isolated DL-fructosazone (DL-glucosazone), and making use of reduction, oxidation and epimerisation, he converted DL-fructosazone into D(–)-fructose, D(+)-glucose, D(+)-mannose and other aldohexoses.

(iv) When hydrolysed with barium hydroxide solution, dibromoacraldehyde forms a mixture a- and 8-acrose (E. Fischer, 1887):

$$BrCH_2CHBrCHO \xrightarrow{Ba(OH)_2} HOCH_2CHOHCHO \xrightarrow{\text{dimerises}}$$

$$HOCH_2CO(CHOH)_3CH_2OH$$

Aldotrioses. Glyceraldehyde is the only aldotrios

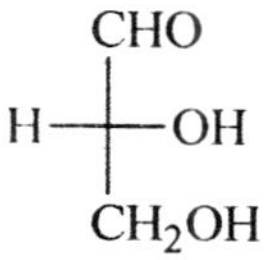

$$\begin{array}{c} CHO \\ H-\!\!\!\!\!-\!\!\!\!\!-OH \\ CH_2OH \end{array} \qquad \begin{array}{c} CHO \\ HO-\!\!\!\!\!-\!\!\!\!\!-H \\ CH_2OH \end{array}$$

D(+)-glyceraldehyde 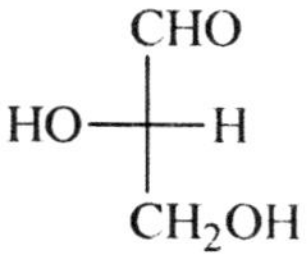 L(–)-glyceraldehyde

Aldotetroses, $HOCH_2CHOHCHOHCHO$. This structure contains two asymmetric carbon atoms, and so there are four optical isomers (two pairs of enantiomers). All are known, and correspond to D- and L-threose and D- and L-erythrose.

D(+)-Glyceraldehyde may be stepped up by the Kiliani reaction to give D(–)-threose and D(–)-erythrose. The question now is: which is which ?

On oxidation, D-erythrose forms mese-tartaric acid. Therefore D-erythrose must be (I), and consequently (II) must be D-threose.

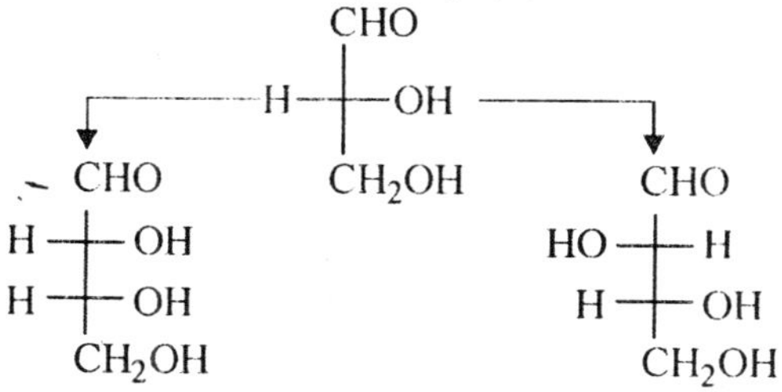

$$\begin{array}{c} CHO \\ H-\!\!\!\!\!-\!\!\!\!\!-OH \\ CH_2OH \end{array}$$

$$\begin{array}{c} CHO \\ H-\!\!\!\!\!-\!\!\!\!\!-OH \\ H-\!\!\!\!\!-\!\!\!\!\!-OH \\ CH_2OH \end{array} \qquad \begin{array}{c} CHO \\ HO-\!\!\!\!\!-\!\!\!\!\!-H \\ H-\!\!\!\!\!-\!\!\!\!\!-OH \\ CH_2OH \end{array}$$

D(–)-erythrose (I) D(–)-threose (II)

Aldohexoses, $HOCH_2CHOHCHOHCHOHCHOHCHO$. This structure contains four asymmetric carbon atoms, and so there are sixteen optical isomers (eight pairs of enantiomers). All are known, and may be prepared by stepping up the aldopentoses. D-ribose gives D(+)-allose and D(+)-altrose; D-arabinose gives D(+)-glucose and D(+)-mannose; D-xylose gives D(–)-gulose and D(–)-idose, and D-lyxose gives D(+)-galactose and D(+)-talose.

(VII) and (VIII) must be allose and altruse, but which is which?

On oxidation with nitric acid, the former gives an optically inactive (allaric) and the latter an optically active (talaric) dicarboxylic acid.

Therefore (VII) is allose and (VIII) is altrose.

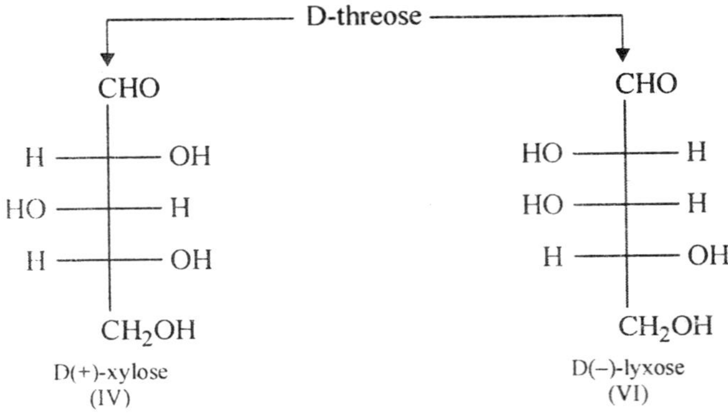

Fig. 16

(XIII) and (XIV) must be galactose and talose, but which is which? On oxidation with nitric acid, the former gives an optically inactive (galactaric) and the latter an optically active (talaric) dicarboxylic acid. Therefore (XIII) is galactose and (XIV) is talose.

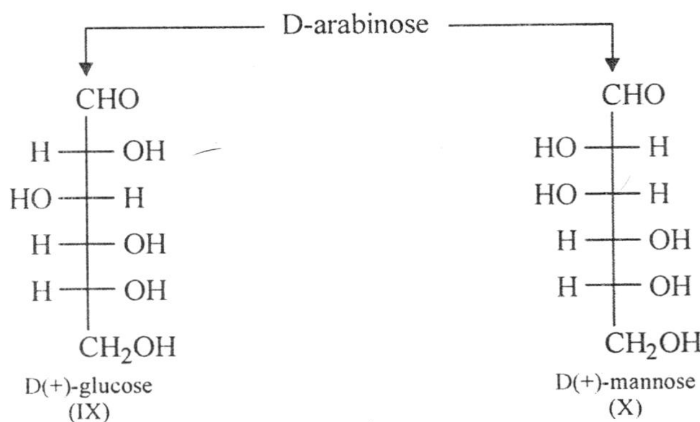

Fig. 17

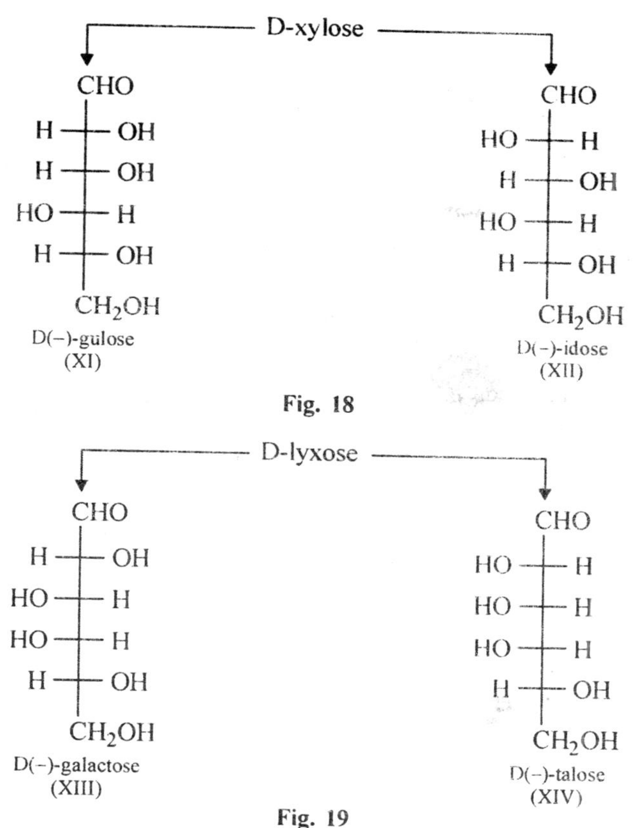

Fig. 18

Fig. 19

The elucidation of the configurations of the remaining four aldohexoses is not quite so simple since, on oxidation with nitric acid, glucose and mannose *both* give optically active dicarboxylic acids, as also do gulose and idose; in all four configurations (IX, X, XI, XII), replacement of the two terminal groups (CHO and CH_2OH) by carboxyl groups leads to dicarboxylic acid whose structures have no plane (or any other element) of symmetry.

It has been found, however, that the dicarboxylic acid from glucose (glucaric acid) is the sape as that produced from gulose. Actually the two glucaric acids obtained are enantiomers, D-glucose giving D-glucaric acid and D-gulose L-glucaric acid. Since glucaric acid, $HO_2C(CHOH)_4$-CO_2H, is produced by the oxidation of the terminal groups with the rest of the molecule unaffected, it therefore follows that the 'rest of the molecule' must be the same for both glucose and gulose.

Inspection of formulae (IX, X, XI and XII) shows that only (IX) and (XI) have the 'rest of the molecule' the same; by interchanging the CHO and CH,OH groups of (IX), (XI) is obtained. Therefore (IX) must be glucose (since we know that glucose is obtained from arabinose), and (XI) must be gulose. Consequently (X) is mannose and (XII) is idose.

Ketohexoses. Fructose is a 2-ketohexose, and natural fructose is laevorotatory. Since D-glucose gives the same osazone as natural fructose, the latter must be D(–)-fructose.

Furthermore, since osazone formation involves only the first two carbon atoms in a sugar, it therefore follows that the configuration of the rest of the molecules in glucose and fructose must be the same. Hence the configuration of D(–)-fructose is (XV).

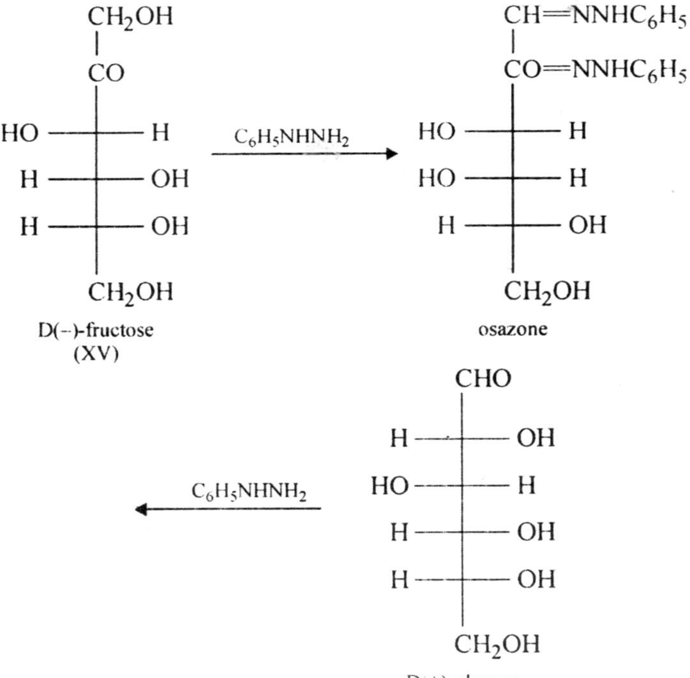

The configurations of the other 2-ketohexoses are:

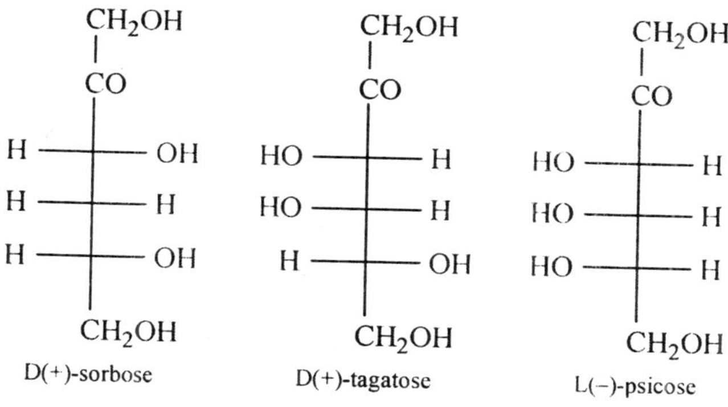

Fig. 20

DETERMINATION OF THE SIZE OF SUGAR-RINGS

E. Fischer (1893) refluxed glucose with excess of methanol in the presence of a small amount of hydrochloric acid and obtained a white crystalline solid containing *one* methyl group. This compound was no longer reducing, and also did not form an osazone. Thus Fischer had prepared methyl glucoside.

Ekenstein (1894) isolated a second isomer from the same reaction, and Fischer explained the existence of these two isomers by suggesting a *ring* structure for them, and followed Tollens in believing that they were five-membered rings.

There was no experimental evidence for the existence of a five-membered ring, and it was shown to be incorrect by Haworth, Hirst and their co-workers (1926 onwards).

They proved that the above glucosides were *six-membered* rings, *i.e.*, glucopyranosides. Their method was to fully methylate the glucoside (I), hydrolyse the product, methyl tetramethyl-α-D-ghicoside (II) to tetramethyl-α-D-glucose (III), oxidise this with bromine-water at 90°C to the lactone (IV), and then to oxidise this with nitric acid.

The product obtained was shown to be xylotrimethoxy-glutaric acid (V) this can be obtained directly by the oxidation of methylated xylose). The most reasonable interpretation of these results is in accordance with the assumption that methyl glucoside is a six- and *not* a five-membered ring. Thus:

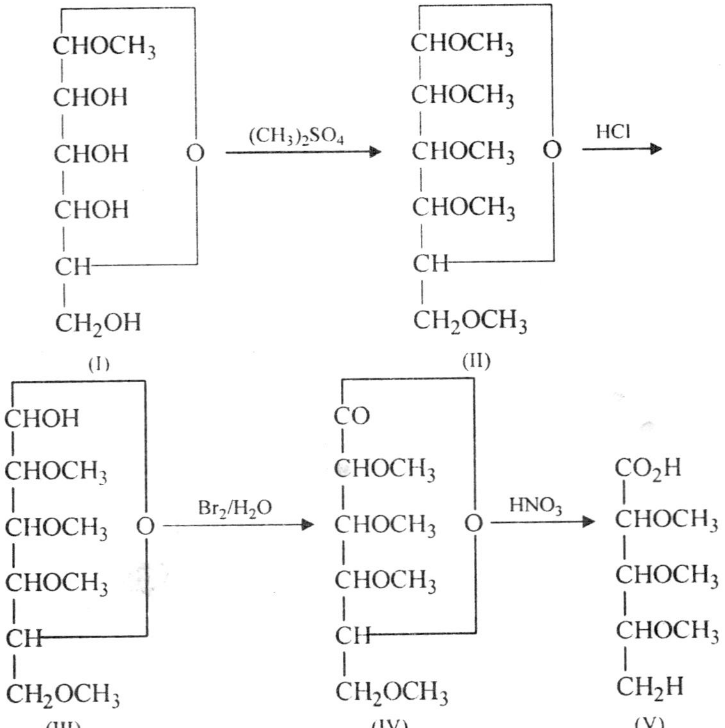

$$\begin{array}{l} CHOCH_3 \\ | \\ CHOH \\ | \\ CHOH \quad O \\ | \\ CHOH \\ | \\ CH \\ | \\ CH_2OH \\ \text{(I)} \end{array} \xrightarrow{(CH_3)_2SO_4} \begin{array}{l} CHOCH_3 \\ | \\ CHOCH_3 \\ | \\ CHOCH_3 \quad O \\ | \\ CHOCH_3 \\ | \\ CH \\ | \\ CH_2OCH_3 \\ \text{(II)} \end{array} \xrightarrow{HCl}$$

$$\begin{array}{l} CHOH \\ | \\ CHOCH_3 \\ | \\ CHOCH_3 \quad O \\ | \\ CHOCH_3 \\ | \\ CH \\ | \\ CH_2OCH_3 \\ \text{(III)} \end{array} \xrightarrow{Br_2/H_2O} \begin{array}{l} CO \\ | \\ CHOCH_3 \\ | \\ CHOCH_3 \quad O \\ | \\ CHOCH_3 \\ | \\ CH \\ | \\ CH_2OCH_3 \\ \text{(IV)} \end{array} \xrightarrow{HNO_3} \begin{array}{l} CO_2H \\ | \\ CHOCH_3 \\ | \\ CHOCH_3 \\ | \\ CHOCH_3 \\ | \\ CH_2H \\ \text{(V)} \end{array}$$

Fig. 21

Fischer (1914) also prepared methyl glucoside by dissolving glucose in methanol and letting it stand at room temperature in the presence of hydrochloric acid. He now obtained compounds which Haworth *et al.*, using the above method, showed to be iue-membered rings, *i.e.*, methyl D-glucofuranosides. Hudson (1937, 1939) has also determined the size of sugar-rings, but his method was to oxidise methyl glycosides with periodic acid. As we have seen, periodic acid splits 1,2-glycols, and *one molecule of acid* is used for *each pair of adjacent alcoholic groups*, *e.g.*,

$$R^1CHOHCHOHR^2 \xrightarrow{1\,HIO_4} R^1CHO + R^2CHO$$

$$R^1CHOHCHOHCHOHR^2 \xrightarrow{2\,HIO_4} R^1CHO + HCO_2H + R^2CHO$$

Compounds containing an aldehyde or ketonic group adjacent to an alcoholic group are disc attacked by periodic acid in a similar manner to glycols:

$$\text{RCHOHCHO} \xrightarrow{1\,\text{HIO}_4} \text{RCHO} + \text{HCO}_2\text{H}$$

$$\text{R}^1\text{COCOHR}_2 \xrightarrow{1\,\text{HIO}_4} \text{R}^1\text{CO}_2\text{H} + \text{R}^2\text{CHO}$$

Aldoses and pentoses are *completely* broken down by periodic acid, *i.e.*, *free* sugars behave as open-chain compounds in this reaction, *e.g.*, an aldopentose consumes four molecules of periodic acid, and the products are four molecules of formic acid and one molecule of formaldehyde (from the terminal CH_2OH group).

$$\text{OCHCHOHCHOHCHOHCH}_2\text{OH} \xrightarrow{4\,\text{HIO}_4} 4\text{HCO}_2\text{H} + \text{CH}_2\text{O}$$

Thus, estimating the periodic acid used and the formic acid and/or formaldehyde produced will indicate the number of free adjacent oxidisable groups (CHOH, CHO or CO). On the other hand, the ring in *glycosides* is stable in the presence of periodic acid. Thus, the oxidation of methyl glucoside (produced under reflux conditions; see above) uses two molecules of periodic acid and produces *one* molecule of formic acid. Only one structure fits these facts, namely, a *six-membered* ring:

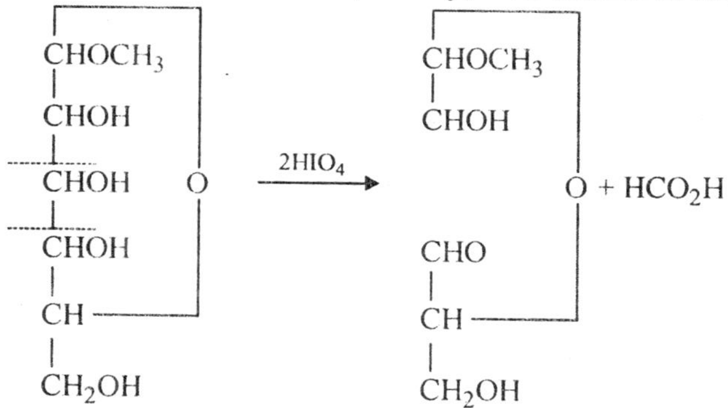

The other product of the oxidation has also been isolated and characterised by Hudson. This periodic acid oxidation method is now very widely used for structure investigation of carbohydrates.

Since methylation of sugars under different conditions produces different-sized rings, the question is: what is the size of the ring in the original sugar? Various sources of evidence, *e.g.*, X-ray analysis, show that the normal sugars are *pyranoses*.

One other point will be mentioned here, and that is the configuration of the *first* carbon atom. When the open-chain structure is closed to form the ring, two configurations of the new asymmetric carbon atom are possible, the one with the hydrogen to the left (the α-anomer), and the one with the hydrogen to the right (the β-). This is an arbitrary arrangement, but it has now been shown that the α- and β-anomers actually have the configurations originally arbitrarily assigned to them, *e.g.*. X-ray analysis of α-D-glucose has shown that the 1,2-hydroxyl groups are in the *cis*-position (McDonald *et al.*, 1950).

Since the L-series are the enantiomers of the D-series, in the former the a-anomer has the hydrogen on the first carbon atom to the right and the β-anomer has the hydrogen to the *left*. These are *opposite* to those of the D-series, *i.e.*, the configuration of C-l in an α-D-aldose is identical with that of the P-L-aldose.

GENERAL NATURE OF PROTEINS

Thename protein was introduced by Mulder (1839), who derived it from the Greek wordproteios (meaning, first). Proteins are nitrogenous substances which occur in the protoplasm of all animal and plant cells. Their composition varies with the source: carbon, 46-55 per cent; hydrogen, 6-9 per cent; oxygen, 12-30 per cent; nitrogen, 10-32 per cent; sulphur, 0.2-0.3 per cent. Other elements may also be present, *e.g.*, phosphorus (nucleoproteins), iron (haemoglobin). As we have seen proteins can be broken down into smaller and smaller fragments until the final products are the amino-acids. This sequence may be written:

protein → polypeptides → peptides → amino-acids

There is no sharp dividing line between peptides, polypeptides and proteins. One arbitrary convention designates proteins as those molecules with a molecular weight above ~ 10 000 and peptides (poly-peptides) as those molecules with a molecular weight below ~ 10 000. In general, proteins and peptides differ in physical and chemical properties which can be correlated with the differences in molecular size. Both groups often exhibit physiological activity, behaving as, *e.g.*, enzymes, hormones, growth factors, etc.

Synthetic peptides of very high molecular weight are often referred to as polypeptides, and their methods of preparation and the study of their properties have provided a great deal of information on the structure and properties of proteins.

Proteins are amphoteric, their behaviour as an anion or a cation depending on the pH of the solution. At some definite pH, characteristic for each protein, the positive and negative charges are exactly balanced, *i.e.*, there is no net charge on the protein molecule, and the molecules will not migrate in an electric field. In this condition the protein is said to be at its isoelectric point, and at this pH the protein has its least solubility, *i.e.*, it is most readily precipitated (cf. amino-acids). The osmotic pressure and viscosity of the protein solution are also a minimum at the iso-electric point. The amphoteric nature of proteins is due to the presence of a large number of free acidic and basic groups arising from the amino-acid units in the molecule. These groups can be titrated with alkali or acid, and by this means it has been possible to identify acidic and basic groups belonging to the various amino-acid units.

All proteins are optically active, and may be coagulated and precipitated from aqueous solution by heat, the addition of acids, alkalis, salts, organic solvents miscible with water, etc. Proteins in this precipitated state are said to be denatured, and the process of reaching this state, denaturation, occurs most readily near the isoelectric point. Denaturation is now believed to be the result of changes in conformation or unfolding of the protein molecule. Associated with denaturation are changes in optical rotation and (usually) the loss of biological activity, *e.g.*, enzymes (all are proteins) become inactive when denatured.

Denaturation is generally irreversible, but many examples are now known where the process has been reversed. This reversal of denaturation has been called renaturation or refolding. When denaturation is effected by heat, renaturation does not usually result on rapid cooling. If, however, cooling is carried out very slowly, renaturation often occurs. In these circumstances the process of renaturation has been referred to as annealing.

Proteins exhibit a variety of colour reactions, *e.g.*,

(i) **Biuret Reaction:** Addition of a very dilute solution of copper sulphate to an alkaline solution of a protein produces a red or violet colour. This reaction is due to the presence of the grouping —CO—NH—CHR—CO—NH—. At least two peptide linkages (—CONH—) must be present (dipeptides do not give the test).

(ii) **Xanthoproteic Reaction:** Proteins usually produce a yellow colour when warmed with concentrated nitric acid, and the colour becomes orange when the solution is made alkaline. This reaction is due to the nitration of the benzene ring in phenylalanine, tyrosine and tryptophan.

(iii) **Milton's Reaction:** Millon's reagent (mercuric nitrate in nitric acid containing a trace of nitrous acid) usually produces on addition to a protein solution a white precipitate which turns red on heating. This reaction is characteristic of phenols, and so is given by proteins containing tyrosine (this is the only phenolic amino-acid that occurs in proteins).

(iv) **Ninhydrin Test:** Proteins (and peptides) give this test, but the colours are different from that of the amino-acids.

The molecular weights of proteins have been determined by means of ultracentrifugal sedimentation, osmotic pressure measurements. X-ray diffraction, light scattering effects, molecular sieves (gel fihration), and by chemical analysis. Chemical methods are based on the estimation of a particular amino-acid. Thus, suppose the percentage composition of amino-acids in a protein has been determined. From these values it is possible to calculate the mole proportions of each amino-acid by dividing its percentage weight by its molecular weight. We now choose the amino-acid present in the least molar amount and on the assumption that only one of these amino-acid residues is present in the protein, the molecular weight, Mi of the protein, is given by

$$M_1 = \frac{100}{x} \times m \quad \text{or} \quad x = \frac{m}{M_1} \times 100$$

where x is the percentage weight and m is the molecular weight of the amino-acid. If two molecules of the amino-acid are present per molecule of protein, the percentage weight is still x, but now we have

$$x = \frac{2m}{M_2} \times 100 = \frac{2m}{2M_1} \times 100$$

i.e., the molecular weight M_2 is $2M_1$. Hence, if n molecules (where n must be an integral number) of the amino-acid are present, the molecular weight of the protein is nM_1. Therefore M_1 is the minimum molecular weight of the protein and nM_1 is its true molecular weight. As an example, let us consider the protein bovine insulin. The amino-acid that occurs in the smallest molar amount is threonine: 2 per cent, w = 119.

$$\therefore\ M_1 = \frac{100}{2} \times 119 = 5950$$

Now, bovine insulin has been shown to contain one molecule of threonine and hence its true molecular weight is also 5950.

Since the modern methods of estimating amino-acids have a high degree of accuracy, a knowledge of the minimum molecular weight is extremely valuable. This is because many of the methods used for the determination of molecular weights of proteins (and peptides) are apparently accurate only within the limits of about 3-5 per cent. It should be noted that elemental analysis as a means of obtaining molecular formulae of proteins is unsatisfactory because of their very high molecular weights.

The average molecular weight of the common amino-acids is about 141.5, and since one molecule of water is lost in the formation of the peptide bond, a peptide containing n amino-acid residues has an approximate molecular weight 141.5n − 18n == 123.5n. Hence, n = M/123.5. For the purpose of simplifying the calculation (and with little effect on the approximation), we may replace 123.5 by 125. The values of molecular weights recorded for proteins vary considerably, ranging from about 5000 to many millions.

One of the difficulties in protein chemistry (including peptides) is to be able to decide whether the specimen being investigated is pure. Although many proteins (and peptides) have been obtained crystalline, these have no characteristic melting points. Various criteria are therefore used to show homogeneity, *e.g.*, constant solubility, chromatography (column, paper, and ion-exchange), paper electrophoresis, etc.

Since the solubility of a protein depends on pH, the presence or absence of salts, etc. (cf. above), by controlling these factors it is possible to separate proteins. Thus, for example, by adjusting the pH of a solution containing a mixture of proteins (or peptides) to the isoelectric point of each protein in turn, each of these will be precipitated in turn. Alternatively, the salting out method may be used to separate proteins. The solubility of many proteins is increased in the presence of small concentrations of various neutral salts. This is referred to as salting in, and bivalent cations are more effective than univalent cations. As the concentration of the ion is increased, the solubility of the protein passes through a maximum, then begins to decrease and at a sufficiently high concentration (of ion) the protein is precipitated, *i.e.*, salted out. Not only can cations precipitate proteins (as salts), but so can suitable anions, *e.g.*, tungstic acid, phosphotungstic acid, trichloroacetic acid, etc.

A third method of separation of proteins based on solubility is the controlled precipitation by organic solvents miscible with water. From the foregoing account it can be seen that, in general, methods used for isolating proteins are also used for their separation and purification. Many proteins are not composed of a single peptide chain but consist of a number of subunits. Furthermore, these subunits may or may not be identical.

Peptides have been classified as homeomeric when the products of hydrolysis are amino-acids only, and as heteromeric when other products in addition to amino-acids are obtained. A large number of peptides are linear but many are cyclic. Cyclic peptides have been classified as homodetic when their structures contain only peptide linkages. On the other hand, when the rings contain both amide (peptide) and other types of linkages, *e.g.*, disulphide, the cyclic peptides are classified as heterodetk, *e.g.*, oxytocin. This latter class has also been called cyclodepsipeptides, but apparently some authors restrict this term to those heteromeric cyclic peptides composed of amino-acids and hydroxyacids linked by amide and ester bonds. In this case, the compounds have been referred to as the peptolides. They have been isolated from bacteria, fungi, etc., and many show biological activity.

CLASSIFICATION OF PROTEINS

Several arbitrary classifications of the proteins are in use. One method divides the proteins into two groups, fibrous proteins, which are insoluble in common solvents, but are soluble in concentrated acids and alkalis, and globular proteins, which are soluble in water and in dilute acids, alkalis and salts.

A more common method of classification is the division of proteins into the three main groups: simple, conjugated, and derived proteins. Each group is subdivided into a number of classes designated by general names. Each class contains sub-classes of proteins of similar but not identical physical and chemical properties, *e.g.*, in A(i), below, one sub-class of albumin is serum albumin. The term serum indicates that this group of albumins occurs in the blood serum of vertebrates, e.g; man, horse, sheep, dog, etc. However, all these serum albumins differ from each other.

(A) **Simple Proteins:** These give only amino-acids or their derivatives on hydrolysis.

(i) **Albumins:** These are soluble in water (and in acids and alkalis), and are coagulated by heat. They are precipitated by saturating their solutions with ammonium sulphate.

Albumins are usually low or deficient in glycine; some albumins are serum albumin, egg albumin and lactalbumin.

(ii) **Globulins:** These are insoluble in water, but are soluble in dilute salt solution and in dilute solutions of strong inorganic acids and alkalis. They are precipitated by half saturating their solutions with ammonium sulphate, and they are coagulated by heat.

Globulins usually contain glycine; some typical globulins are serum globulin, tissue globulin and vegetable globulin.

(iii) **Prolamins :** These are insoluble in water or salt solution, but are soluble in dilute acids and alkalis, and in 70-90 per cent ethanol.

Prolamins are deficient in lysine, and contain large amounts of proline; some prolamins are zein (from maize), gliadin (from wheat) and hordein (from barley).

(iv) **Glutelins :** These are insoluble in water or dilute salt solution, but are soluble in dilute acids and alkalis; they are coagulated by heat. They are comparatively rich in arginine, proline and glutamic acid.

Some glutelins are glutenin (from wheat) and oyrzenin (from rice).

(v) **Scleroproteins (Albuminoids) :** These are insoluble in water or salt solution, but are soluble in concentrated acids or alkalis.

Examples: keratin (from hair, hoof), fibroin (from silk); these are not attacked by enzymes.

Submembers of the scleroproteins are:

(a) Collagens (in skin, tendons and bones); these form gelatin (a water-soluble protein) when boiled with water. Collagens are attacked by pepsin or trypsin.

(b) Elastins (in tendons and arteries); these are not converted into gelatin, and are attacked slowly by trypsin.

(vi) **Basic Proteins :** These are strongly basic, and fall into two groups.

(a) **Histones :** These are soluble in water or dilute acids, but are insoluble in dilute ammonia. They are not coagulated by heat, and contain large amounts of histidine and arginine, but contain no tryptophan and very little cystine or methionine; they are hydrolysed by pepsin and trypsin.

Histones are the proteins of the nucleic acids, haemoglobin, etc.

(b) **Protamins :** These are more basic than the histones and have a simpler structure. They are soluble in water, dilute acids and dilute ammonia; they are not coagulated by heat, and are precipitated from solution by ethanol. They contain large amounts of arginine, and occur in various nucleic acids. They do not contain sulphur, and are hydrolysed by various enzymes, *e.g.*, trypsin, papain, but not by pepsin.

B. Conjugated Proteins: These are proteins which contain a non-protein group (*i.e.*, a compound not containing amino-acid residues) attached to the protein part. The non-protein group is known as the prosthetic group, and it may be separated from the protein part by careful hydrolysis.

(i) **Nucleoproleins:** The prosthetic group is a nucleic acid.

(ii) **Chromoproteins:** These are characterised by the presence of a coloured prosthetic group. Examples: chlorophyll and haemoglobin, but in many cases the prosthetic group is organic, *e.g.*, in visual purple the prosthetic group is a carotenoid pigment.

(iii) **Glycoproteins:** In these the prosthetic group contains a carbohydrate or a derivative of the carbohydrates. They are also known as mucoproteins.

(iv) **Phosphoproteins:** These are conjugated proteins in which the prosthetic group contains phosphoric acid in some form other than in the nucleic acids or in the lipoproteins.

(v) **Lipoproteins:** In these the prosthetic group is lecithin, kephalin, etc.

(vi) **Metalloproteins:** These all contain a metal which is an integral part of the structure. Many metals occur, *e.g.*, iron, magnesium, copper, manganese. Examples are haemoglobin and chlorophyll which, as we have seen, may also be classed as chromoproteins.

OXYTOCIN, INSULIN, THYROTROPIN RELEASING HORMONE, AND ANTAMANIDE

Oxytocin : As an illustration of the principles involved in the determination of the sequence of amino-acids, we shall first consider oxytocin, the hormone which occurs in the posterior pituitary gland and is responsible for uterine contraction. The structure was established independently by du Vigneaud *et al.* (1953, 1954) and Tuppy *et al.* (1953). Oxytocin is extracted from the gland by acetic acid and is purified by chromatography or electrophoresis.

The isoelectric point of oxytocin is 7.7, a value which suggests the presence of a free amino group and no free carboxyl group. Complete hydrolysis with acid and the quantitative estimation of the amino-acids (chromatography on starch) showed the presence of an equimolecular mixture of eight acids: cystine, glycine, leucine, isoleucine, proline, aspartic acid, glutamic acid and tyrosine. Ammonia was also obtained, the ratio of this to any one amino-acid being 3:1. The production of ammonia in this proportion suggests the presence of three carbonamido groups. Also, the molecular weight of oxytocin (determined by physical methods) was about 1 000, a value which indicates that the molecule is an octapeptide.

Tuppy's procedure was as follows. Since oxidation of oxytocin with performic acid gives a di-sulphonic acid with a molecular weight corresponding to an oxytocin disulphonic acid, this suggests that oxytocin is a ring compound, the ring therefore including the S—S bond of cystine. On controlled hydrolysis of oxidised oxytocin with hydrochloric acid, four dipeptides and two tri-peptides were isolated, together with two molecules of cysteic acid.

(I) Asp $\rightarrow$ $CySO_3H$	(II) $CySO_3H$ $\rightarrow$ Tyr
(III) Leu $\rightarrow$ Gly	(IV) Ileu $\rightarrow$ Glu
(V) Tyr-(Glu, lieu)	(VI) $CySO_3H$.(Leu, Pro)

The sequence in each dipeptide (I)-(IV) was established by the DNP method, *i.e.*, treatment of the dipeptide with FDNB followed by hydrolysis with acid, and identification of the dinitrophenyl derivative (the 'end group') by chromatography. End-group analysis of (V) showed tyrosine was the amino-terminal residue, and it therefore follows from the sequence in (IV) that the sequence in (V) is Tyr $\rightarrow$ Ileu $\rightarrow$ Glu (this must be so, since only one amino-acid residue of each kind is present in oxytocin). Furthermore, from the sequence in (II) it follows that in oxytocin the sequence of four amino-acids is $CySO_3H$ $\rightarrow$ Tyr $\rightarrow$ lieu $\rightarrow$ Glu. Also, from the sequence in (III), the sequence in (VI) must be $CySO_3H$ $\rightarrow$ Pro $\rightarrow$ Leu, since Leu must be the terminal residue in order that (III) may be obtained. Hence the sequence of these four amino-acids is $CySO_3H$ → Pro → Leu → Gly.

Partial hydrolysis of oxidised oxytocin with the proteinase isolated from Bacillus subtilis gave glycine amide and tetrapeptides (VII) and (VIII), the amino-acids in which were identified by

(VII) $CySO_3H$.(Glu, Tyr, Ileu) (VIII) Asp.($CySOa_3$, Leu, Pro)

hydrolysis and chromatography, and also the end-group was determined. The sequence in (VII) has already been established to be $CySO_3H$ → Tyr → Ileu → Glu (see above), and since the addition of Asp to (VI) gives (VIII), it follows that the sequence in (VIII) is Asp → $CySO_3H$ → Pro → Leu (Leu shown to be the end-group).

Isolation of glycine amide (from the enzyme hydrolysis) shows that it is an end-group, and from (III) and (VIII) it follows that there is the following sequence:

Asp → $CySO_3H$ → Pro → Leu → $GlyNH_2$

Since the amino-terminal group in (VII) is $CySC_3H$, combining the two sequences now established, the sequence in oxidised oxytocin that accounts for all the facts is:

$$
\begin{array}{l}
\ NH_2 \quad\ NH_2 \\
\quad | \qquad\quad | \\
CySO_3H \rightarrow Tyr \rightarrow Ileu \rightarrow Glu \rightarrow Asp \rightarrow CySO_3H \\
\qquad\qquad\qquad\qquad\qquad\qquad \rightarrow Pro \rightarrow Leu \rightarrow GlyNH_3
\end{array}
$$

The carbonamido groups have been placed as shown because (a) oxytocin contains three such groups (see above); (b) the terminal glycine amide accounts for one (see above); (c) glutamic and aspartic acids are the only two acids which each possess two carboxyl groups, and since in the others, all monocarboxylic acids, the carboxyl group must be involved in the peptide link, then only these two dicarboxylic acids can have carbonamido groups. There is, however, the problem of deciding which carboxyl group is the carbonamido group, *i.e.*, whether it is the α or γ one of Glu and the α or β one of Asp.

Finally, since oxidised oxytocin is produced without chain fission, it suggests the presence of the S—S ring (see above). Assuming that α-carboxyl groups (of Glu and Asp) are involved in the peptide linkages, then the structure of oxytocin is:

```
                          NH2     NH2
                          |       |
Cy → Tyr → Ileu → Glu → Asp → Cy → Pro → Leu → GlyNH2
|                               |
S-------------------------------S
```

Du Vigneaud's procedure is different from Tuppy's in that the structure of oxytocin was determined mainly as the result of the examination of many fragments obtained by the partial hydrolysis of oxytocin, performic acid-oxidised oxytocin oxidised with bromine-water, and desulphurised oxytocin. The resulting peptides were separated into acidic and neutral components by means of ion-exchange resins, and were then further separated by paper chromatography. It was shown that oxidised oxytocin had only one ^v-terminal group (DNP method) and that this was cystine. When oxidised oxytocin was treated with bromine-water, a dibromopeptide and a heptapeptide were obtained. Hydrolysis and end-group analysis of the dipeptide showed it to be $CySO_3H \rightarrow TyrBr_2$ (3,5-dibromo derivative). Hydrolysis of the heptapeptide gave $CySO_3H$, Leu, Ileu. Pro, Glu, Asp, Gly and ammonia, and end-group analysis showed that the N-terminal residue was isoleucine. Since oxytocin has only one terminal amino group (see above), the amino group in isoleucine must have formed the peptide link with tyrosine. Thus, the sequence of three residues is established: $CySO_3H \rightarrow$ Tyr $\rightarrow$ Ileu.

Controlled hydrolysis of the heptapeptide produced four fragments, (XIII)-(XVII), and hydrolysis of desulphurised oxytocin (by means of Raney nickel) gave four fragments, (XVHI)-(XXI).

(IX) Asp, $CySO_3H$ (X) $CySO_3H$, Pro (XI) $CySO_3H$, Pro, Leu (XIII) $CySO_3H$, Pro, Leu, Gly (XIII) $CySO_3H$, Asp, Glu (XIV) Leu, Gly, Pro (XV) CySSCy, Asp, Glu (XVI) Tyr, CySSCy, Asp, Glu (XVII) Tyr, CySSCy, Asp, Glu, Leu, Ileu (XVIII) Ala, Asp (XIX) Glu, Ileu (XX) Ala, Asp, Glu (XXI) Ala, Asp, Glu, Leu, Ileu

In peptides (XVII) and (XXI), differentiation between Leu and Ileu was not made, *i.e.*, these peptides contain only one of these acids, but which one was not determined (both acids appeared together on the chromatogram). Application of the DNPmethod to (IX) showed that its sequence was Asp $\rightarrow CySO_3H$. Consideration of the acids in (IX)-(XII) shows that the sequence of five residues in oxidised oxytocin is therefore (XXII):

(XXII) Asp → $CySO_3H$ → Pro → Leu → Gly

This accounts for (XIV) and at the same time shows its sequence. On the other hand, since (XIII) contains (IX), it follows that Glu may be added to form the sequence (XXIII).

(XXIII) Glu → Asp → $CySC_33H$ → Pro → Leu → Gly

Now, in desulphurisation, the —CH_2S— group is converted into the —CH_3 group. Thus, instead of cystine, two molecules of alanine (which is not present in oxytocin) will be produced. Hence, (XVIII) corresponds to (IX) and (XX) to (XIII). Also, the isolation of (XIX) shows that Glu is linked to lieu, and since Glu is linked to Asp as shown in (XXIII), lieu must be in the sequence (XXIV).

(XXIV) Ileu → Glu → Asp → $CySC_3H$ → Pro → Leu → Gly

Since lieu is now assigned, it follows that (XVII) is Tyr, CySSCy, Asp, Glu, Ileu, and (XXI) is Ala, Asp, Glu, Ileu. If Tyr is joined to one half of the cystine residue, with Asp joined to the other half, then (XVI) is accounted for, *i.e.*, oxytocin contains the sequence

Tyr → CyS

|

Ileu → Glu → Asp → CyS → Pro → Leu → Gly

This accounts for the eight amino-acids, and since the only free amino group present is in cystine (see above), and since oxidation does not bring about fission, oxytocin must be cyclic, and this is satisfied by joining Tyr to lieu. The Gly end is not satisfactory, since this residue is present as carbon-amide. This was confirmed by application of the Edman method of end-group analysis to oxidised oxytocin. The first four acids were removed, the order of removal being: $CySC_3H$, Tyr, lieu and Glu. The carbonamido groups were placed as described above (in Tuppy's method), and the structure for oxytocin was therefore the same as that established by Tuppy. The structure of oxytocin has been confirmed by a number of syntheses. The one described here is that of du Vigneaud *et al.* (1959). In the following equations, the symbols used are OEt = ethyl ester, NP = p-nitrophenyl ester, Bzl = benzyl, Z = benzyloxycarbonyl:

Insulin

This is the hormone which occurs in the pancreas and was the first protein whose amino-acid sequence was worked out (Sanger et al., 1951-1955). Measurement of the molecular weight of insulin gave values which varied with the concentration. The values were multiples of 12000, viz. 12000, 24000, and 36000. The minimum molecular weight of insulin, determined from amino-acid analysis, was about 6 000. It was originally believed that 12000 was the true molecular weight, but later work based on osmotic pressure and sedimentation measurements in organic solvents showed that the 'monomer' of insulin actually did have the molecular weight 6000.

N-Terminal amino-acid determination (DNP method) showed the presence of one glycine residue and one phenylalanine residue. Hence, insulin contains two peptide chains linked together. Since amino-acid analysis had shown the presence of cystine, it was assumed that the two chains were joined by disulphide bonds. Insulin was therefore oxidised with performic acid. This produced two peptides, which were separated (by electrophoresis or by chromatography) and examined individually. The peptide with the N-tenninal glycine residue was called the A-chain and that with the N-terminal phenylalanine residue was called the B-chain.

Each chain was subjected to partial hydrolysis with acids and with enzymes. The A-chain gave about 35 fragments with acid hydrolysis and about 10 fragments with enzymic hydrolysis; the B-chain gave respectively about 50 and about 10. The fragments were separated (by electrophoresis or by chromatography) and examined by end-group analysis (DNP method) and for amino-acid residues. Then, by means of the overlapping procedure, the primary structure of each chain was deduced. The A-chain was shown to contain 21 amino-acid residues and the B-chain 30. The A-chain contained four cysteic acid residues and the B-chain two. Hence, the A-chain contains a disulphide ring and is linked to the B-chain by two disulphide bonds.

Insulins from different sources, *e.g.*, cattle, sheep, horses, etc., differ slightly, but all show identical hormonal activity. The formula shown is that of bovine insulin. It will be seen that there is a small ring system containing cystine, alanine, serine and valine.

The differences in the insulins from various sources appear to concern this ring only. Thus, the sequence Ala → Ser → Val in bovine insulin is replaced by Ala → Gly → Val in sheep insulin, and by Thr → Gly → Ileu in horse insulin (Brown *et al.*, 1955; Harris *et al.*, 1956). Katsoyannis *et al.* (1966) have synthesised human and sheep insulins; Merrifield *et al.* (1968) have synthesised the A- and B-chains by the solid phase method.

THYROTROPIN RELEASING HORMONE (TRH)

Schally *et al.* (1966, 1969) isolated (by means of electro-phoresis and chromatography) a few milligrams of this hormone in an impure state from about a quarter of a million porcine hypothalami. Because of the small amount available, the general strategy of determining primary structures of proteins was very much restricted. Acid hydrolysis of the hormone gave three amino-acids, histidine, glutamic acid and proline, in essentially equi-molecular proportions. It was also shown that these three amino-acids were derived from the sequence Glu-His-Pro (I) as a probable part of TRH. (I) and other synthetic tripeptides of these three amino-acids in several alternative sequences, however, showed no hormonal activity (Schally *et al.*, 1968). It was also shown that TRH had neither a free amino nor a free carboxyl group. One possibility is that the tripeptide is cyclic.

Folkers *et al.* (1969) carried out synthetic experiments on (I), modifying both the amino and carboxyl groups and tested the hormonal activity of each product. In this way it was shown that methylation (carboxyl groups) and ammonation of (I) gave a compound which was identical with TRH biologically and chroma-tographically. The structure of TRH was thought to be (II), L-pyroglutamyl-L-histidyl-L-proline amide, (pyroGlu-His-ProNH_2), and this was confirmed by (i) hydrolysis to the three amino-acids (Glu, His and Pro); (ii) its NMR spectrum. Assignable signals were (τ-values): 2-39 bs (2-H His); 3-10 bs (4-H His); 5.55 m (α-H His); 5.80 m (α-H pyroGlu); 6.35 m (α-H Pro); 7.0 m (CH_2 His); 7.0 m (5-CH_2 Pro); 7.7 m (CH_2CH, pyroGlu); 8.1 bm (CH_2CH_2 Pro) [b = broad]; (iii) seventeen R_F values obtained in different solvent systems were identical with the R_F values of TRH under the same conditions. Antamanide. This is an antitoxin which has been isolated from the fungus Amanita phalloides by chromatographic methods. Its structure has been elucidated by Wieland *et al.* (1968) who used a combination of gas chromatography and mass spectrometry.

Hydrolysis of antamanide (HCl—AcOH) gave a mixture of the amino-acids alanine, phenyl-alanine, proline, and valine in the molar ratio 1:4:4:1. Since it did not give the ninhydrin test and also was electrophoretically neutral, antamanide was assumed to be a cyclic peptide. The mass spectrum showed the presence of molecular ion at At = 1146. This corresponds to ~ 9-2 amino-acid residues and taking this in conjunction with the molar ratio given above, antamanide is most probably a cyclic decapeptide. This cyclic structure was supported by the fact that the molecular ion had a very high intensity. It was also established that all the amino-acids had the L-configuration.

The first approach to the determination of the primary structure was purely chemical. Hydrolysis of antamanide with HC1—AcOH under controlled conditions gave a mixture of linear peptides and unchanged cyclopeptide. These were separated by preparative TLC. Amino-acid analysis of the quickest migrating component showed it was composed of two Phe, one Ala, one Val, and four Pro residues. Application of the Edman method showed the sequence:

Phe . Phe . Ala . Val . Pro . Pro . (Pro . Pro)

This leaves two Phe residues to be accounted for, and these were originally placed at the beginning of the chain to give (—Phe_4.Ala.Val.Pro_4—) as the sequence in the cyclic decapeptide. The octapeptide: Phe.Phe.Ala.Val.Pro.Pro.Pro.Pro was synthesised (standard methods were used), but it was found to be chromatographically different from the octapeptide obtained by fragmentation of antamanide (see above).

Since further chemical work failed to elucidate the amino-acid sequence, mass spectrometry was used. As already mentioned, M = 1146, and an unusual feature of the mass spectrum was the absence of intense fragment ions above m/e 588 and m/e 560, the only exception being an ion at M — 91 (91 = benzyl from Phe). On the basis of the ions observed, it was possible to deduce two probable pentapeptide sequences. The absence of characteristic fragment ions between m/e 588 and m/e 1055 (M—91) may be readily explained by the immediate decom-position of the antamanide molecule into two pentapeptides, each of which then decomposes stepwise.

On the basis of the above observations and deductions, it was concluded that the structure of antamanide is derived by coupling the two sequences Pro.Pro.Phe.Phe.Val and Pro.Pro.Ala.Phe.Phe. These sequences were confirmed by a combination of gas chromatography and mass spectrometry.

MOLECULES OF PROTEIN

The spatial arrangement of the polypeptide chain in a protein molecule is determined by the primary structure (amino-acid sequence) of that protein. The conformation that the polypeptide 'backbone' assumes is called the secondary structure, and the way in which the entire molecule folds to produce a specific shape is called the tertiary structure of the protein. Finally, there is the quaternary structure of a protein. This is concerned with those proteins which contain subunits and is a description of the arrangement and ways in which the subunits are held together. Various types of bonds and/or forces are responsible for the stabilisation of these protein' structures' other than the primary structure.

The Hydrogen Bond : This is an extremely important factor in the stabilisation of protein conformations. The most common type of hydrogen bonding occurs between the carbonyl oxygen of one peptide bond and the hydrogen atom of the amino group of another peptide bond. Although this bond is weak, because a large number of them are involved, the overall stabilisation is considerable. The strength of the hydrogen bond is a maximum when the three atoms (O, H, and N) are collinear, and a result of hydrogen bonding is a shortening of the distance between O and N (*i.e.*, less than that calculated from the sum of the van der Waals radii). This shortening has been used as a means of establishing the presence of these bonds in crystalline proteins (X-ray analysis). Infrared spectroscopy is also a means of showing the presence of hydrogen bonds.

$$>C=O..........H-N<$$

ELECTROSTATIC FORCES

These may be forces of repulsion or attraction, depending on the charge on the polar groups. The two polar groups are carboxylate, $-CO_2^-$, and ammonium, $-NH_3^+$. When like-charged polar groups are close together, the repulsive forces will destabilise the protein conformation. When the polar groups carry unlike charges, the attractive forces will stabilise the conformation; these attractive forces are often referred to as 'salt linkages' or ionic bonds. Unlike hydrogen bonds, these electrostatic forces are independent of the orientation of the two polar groups with respect to each other. In aqueous solution, the polar groups will be hydrated and this will considerably decrease the electrostatic forces. Also, since the charges ($-CO_2^-$, and $-NH_3^+$) depend on the pH of the solution the conformation can also depend on the pH.

Other forces that may operat are dipole-dipole interactions and van der Waals forces.

1. Compound A, $C_5H_{10}O_4$, is oxidised by Br_2-H_2O to the acid, $C_5H_{10}O_5$, which readily forms a lactone. A forms a triacetate with

Ac_2O and an osazone with $PhNHNH_2$. A is oxidised by HIO_4, only one molecule of which is consumed. What is the structure of A?

2. Compound A, $C_5H_{10}O_4$, on oxidation with Br_2—H_2O; gives acid B, $C_5H_{10}O_5$. A forms a triacetate (Ac_2O) and is reduced by HI to n-pentane. Oxidation of A with HIO_4 gives, among other products, 1 molecule of CH_2O and 1 molecule of HCO_2H. What are the possible structures of A and how could you distinguish between them?
3. Designate, using R and S symbols, D (+) glucose.
4. Ruthenium tetroxide is known is oxidise a single CHOH group to ketone. How could you convert $HOCH_2(CHOH)_4CHO$ into $HOCH_2(CHOH)_2COCHOHCHO$?

Inter-and Intramolecular Chemical Bonds

The disulphide bond has already been discussed in connection with the primary structure of a protein. Since, however, the primary structure influences the other types of 'structure', the presence of disulphide bonds will affect the conformation of the protein.

THE HYDROPHOBIC BOND

The exact nature of this 'bond' is still a matter of debate and, in any case, the term 'bond' is misleading since no type of bond (in the usual meaning of the term) is involved. According to one theory, hydrophobic bonds are a consequence of the hydrophobic character (*i.e.*, non-polar and have little attraction for water) of alkyl side-chains of the amino-acids and also a consequence of the structure of water.

Water molecules form hydrogen bonds among themselves, but this produces only three-dimensional clusters. This cluster-formation results in a decrease in the entropy of the system. Furthermore, since hydrophobic side-chains can fit into suitable types of clusters of water molecules, the clusters are further stabilised and so the entropy of the system is further decreased.

On the other hand, if different hydrophobic side-chains in the protein molecule come into contact with one another (through folding of the chain), the area now exposed to water is decreased and so the number of 'cages' is reduced.

This results in an increase in entropy, *i.e.*, the formation of hydrophobic bonds, viz. contact between hydrophobic side-chains, increases the stability of the system. Hence, maximum stability is achieved by the hydrophobic side-chains lying inside the protein molecule. At the same time, hydrophilic groups (*i.e.*, polar groups, and have a high attraction for water) will tend to lie on the surface of the protein molecule.

The orders of hydrophobic and hydrophilic character are:

Hydrophobic: Phe > Ala > Val > Gly > Leu > Cys.

Hydrophilic: Tyr > Ser > Asp > Glu > $AspNH_2$ > $GluNH_2$ > Arg.

Secondary Structure of Proteins

The α-Helix. The α-helix model for the conformation of proteins was proposed by Pauling *et al.* (1951). This was suggested on theoretical grounds, and its presence was subsequently verified by experimental evidence. Some of the arguments on which this model was based were:

(i) The peptide group is planar.

(ii) Hydrogen bonding stabilises the conformation, and the strength of this bond is a maximum when the atoms concerned (C=O—H—N) are collinear or, failing this ideal situation, do not deviate by more than 30° (from collinearity).

(iii) The model to be chosen permits the maximum number of hydrogen bonds (of the type in (iii)).

The model which best satisfied these requirements was the α-helix. Pauling then proposed a helix in which each turn contained either 3.7 or 5.1 amino-acid residues. Further considerations (largely stereochemical) showed that a helix with 3.7 residues per turn was more stable than any other. This α-helix is represented by (I) and it should be noted that each hydrogen bond is formed between the CO group of one residue and the NH group of the fourth residue in the chain. This hydrogen bonding prevents free rotation and so the helix is rigid. Furthermore, at least three adjacent hydrogen bonds must be broken before free rotation can occur in a segment of the helix.

The α-helix may be left- or right-handed. The common amino-acids, except glycine, are optically active and all have the L-configuration. Moffitt (1956) deduced theoretically that the right-handed helix (for

L-amino-acids) is more stable than the left-handed helix. Hence, the right-handed helix is the one that would be expected to occur naturally.

The β-conformation: Pauling *et al.* (1951) also proposed another conformation, the β–conformation or pleated sheet. In this the polypeptide chain is extended and chains are held together by inter-molecular hydrogen bonds. Two types of pleated sheets are possible, parallel (IIa) in which all the chains run in the same direction, and anti-parallel (IIb) in which the chains run alternately in opposite directions.

The existence of the a-helix in proteins in the solid state has been established by X-ray analysis. Also, X-ray data have shown that the a-helix has two types of repeat units, one being the pitch or distance between two successive turns (~50-5.5Å), and the other being the distance, in the direction of the helical axis, between two like atoms in the chain (1.5 Å), *e.g.*, the 'rise' from the first N to the second N in NHCOCHRNH. The diameter of the helix has been estimated to be about 10Å. ·

Not all polypeptide chains are capable of forming the α-helix, since the stability of this helix depends on the nature and sequence of the side-chains (R groups) in the polypeptide chain, *e.g.*, proline inhibits the formation of a helical conformation. This is due to the fact that the peptide bond is of the type —CO—N<, *i.e.*, there is no hydrogen on the nitrogen atom and hence hydrogen bonding is not possible (for proline). Thus the α-helix ends at this point (the proline molecule), proceeds in some conformation other than the helix and can then start again as a helix. The amount of the helix form varies in proteins, ranging from zero to about 100 per cent.

X-ray analysis has also established the existence of the pleated sheet structure in solid proteins. The chains are parallel in, *e.g.*, keratin, and are anti-parallel in, *e.g.*, fibroin. The calculated distance between two CHR groups on the same side of a chain is 7.2 Å, but the actually measured repeat unit is 7.0 Å. This shortening has been attributed to crowding caused by the side-chains (R), thereby preventing the chain from being fully extended.

Although, X-ray analysis may be used to determine the conformations of proteins in the solid state, other methods must be used to investigate the conformations in solution. As we have seen, the stability of the α-helix is largely dependent on hydrogen bonding. Thus, in solvents which cannot form hydrogen bonds with the protein solute, *e.g.*, chloroform, there will be no competition and so a helical protein will tend to retain this conformation. On the other hand, if the solvent can form hydrogen bonds with the protein solute, *e.g.*, water, acetic acid, etc., the conformation of the protein will now depend on the relative strengths of the 'chain' hydrogen bonds, those between the protein and solvent, and those between solvent and solvent. With water as solvent, all these hydrogen bonds have about the same strength and so the a-helix is relatively stable. On the other hand, with dichloroacetic acid as solvent, hydrogen bonding between protein and solvent occurs to a large extent, resulting in extensive loss of the helical conformation and the appearance of the randomly coiled configuration. Complete loss of the α-helix occurs when the helical protein is dissolved in water containing urea. Urea is a resonance hybrid and the negatively charged oxygen atom can therefore form hydrogen bonds with the protein molecule which are stronger than the 'chain' hydrogen bonds.

$$(H_2\ddot{N})_2C=O \longleftrightarrow (H_2\overset{+}{N})(H_2\ddot{N})C{-}\bar{O} \longleftrightarrow (H_2\ddot{N})(H_2\overset{+}{N})C{-}\bar{O}$$

The random coil conformation is, unlike the α-helix, very flexible and the change from one form into the other, *i.e.*, the helix-coil transition, can be effected by changes in temperature or pH.

Optical rotatory studies have provided a valuable means of estimating the helical content of proteins in solution. As we have seen the common amino-acids have the L-configuration and so it can be anticipated that proteins will exhibit optical rotations which will have some relationship to the constituent amino-acid residues (the value of the rotation for proteins is always negative). In relatively short peptides, the optical rotation is approximately the sum of the contributions of the amino-acid residues present (see Rule of optical superposition). As the peptide chain increases in length, the deviations from this 'addition rule' may become considerable. In the randomly coiled configuration, the addition rule holds reasonably well, but when the chain assumes a helical conformation, because this structure is chiral the optical rotation of the protein will consist of the sum of two contributions, that due to the rotation of the amino-acid residues (negative) and that due to the helix. Furthermore, the contribution by the helix will depend on its length. Thus, it may be possible to estimate approximately the amount of helical content present in the protein from the optical rotations (observed and estimated from the addition rule).

Better quantitative results have been achieved by means of optical rotatory dispersion" studies. Both the amplitude and the shape of ORD curves are very sensitive to changes in the helical content of proteins.

Other methods are also available for the determination of helical content. Infrared spectroscopic studies have shown that the frequency of the absorption band due to hydrogen bonding depends on the C=O--H—N angle. Since these are quite different in the a-helix and pleated sheet, these two conformations can be distinguished.

Deuterium exchange experiments have been used to estimate helical content. When dissolved in deuterium oxide, rapid exchange would be expected to occur between deuterium (of D_2O) and protein hydrogen atoms attached to oxygen or nitrogen. Experimental results, however, have shown that for most proteins the exchange rates are much lower than expected. This slow exchange has been attributed to the hydrogen atoms of the peptide groups (CONH) in the helical regions of the protein molecule. The reason for this is uncertain, but it is possible to estimate helical contents, the values of which are in reasonable agreement with those obtained by ORD studies.

NMR spectroscopy has recently been used to distinguish between the α-helix and random coil configurations. The NMR spectrum of a single 'straight-chain' polypeptide is essentially that derived from the superposition of its constituent amino-acid residues. As a result of experimental work, it has been found that the signal of an α-proton (α-CH—NH—) in a helix occurs upfield with respect to that in the random coil and that the NH signal in the helix occurs downfield with respect to that in the random coil. Hence, it is possible to detect helix-coil transitions by NMR spectroscopy.

TERTIARY STRUCTURE OF PROTEINS

As we have seen hydrogen bonding is of extreme importance in the stabilisation of the secondary structures, the a-helix and pleated sheets. On the other hand, folding of the entire molecule, *i.e.*, the tertiary structure, involves hydrogen bonding, ionic, chemical, and hydrophobic bonds. The tertiary structure that a protein assumes under normal conditions of temperature and pH will be its most stable arrangement. This has been referred to as the native conformation of that protein. Two major molecular shapes occur naturally, fibrous and globular. Fibrous proteins have a large helical content and are essentially rigid molecules of rod-like shape.

On the other hand, globular proteins have a polypeptide chain which consists partly of helical sections and folded about the random coil sections to give a 'spherical' shape. In globular proteins, most polar groups lie on the surface of the molecule and most hydro-phobic side-chains lie inside the molecule.

The tertiary structures of proteins have been elucidated by methods which give information on the shapes of molecules, *e.g.*. X-ray analysis, viscosity measurements, diffusion, light-scattering, ultracentrifuge method, electron microscopy (cf. molecular weights of proteins).

When a protein undergoes denaturation the changes that occur involve changes in secondary and/or tertiary structures of proteins. This has been established by, *e.g.*, large changes in optical rotation and in the ORD curves of the protein.

α-keratin, which is found in hair and wool, consists of three (or seven) α-helices wound round each other like strands in a rope. On the other hand, silk fibroin consists of pleated sheats in which the polypeptide chains are anti-parallel. These two are fibrous proteins. Two globular proteins are mycoglobin and haemoglobin, both of which contain haem (they are chromoproteins).

QUATERNARY STRUCTURE OF PROTEINS

Both fibrous and globular proteins may consist of only one polypeptide chain or of severarchains. In the latter case, the protein is said to be oligomeric, the individual chains being known as protomers or subunits. These subunits may or may not be identical, and when they are held together by hydrogen bonds, they may be separated by, *e.g.*, dissolving in water containing urea. Mycoglobin consists of a single polypeptide chain which contains about eight straight segments (α-helices) which are folded in an irregular manner at the random-coil sections. Haemoglobin, however, contains four subunits, two identical α-chains and two identical β-chains. Each subunit has a tertiary structure similar to that of mycoglobin.

The Peptide Linkage

As we have seen proteins are hydrolysed by acids, alkalis, or enzymes to a mixture of amino-acids. Fischer (1902) and Hofmeister (1902) suggested that amino-acids in proteins are joined in a linear fashion by peptide linkages, *i.e.*, by the —CONH— group, the carboxyl group of one amino-acid molecule forming an amide by combination with the amino-group of the next amino-acid molecule, etc. Thus, on this basis, a protein molecule may be represented as a linear polymer of amino-acid molecules. The infrared spectra of amides have been extensively studied and many assignments associated with the —CONH— group have been made. Thus, for example, polypeptides and proteins show bands near 3300 and 3100 cm^{-1}, which are characteristic of the hydrogen-bonded N—H group (str.) in secondary amides (RCONHR). Also shown are bands near 1650 and 1550 cm^{-1}, which are characteristic of the C=O (str.) in secondary amides. Ultraviolet spectra studies have shown that the peptide bond absorbs in the region 180-220 nm.

Pauling *et al.* (1953) have carried out X-ray studies on a number of crystalline peptides and deduced the various bond lengths (in A) and bond angles in these compounds.

The conclusions reached from these results were:

(i) The atoms in the group, —CONH—, are planar and the O and H are trans.

(ii) Since the peptide C—N bend length (*i.e.*, C—N of —CONH—), 1.32 Å is shorter than the usual C—N bond length (~ 1.47 Å), this bond has some double bond character.

We can explain the double bond character as being due to resonance, and hence hindered rotation about this C—N bond permits the possibility of geometrical isomerism, the (rans-isomer being the more likely because of the much larger steric repulsion operating in the cis-isomer.

On the other hand, rotation can occur about the R^1CH—CO and the R^2H—NH bond. It is therefore possible to describe the conformation of the protein molecule in terms of rotations about the C^α—C' bond (ψ) and about the N—C^α bond (ϕ). Thus, the conformation of a protein molecule containing n amino-acid residues is described by the parameters:

$$(\phi_1\psi_1)\ (\phi_2\psi_2)\ \ldots\ (\phi_n\psi_n)$$

This problem of conformation is considered further.

The Primary Structure of Peptides

The primary structure of a peptide (or protein) is the sequence of the amino-acid residues in the molecule. First, let us consider a dipeptide composed of two different amino-acids, **A** and **B**. These may be combined in two different ways:

H_2N—**A**—CONH—**B**—CO—H HO_2C—**A**—NHCO—**B**—NH_2
(I)   (II)

H_2N—**B**—CONH—**A**—CO_2H

(IIa)

Inspection of either (I) or (II) shows that the two ends of each molecule are different. The 'amino-end' is said to be N-tenninal and the 'carboxyl-end' is said to be C-tenninal. The general method of writing the sequence of amino-acids in a peptide (polypeptide or protein) is with the terminal amino-group on the left. (I) is in accordance with this convention, but not (II), which should therefore be written as (IIa). The peptides are then named as acylated derivatives of the terminal amino-add residue on the right hand side.

Table gives a list of symbols used for amino-acids; they are usually the first three letters of the name. When the sequence of the amino-acids is not known, the symbols are enclosed in brackets and are separated by commas. When the sequence is known, the units are separated by dots or dashes, or by arrows which indicate the direction of linkage from carboxyl to amino. Since the conventional way of writing peptide formulae has the terminal amino-group on the left (see (I) and (IIa)), the arrow will point left-to-right. A terminal amino-group may be indicated by H and a terminal carboxyl group by OH. Finally, carbonamido-groups may be indicated by the addition of NH_2 to the symbol, *e.g.*, asparagine and glutamine (these have proposed alternatives): AspNH; (Asn) and $GluNH_2$ (Gin). Also, a proposed alternative for tryptophan is Trp. The following formulae illustrate the conventions:

(Ala, Gly, Tyr) Thr.(Val, Arg) Ala.Glu.Val or

H.Ala.Glu.Val.OH or Ala → Glu → Val or Ala—Glu—Val

Returning now to the problem of amino-acid sequence, it can be shown that, in general, for n different acids n different combinations are possible. Furthermore, had not the common amino-acids been all L-acids, the total number of possible combinations would have been very much larger. As a simple example let us consider a tripeptide. The first thing to do in this case is to determine the nature of the amino-acid residues. This is usually carried out by acid hydrolysis and chromatography. Suppose the amino-acids are shown to be A, B and C. These can be written in six (3!) different combinations:

A.B.C   A.C.B   B.A.C   B.C.A   C.A.B   C.B.A

The problem then is: how do we ascertain which is the actual combination of the tripeptide under investigation? Inspection of the formulae shows that if we were able to determine the N-terminal amino-acid (N—T—AA), we would be able to group the six possibilities into three pairs, *i.e.*, we would now know that our tripeptide was either one or other of a pair. We would also be in this situation if we were able to determine the C-tenninal amino-acid (C—T—AA). Thus:

N-Terminal	**C-Terminal**
(a) A.B.C and A.C.B	(d) B.C.A and C.B.A
(b) B.A.C and B.C.A	(e) A.C.B and C.A.B
(c) C.A.B and C.B.A	(d) A.B.C and B.A.C

Since this results in different pairing, the determination of both N- and C-terminal groups will give the amino-acid sequence of the tripeptide, *e.g.*, if the N—T—AA determination showed that the tripeptide was in group (b) and the C—T—AA determination showed that the tripeptide was in group (d), the tripeptide is therefore B.C.A.

Now let us assume that the N— and C—T—AA methods are such that their application results in the removal of the respective terminal amino-acid. In these circumstances we would be left with different fragments according to the order of application (of the methods), *e.g.*, for B.C.A:

(i) N—T—AA first: fragment C.A.

(ii) C—T—AA first: fragment B.C.

By repeating either of these determinations, the amino-acid sequence is solved. Thus, the sequence may be determined by use of one method twice or by use of each method once.

Now let us consider the tetrapeptide whose acids have been shown to be (A, B, C, D). There are 24 (4!) possible combinations (as shown). Suppose the N— and C—T—AA determinations

A.B.C.D	B.A.C.D	C.A.B.D	D.A.B.C
A.B.D.C	B.A.D.C	C.A.D.B	D.A.C.B
A.C.B.D	B.C.A.D	C.B.A.D	D.B.A.C
A.C.D.B	B.C.D.A	C.B.D.A	D.B.C.A

A.D.B.C	B.D.A.C	C.D.A.B	D.C.A.B
A.D.C.B	B.D.C.A	C.D.B.A	D.C.B.A

showed respectively B and D. The tetrapeptide is therefore B.A.C.D or B.C.A.D. As for the tripeptide, the fragments obtained will depend, on the order of application:

(i) N—T—AA first: fragments A.C.D or C.A.D.

(ii) C—T—AA first: fragments B.A.C or B.C.A.

(iii) Both N— and C—T—AA (irrespective of order): fragments A.C or C.A.

Hence, the amino-acid sequence of the tetrapeptide may be determined by using either the N—T—AA or the C—T—AA method three times, or by a combination of these methods (also three operations). The general method of amino-acid sequence analysis, however, does not use both end-group analyses on the original peptide. Only one end-group is determined and this is then followed by fragmentation of the peptide in at least two different ways. The smaller peptides are then subjected to amino-acid sequence determination by end-group analysis. In this way, the various small peptides 'overlap' and so it becomes possible to deduce the complete sequence of the amino-acids in the original peptide.

C-TENNINAL AMINO-ACID DETERMINATION

The most widely used method is that of hydrazinolysis (Akabori *et al.*, 1956). The peptide (or protein) is heated with anhydrous hydrazine at 100°C. This converts all amino-acid residues except the C-terminal one into amino-acid hydrazides.

$$\text{---NHCHR}^1\text{CONHCHR}^2\text{CONHCHR}^3\text{CO}_2\text{H}$$

$$\downarrow \text{N}_2\text{H}_4$$

$$\text{---+ H}_2\text{NCHR}^1\text{CONHNH}_2 + \text{H}_2\text{NCHR}^2\text{CONHNH}_2 + \text{H}_2\text{NCHR}^3\text{CO}_2\text{H}$$

The mixture of products is subjected to chromatography on a column of a strong cation-exchange resin. On elution the strongly basic hydrazides are retained, but the free amino-acid is eluted and can be identified.

Another chemical method involves the reduction of the peptide (or protein) with lithium boro-hydride or lithium aluminium hydride. This converts the free terminal carboxyl group to a primary alcoholic group. Hydrolysis produces a mixture of amino-acids and an amino- alcohol, the latter being separated and identified by paper chromatography.

A widely used method makes use of the enzyme carboxypeptidase. This enzyme attacks peptides (or proteins) only at the end which contains the free α-carboxyl group. When this terminal amino-acid residue is liberated, the new terminal free carboxyl group is attacked by the enzyme. Thus, in the peptide --- X.Y.Z, after a given time of hydrolysis, a number of 'successive' terminal amino-acids will have been liberated, but in amounts Z > Y > X > ---. Hence, by identification and quantitative determination of the amino-acids, their sequence can be established.

N-Terminal Amino-Acid Determination

The Edman method (1950) is very widely used and its basis is the reaction between phenyl isothiocyanate and the peptide (or protein) to form the phenyl-thiocarbamyi (PTC)-peptide (or protein) in the presence of dilute alkali. When treated with dilute acid (hydrochloric or trichloroacetic acid), the PTC-peptide is converted into a phenyl-thiohydantoin (PTH) and a peptide which now has tost the N—T—AA of the original of the peptide. The mechanism of this step is uncertain. The PTH may be separated and identified by paper chromatography,

$$C_6H_5NCS + H_2NCHR^1CONHCHR^2CONHCHR^3CO_2H \xrightarrow{OH^-}$$

$$C_6H_5NHCSNHCHR^1CONHCHR^2CONHCHR^3CO_2H \xrightarrow{H^+}$$

$$\begin{array}{c} NH - CHR^1 \\ | \quad\quad | \\ SC \quad\quad CO \\ \diagdown \; \diagup \\ N \\ | \\ C_6H_5 \end{array} + H_2NCHR^2CONHCHR^3CO_2H$$

and the process can now be repeated on the degraded peptide. The Edman N-terminal amino-acid determination has now been automated, and because of this, can be used to determine the amino-acid sequence in polypeptides, *i.e.*, the step involving the splitting of peptides (or proteins) into smaller fragments may be unnecessary in many cases, particularly where the peptide is relatively small.

Another widely used method is the DNP method (Sanger, 1945). l-Fluoro-2,4-dinitrobenzeae (FDNB) very readily reacts with amino-groups in the presence of sodium hydrogen carbonate solution (mildly alkaline) at room temperature to form 2,4-dinitrophenyl (DNP) derivatives which are stable to acids. Hence, hydrolysis with acid of the DNP-peptide produces the DNP-amino-acid and a mixture of free amino-acids. DNP-derivatives are formed with any free amino-group. Thus the basic amino-acid, lysine, will react even if it is not an Merminal group (see Table). The hydroxyl group of tyrosine, the thiol group of cysteine, and the imidazole nucleus of histidine also react (although, more slowly than an amino-group). Hence, the DNP method may give rise to a number of DNP derivatives. These, however, may be readily isolated and identified by chromatography (particularly TLC). If the basic amino-acid is not N-terminal, then it will form the mono-DNP derivative; if N-terminal, the di-DNP derivative. The DNP derivatives of most of the amino-acids have been prepared and characterised. The DNP method cannot be used repetitively, since its use requires complete hydrolysis of the DNP derivative (cf. the Edman method).

A recent modification of the DNP method is the use of 5-dimethylaminonaphthalene-1-sulphonyl chloride, 'dansyi' chloride (DNS-C1), in place of FDNB. This modification is called the 'dansyi method', and its use is similar to that of the DNP method (replace FDNP by DNS-C1 in the above equation). This dansyi method is now being widely used because the dansyi group, being highly fluorescent, permits the detection and estimation of dansyi amino-acids in minute amounts by fluorimetric methods. Apart from some other chemical methods, an enzymic method is also available for N-terminal amino-acid determination. The enzyme leucine aminopeptidase attacks peptides (or proteins) only at the end which contains the free amino-group and proceeds to liberate, in succession, each new terminal amino-acid. Hence, after a given time of hydrolysis, estimation of the amounts of free amino-acids will give their sequence.

PARTIAL HYDROLYSIS OF PEPTIDES

Since the 'overlapping procedure' is generally used for the elucidation of the amino-acid sequence, methods are necessary to bring about partial hydrolysis of peptides (or proteins). Different hydrolytic paths are possible because of the different susceptibilities of the various peptide bonds to attack by hydrolytic reagents. First, let us consider the application of the overlapping procedure. Suppose we have a hexapeptide whose amino-acids have been shown to be (A, B, C, D, E, F) and whose N-terminal amino-acid has been shown to be C. The hexapeptide may therefore be written as C(A, B, D, E, F). Now suppose that on partial hydrolysis the small peptides obtained were (as shown by amino-acid analysis):

C—A, (B, E), (B, D), C.(A, E), (B, D, F) and (B, D, E, F)

Since we have C—A, C.(A, E) must be C—A—E. Hence (B, E) is E—B. and we now have C—A—E—B. Therefore (B, D) is B—D, and so we have C—A—E—B—D. Finally, since only F is missing, the hexapeptide is C—A—E—B—D—F; this explains fragment (B, D, E, F). These results may be tabulated as shown.

As the peptide becomes more complex, the amino-acids usually occur more than once. This will increase the difficulty in elucidating the amino-acid sequence, *e.g.*, both hexapeptides C—A—E—B—D—E and C—A—E—E—B—D might possibly give rise to fragments C—A, (B, E), (B, D), C(A, E), (B, D, E) and (B, D, E, E). Since two (B, D, E) are possible, it will be necessary to use end-group analysis to decide which is correct; overlapping is of no help here.

N—T—AA			C—T—AA		
C	A				
C	A	E			
		E	B		
			B	D	
		E	B	D	F
C	A	E	B	D	F

Suppose we found it to be E(B, D). This could still have been derived from either hexapeptide. If, however, we found it to be B(D, E), then the hexapeptide cannot be C—A—E—E—B—D. It also follows that (B, E) is E—B and so B(D, E) is B—D—E. The order in (B, E) would be confirmed by end-group analysis, and this would also show (B, D, E, E) is E(B, D, E). Hence, the hexapeptide is C—A—E—B—D—E. From this example it can be seen that it will usually be necessary to carry out end-group analysis on each fragment-peptide obtained. It can also be seen that had the fragment (E, E) appeared in the partial hydrolysate, solution to the problem would have been easier. Its absence suggests that the particular method of hydrolysis used readily splits the E—E bond. Hence, it is better to use different hydrolytic reagents which can selectively break peptide bonds.

Partial hydrolysis with acids is generally unsatisfactory, since bond-breaking tends to occur randomly and also results in a large number of small peptides which may be very difficult to separate. Even so, this approach is often successful for relatively small peptides. On the other hand, enzymic hydrolysis is extremely useful because each enzyme hydrolyses only certain types of peptide bond. Trypsin splits peptide bonds in which the carbonyl group is part of a lysine or an arginine residue. Chymotrypsin splits peptide bonds in which the carbonyl group is part of a phenylalanine, tyrosine, or tryptophan residue. Hence, the separate use of these two enzymes will result in splitting of the peptide (or protein) in different ways to give relatively large fragments.

Other enzymes are also available, *e.g.*, carboxypeptidase leucine aminopeptidase pepsin (NH group of Leu, Asp, Glu, etc.), papain (CO group of Gly, Arg, Lys, etc.), etc. Since the specificity of the various enzymes differs considerably, the less specific enzymes are generally more satisfactory when used with relatively small peptides. Large peptides require the use of the specific enzymes; otherwise a large number of fragment-peptides will be obtained (see acidic hydrolysis, above).

Chemical methods have also been introduced to split peptides (or proteins) at specific peptide bonds. One of the most successful is the reaction between cyanogen bromide and the peptide in aqueous formic acid at room temperature. Only peptides in which the CO group is that of a methionme residue are split, the products being a homoserine lactone and the 'rest' of the peptide.

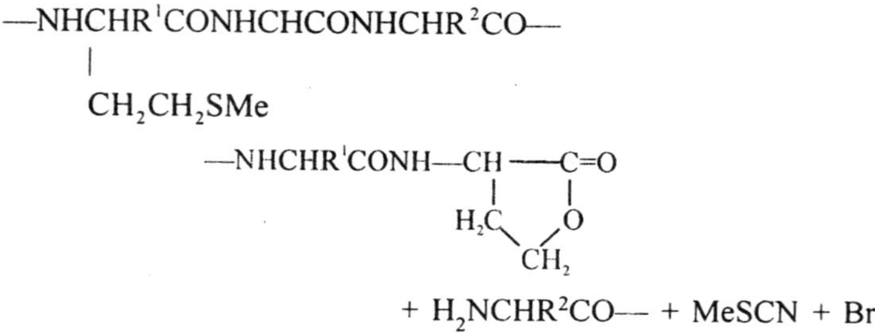

$$\text{—NHCHR}^1\text{CONH}\underset{\displaystyle |\atop \displaystyle CH_2CH_2SMe}{\text{CH}}\text{CONHCHR}^2\text{CO—}$$

$$\text{—NHCHR}^1\text{CONH—CH—C=O (ring: } H_2C\text{—}CH_2\text{—O)}$$

$$+ H_2NCHR^2CO\text{—} + MeSCN + Br^-$$

It should also be remembered that this step of cleaving peptides into smaller fragments may be avoided in many cases.

Protein Subunits, Cyclic Structures, and Disulphide Bonds

In this section we shall deal with some further problems involved in the determination of amino-acid sequence. If the protein consists of a single linear peptide chain, then the methods described above may be readily applied. Thus, end-group analysis will show the presence of a single N—T—AA or of a single C—T—AA (or one of each if both end-group analyses are performed). In many cases, however, the results indicate the presence of n end groups (N or C). This usually means that the protein is composed of n peptide chains. If there are no end groups (*i.e.*, no free α-NH_2 or α-CO_2H), this is strong evidence that the protein has a cyclic structure. Other evidence that may be used to show the presence of a cyclic peptide is the neutrality towards electrophoresis and the failure to give the ninhydrin test.

When the subunits in a protein are not cross-linked by covalent bonds, they are readily dissociated into the individual units by, *e.g.*, dissolving in a urea solution. These units are then separated, purified and examined by the methods described above. On the other hand, if cystine is present the subunits may be held together by the disulphide bond and/or a single chain may contain an intramolecular disulphide ring. These bonds are usually split before the primary structure is determined. One common method is oxidation of the molecule with performic acid to give cysteic acid (the sulphonic acid). Alternatively, the disulphide bond may be reduced to thiol by means of, *e.g.*, sodium borohydride? and the products treated with, *e.g.*, iodoacetic acid. The primary structures of these products are then determined and the positions of the disulphide linkages are deduced from the positions (in the sequence) of the cysteic acid residues or of the carboxy-methylcysteine residues. It should also be noted that performic acid oxidises methionine to the corresponding sulphone and also destroys tryptophan.

More recently, the primary structures of peptides and proteins have been elucidated by means of mass spectrometry. Because the volatility of the amino-acids is so low, special techniques were developed to examine their mass spectra. Even so, this led to a number of inaccuracies and so more volatile derivatives have been used. Me and Et esters have been examined in great detail and one fragment pattern usually observed is that due to fissions at a and b. The molecular ion is generally observed, but its intensity is usually weak. On the other hand, the intensity of the ion $[\overset{+}{N}H_2 = CHCO_2Et]$ m/e 102, is usually strong or medium. Also, the intensity of the 'amine ion' $[RCH = \overset{+}{N}H_2]$ (M – 73) is generally strong or medium and so it is often possible to deduce the size of R (subtract 29 mass units ($CHNH_2$) from M – 73, *i.e.*, M – 102). This means that it is often possible to identify the amino-acid.

The peak at m/e 102 is always accompanied by a peak at m/e 74. This arises from the McLafferty rearrangement. The McLafferty rearrangement can also occur in amine ions when R contains more than two carbon atoms. This gives rise to a strong peak at m/e 30.

This, together with other peaks, may often be a means of identifying valine, leucine and isoleucine.

N-Trifluoroacetylamino-acids, as their methyl esters, are very good derivatives for the separation of amino-acid mixtures by GLC, and these derivatives can be identified by mass spectrometry. Now let us consider the application of mass spectrometry to the determination of the primary structures of peptides (and proteins). Here also, because peptides are involatile, they are first modified chemically. One method is the reduction with lithium aluminium hydride or lithium boro-hydride to give a more volatile product (peptide bonds are reduced, as is also a terminal carboxyl group, to give a polyaminoalcohol).

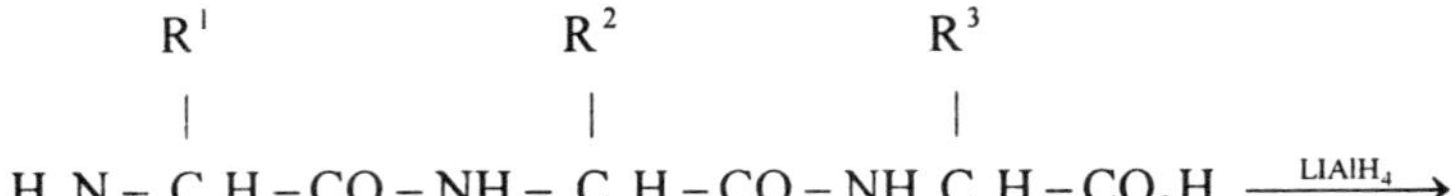

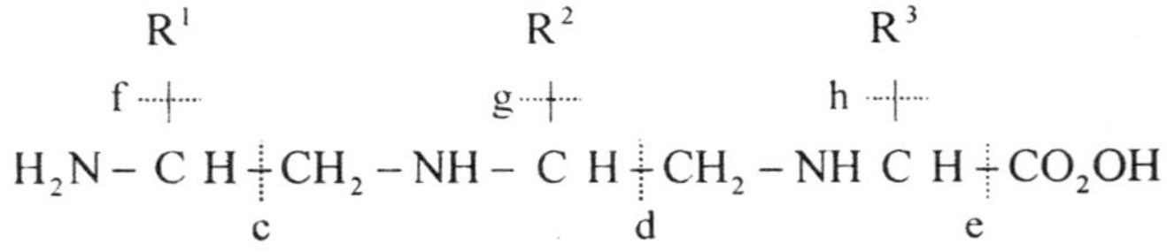

$$H_2N-\underset{|}{\overset{R^1}{C}}H-CO-NH-\overset{R^2}{\underset{|}{C}}H-CO-NH\,\overset{R^3}{\underset{|}{C}}H-CO_2H \xrightarrow{LiAlH_4}$$

$$H_2N-\overset{R^1}{\underset{f}{C}}H \,\vdots_{c}\, CH_2-NH-\overset{R^2}{\underset{g}{C}}H \,\vdots_{d}\, CH_2-NH\,\overset{R^3}{\underset{h}{C}}H \,\vdots_{e}\, CO_2OH$$

When this product is subjected to electron bombardment, it tends to undergo fission at bonds c, d, and e (these are at the most highly branched carbon atoms), *e.g.*, fission at c gives:

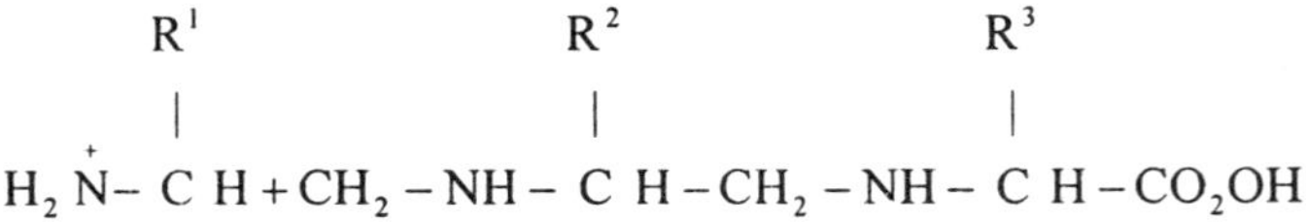

$$H_2\overset{+}{N}-\overset{R^1}{\overset{|}{C}}H+CH_2-NH-\overset{R^2}{\overset{|}{C}}H-CH_2-NH-\overset{R^3}{\overset{|}{C}}H-CO_2OH$$

Thus, the peak for the amine ion will show the size of R^1, and from this can be deduced the size of $(R^2 + R^3)$. Similarly, fission at d will give the sizes of $(R^1 + R^2)$ and R^3. Hence the sizes of R^1, R^2, and R^3, and therefore the corresponding amino-acids are now known, as is also their sequence. With more complicated peptides, peaks due to additional fissions at f, g, h, etc. will also have to be considered.

A later approach is the use of methyl, ethyl, and higher esters of peptides acylated at free amino-groups, *e.g.*, acetyl, trifluoroacetyl, etc.

(cf. amino-acids, above). In these acyi derivatives, the major fragmentations occur at the peptide bond (CO—NH) and also at the RC—CO bond.

It might also be noted that interpretation of the mass spectra of peptides is facilitated if the nature of the constituent amino-acids has been determined first.

SUMMARY OF PRIMARY STRUCTURE DETERMINATION

Here, we shall summarise the strategy adopted to determine the primary structure of peptides and proteins, but it should be realised that there may be variations which depend on the nature of the molecule under consideration.

(i) The peptide or protein must be isolated in a pure state.

(ii) It is first necessary to ascertain whether the protein consists of a single peptide chain or whether it is composed of a number of subunits. If the latter, then the subunits are separated and each chain is examined separately.

(iii) The protein is completely hydrolysed into its constituent amino-acids and their nature and amounts are determined.

(iv) The minimum molecular weight is determined from the amino-acid percentage composition and the molecular weight is also determined by a physical method.

(v) End-group analysis is carried out to determine the nature of the N-and C-terminal groups.

(vi) The amino-acid sequence may be determined by the Edman automated N-terminal method where possible. Alternatively, if the protein is relatively small, it may be subjected to controlled hydrolysis to give a number of simple peptides. These are isolated and purified, and end-group analysis is then applied to these and the amino-acid sequence in the protein may be deduced by the overlapping procedure. When the protein is relatively large, partial hydrolysis is effected in at least two different ways. The amino-acid sequence in each purified fragment is determined and the amino-acid sequence in the protein is deduced by the overlapping procedure.

Mass spectrometry may also be used to determine the amino-acid sequence in the protein or in the various fragments obtained by partial hydrolysis.

SYNTHESIS OF PEPTIDES

The general principles may be illustrated by consideration of the synthesis of a dipeptide. As we have seen two different amino-acids, L—A and L—B, may be combined in two different ways (2!):

$$\underset{\text{(I)}}{H_2NACONHBCO_2H} \qquad \underset{\text{(II)}}{H_2NBCONHACO_2H}$$

To prepare (I), the amino-group of A must be protected (regiospecific control) and the carboxyl group of A must be activated so that it readily reacts with the free amino-group of B. Similarly, to prepare (II), the amino-group of B must be protected and the carboxyl group of B must be activated. Hence, if Y is the protecting group and Z is the activating group, we have:

$$\text{(I) } H_2NACO_2H \xrightarrow{Y} YNHACO_2H \xrightarrow{Z} YNHACOZ$$

$$\xrightarrow{H_2NBCO_2H} YNHACONHBCO_2H \rightarrow \underset{\text{(I)}}{H_2NACONHBCO_2H}$$

$$\text{(II) } H_2NBCO_2H \xrightarrow{Y} YNHBCO_2H \xrightarrow{Z} YNHBCOZ$$

$$\xrightarrow{H_2NACO_2H} YNHBCONHACO_2H \rightarrow \underset{\text{(II)}}{H_2NBCONHACO_2H}$$

In each case, the final step involves the removal of the protecting group Y to give the dipeptide.

Other routes to dipeptides are as follows. The amino-group of the amino-acid which is to be N-terminal is protected and so is the carboxyl group of the amino-acid which is to be C-terminal. These two protected amino-acids may then be combined directly by means of a suitable reagent to give a dipeptide protected at both its N- and C-terminals. Thus (I) may be synthesised as follows (R is the carboxyl protecting group):

$$YNHACO_2H + H_2NBCO_2R \xrightarrow{-H_2O} YNHACONHBCO_2R$$

$$\xrightarrow{2\text{ steps}} \underset{\text{(I)}}{H_2NACONHBCO_2H}$$

Alternatively, the carboxyl group of the M-terminal protected amino-acid is converted into an activated group and, after combination of the two amino-acid derivatives, a dipeptide is obtained which again has both its N- and C-terminals protected, *e.g.*, the synthesis of (II):

$$YNHBCOZ + H_2NACO_2R \rightarrow YNHBCONHACO_2R$$

$$\xrightarrow{2\,steps} H_2NBCONHACO_2H$$

(II)

To extend the length of the peptide chain, one of the protecting groups in the dipeptide is selectively removed and the peptide is built up from this end. Thus, a peptide chain may be extended, one amino-acid residue at a time, from either end of its precursor. On the other hand, a number of suitable simple peptides may be synthesised, these then linked together to give the required protected peptide (or protein), from which the protecting groups are finally removed.

One other point that requires consideration is that if the amino-acid side-chain contains reactive groups, these must be protected. Such reactive groups are, *e.g.*, amino (lysine), carboxyl (aspartic acid), hydroxyl (tyrosine), thiol (cysteine).

Many protecting groups have been introduced and their number is continually increasing. Fischer (1901-1907) introduced methods which, although they led to the synthesis of an octadeca-peptide, are no longer used.

Since peptide synthesis involves protecting N- and C-terminals (and also reactive side-chains), it is necessary to use protecting groups which can be selectively removed one at a time. It is also important that protecting groups should be easily introduced and should be removable under sufficiently mild conditions that the peptide bond is not hydrolysed and that no racemisation or rearrangements occur.

Five useful amino protecting groups are: benzvloxycarbonyl (carbobenzyloxy), t-butyloxy-carbonyl (Boc; carbo-t-butyloxy), trityl (triphenylmethyl), phthaloyl, and tosyi (Ts; p-toluene-sulphonyl). The usual method of protecting a carboxyl group is esterification, the common esters being methyl, ethyl, benzyl, and t-butyl. Reactive side-chain protecting groups are, *e.g.*, benzyl for thiol and hydroxyl, acetyl for hydroxyl, etc.

Activation of the carboxyl group has been carried out in various ways, *e.g.*, by conversion into the acid chloride, acid azide, or p-nitrophenyl ester. Finally, the direct combination between an end amino-group and an end carboxyl group is effected by means of dicyclohexylcarbodi-imide (DCO in an organic solvent (methylene dichloride, THF, etc.). The following account illustrates the applications of these techniques.

Bergmann (1932) introduced benzyloxycarbonyl chloride (also known as carbobenzyloxy chloride) as an amino protecting group, and this appears to be the most widely used method of protection. It is readily prepared by the action of carbonyl chloride on benzyl alcohol in toluene solution.

$$C_5H_5CH_2OH + COCl_2 \rightarrow C_6H_5CH_2OCOCl + HCl$$

The procedure is then as follows:

$$C_6H_5CH_2OCOCl + R^1CH(NH_2)CO_2H \xrightarrow{OH^-}$$

$$C_6H_5CH_2OCONHCHR^1CO_2H \xrightarrow{PCl_5}$$

$$C_6H_5CH_2OCONHCHR^1COCl \xrightarrow[NaOH]{R^2CH(NH_2)CO_2H}$$

$$C_6H_5CH_2OCONHCHR^1CONHCHR^2O_2H \xrightarrow{H_2-Pd}$$

$$C_6H_5CH_3 + CO_2 + NH_2CHR^1CONHCHR^2CO_2H$$

If the amino-acid contains sulphur, then catalytic reduction cannot be used, since the sulphur poisons the catalyst; the removal of the blocking group, however, may be successfully accomplished by means of sodium in liquid ammonia.

A later method of removing this group is to treat the derivative with hydrogen bromide in acetic acid or nitromethane (Ben-Jshai *et al.*, 1952; Anderson *et al.*, 1952):

$$C_6H_5CH_2OCONHCHR^1CONHCHR^2CO_2H \xrightarrow{HBr}$$

$$C_6H_5CH_2Br + CO_2 + \bar{B}r\overset{+}{N}_3\,CHR^1CONHCHR^2CO_2H$$

The use of N-benzyloxycarbonyl derivatives causes no appreciable racemisation.

There are other ways in which this peptide synthesis may be carried out, *e.g.*, the p-nitrophenyl ester method (Z = benzyloxycarbonyl group):

$$ZNHCHR^1CO_2C_6H_4NO_2 \xrightarrow{NH_2CHR^2CO_2H}$$

$$ZNHCHR^1CONHCHR^2CO_2H + HOC_6H_4NO_2 \xrightarrow[CH_3CO_2H]{HBr} \cdots$$

The use of this ester, an activated ester, depends on the fact that it readily reacts with an amino group; the corresponding ethyl ester combines much more slowly. p-Nitrophenyl esters are prepared in high yield by the addition of dicyclohexylcarbodi-imide ($C_6H_{11}N{=}C{=}NC_6H_{11}$), a dehydrating agent, to a solution of the benzyloxycarbonyl derivative of the amino-acid and p-nitrophenol in ethyl acetate (du Vigneaud *et al.*, 1959).

On the other hand, the azide method may be used instead of acid chlorides or esters:

$$ZNHCHR^1CO_2Me \xrightarrow{N_2H_4} ZNHCHR^1CONHNH_2 \xrightarrow{HNO_2}$$

$$ZNHCHR^1CON_3 \xrightarrow{NH_2CHR^2CO_2C_2H_5}$$

$$ZNHCHR^1CONHCHR^2CO_2C_2H_5 \xrightarrow{acid}$$

$$ZNHCHR^1ONHCHR^2CO_2H \xrightarrow{H_2-Pd} NH_2CHR^1ONHCHR^1CO_2H$$

The azide synthesis is not accompanied by racemisation.

The t-butyloxycarbonyl reagent is not used as its chloride, since this is unstable, but is used as its p-nitrophenyl ester:

$$(CH_3)_3COCO_2C_6H_4NO_2 + NH_2CHRCO_2H \rightarrow$$

$$(CH_3)_3COCONHCHRCO_2H + HOC_6H_4NO_2$$

This group is readily removed by HBr—CH_3CO_2H, and also by HCl—CH_3CO_2H, the latter being particularly useful in that the benzyloxycarbonyl group is not removed by this reagent. The best reagent for removing the t-butyloxycarbonyl group appears to be trifluoroacetic acid.

The trityl reagent is simple to use, and may be removed by heating in acetic acid or catalytically (H_2—Pd), *e.g.*,

$$(C_6H_5)_3CCl + NH_2CHRCO_2CH_3 \xrightarrow{Et_3N}$$

$$(C_6H_5)_3CNHCHR^1CO_2CH_3 \xrightarrow[(ii)\,CH_3CO_2H]{(i)\,NaOH}$$

$$(C_6H_5)_3CNHCHR^1CO_2H \xrightarrow[(ii)\,NH_2CHR^2CO_2H]{(i)\,SOCl_2}$$

$$(C_6H_5)_3CNHCHR^1CONHCHR^2CO_2H \xrightarrow{CH_3CO_2H}$$

$$(C_6H_5)_3COOCCH_3 + NH_2CHR^1CONHCHR^2CO_2H$$

Sheehan *et al.* (1940) have used the phthaloyl group as a means of protecting an end amino-group (cf. Gabriel's phthalimide synthesis).

Phthaloylation occurs without racemisation provided the temperature does not exceed about 150°C. On the other hand, Nefkens (1960) has introduced a much milder method of phthaloylation starting from N-carbethoxyphthalimide (prepared from potassium phthalimide and ethyl chloro-formate). This reagent reacts with amino-acids in aqueous sodium hydrogen carbonate solution at room temperature to form the optically pure phthaloyl derivative in excellent yield.

Weygand *et al.* (1961) have also prepared phthaloyl derivatives of amino-acids (without racemisation) by heating the acid with diethyl phthalate and triethylamine in phenol. An example of the use of tosyi chloride as a protecting reagent is (Fischer, 1915):

$$TsCl + NH_2CHR^1CO_2H \xrightarrow[(ii)\,CH_3CO_2H]{(i)\,NaOH}$$

$$TsNHCHR^1CO_2H \xrightarrow{SOCl_2}$$

$$TsNHCHR^1COCl \xrightarrow{NH_2CHR^2CO_2H}$$

$$TsNHCHR^1CONHCHR^2CO_2H \xrightarrow[liq.\ NH_3]{Na}$$

$$NH_2CHR^1CONHCHR^2CO_2H.$$

No racemisation occurs in this synthesis.

Sheehan *et al.* (1956) showed that a protected N-amino-acid combines directly with an amino-acid ester in the presence of dicyclohexylcarbodi-imide in an inert solvent (methylene chloride, THF, etc; Phth = phthaloyl group; see also above):

$$PhthNHCHR^1CO_2H + NH_2CHR^1CO_2C_2H_5 + C_6H_{11}N = C{=}NC_6H_{11} \rightarrow$$

$$PhthNHCHR^1ONHCHR^2CO_2C_2H_5 + C_6H_{11}NHCONHC_6H_{11}$$

The mechanism of this reaction is believed to be:

$$R^1CO_2H + C_6H_{11}N{=}C{=}NC_6H_{11} \rightarrow C_6H_{11}N{=}\underset{}{\overset{OCOR^1}{\overset{|}{C}}}-NHC_6H_{11}$$

$$\xrightarrow{R^2NH_2} R^1CONHR^2 + C_6H_{11}NHCONHC_6H_{11}$$

This reaction occurs with very little racemisation, but there are usually side-reactions which make difficult the purification of the desired product.

A different approach to peptide synthesis is the anhydride method. One application is as follows, the cyclic anhydride, N-carboxyanhydride (NCA) being the unit for polymerisation. The NCA derivative may be prepared in a number of ways. A common method is to convert the amino-acid into its N-benzyloxycarbonyl derivative and then to proceed as shown.

$$PhCH_2OCONHCHRCO_2H \xrightarrow{PCl_5} PhCH_2OCONHCHRCOCl$$

$$\xrightarrow[60°C]{\text{heat in vacuo}} \begin{array}{l} RHC-CO \\ \;| \qquad\quad >O \\ HN-CO \end{array} + PhCH_2Cl$$

(NCA)

Polymerisation is effected by heating the NCA derivative in an organic solvent (dimethylformamide, dioxan, etc.) in the presence of a catalyst, *e.g.*, water, amines, etc.

$$n \begin{array}{l} RHC-CO \\ \;| \qquad\quad >O \\ HN-CO \end{array} + H_2O \rightarrow nCO_2 + H_2NCHRCO-(NHCHRCO)_{n-1}OH$$

If a mixture of different anhydrides is used, the product is a polymer containing different residues in random distribution.

The NCA derivative can also be used to build up a peptide chain, one amino-acid residue at a time. NCA combines with an amino-acid in alkaline solution (pH 10), and after acidification the product is a dipeptide.

$$\begin{array}{l} R^1HC-CO \\ \;| \qquad\quad >O \\ HN-CO \end{array} + H_2NCHR^2CO_2^- \xrightarrow{OH^-} \bar{O}_2CNHCHR^1CONHCHR^2CO_2^-$$

$$\xrightarrow{H^+} H_2NCHR^1CONHCHR^2CO_2H + CO_2$$

This dipeptide may then be coupled with another NCA derivative, and so on.

A different approach to the anhydride method makes use of a mixed anhydride derived from ethyl chloroformate as follows (cf. the activated ester method):

$$ZNHCHR^1CO_2H \xrightarrow[Et_3N]{ClCO_2Et} ZNHCHR^1COOCO_2Et$$

$$\xrightarrow{H_3NCHR^2CO_2Me} ZNHCHR^1CONHCHR^2CO_2Me + EtOH + CO_2$$

Examples have been given above where the carboxyl group has been protected as the methyl or ethyl ester. A difficulty here is that alkaline hydrolysis of the peptide ester may cause racemisation. This difficulty may be avoided by use of benzyl esters, since these can be split by catalytic hydrogeno-lysis (H_2—Pd) to give toluene.

$$RCO_2CH_2Ph \xrightarrow{H_2-Pd} RCO_2H + PhCH_3$$

The thiol group (in cysteine) may be protected by S-benzylation (benzyl chloride in the presence of aqueous ethanolic sodium hydroxide). This group is not removed by HBr—AcOH but is by sodium in liquid ammonia. Benzylation may also be used to protect hydroxyl groups, *e.g.*, in tyrosine, and is removed by HBr—AcOH.

t-Butyl esters are also useful since they may readily be prepared by the action of isobutene on the amino-acid in the presence of a small amount of concentrated sulphuric acid. Furthermore, the t-butyl group is easily removed by treating the ester with anhydrous trifluoroacetic acid or with dry hydrogen chloride.

Since racemisation is always possible in some of the methods described for peptide synthesis, it is desirable to be able to ascertain whether this has happened. One way is to attempt hydrolysis of the synthetic peptide with enzymes (which are highly stereospecific). It may be possible to separate mixtures of diastereoisomeric peptides by paper and thin-layer chromatography, etc. Weinstein *et al.* (1972) have used NMR spectroscopy to determine the amount of racemisation.

Cyclic peptides may be synthesised in various ways. Small peptides may be cyclodimerised, and relatively long peptides cyclised (by self condensation) in dilute solution. A common method starts with peptide active esters. Relatively large rings may be formed by cyclising the peptide under the influence of dicyclohexylcarbodi-imide. The mixed anhydride method has also been used.

SOLID-PHASE PEPTIDE SYNTHESIS

Merrifield (1964) has introduced the 'solid phase' method in which an amino-acid or a peptide is bound chemically to an insoluble synthetic resin and then the chain is built up, one amino-acid residue at a time, at the free end. When the desired peptide has been synthesised, it is liberated from the solid support. The principles used for the peptide synthesis are those which have been described above. The method has been automated, *i.e.*, each addition of the appropriate amino-acid is carried out automatically at a predetermined time (cf. the Edman method). Some outstanding advantages of this solid phase method are:

(i) because of the use of the insoluble solid support, purification of products is not necessary, excess of reagents being removed by thorough washing with suitable solvents;

(ii) high yields;

(iii) the time has been considerably shortened for synthesising peptides (and proteins).

The method may be illustrated with the following example. The resin, which is a copolymer of styrene and divinylbenzene, is chloromethylated. This results in the formation of' benzyl chloride groups' through which the 'first' amino-acid becomes attached as the benzyl ester. This 'first' amino-acid, which is to be the C-terminal end of the peptide, is protected at its amino-group by, *e.g.*, the t-butyloxycarbonyl group, is heated with the resin in the presence of triethylamine in a suitable solvent. The protecting group is selectively removed by HCl—AcOH and the hydrochloride of the amino-group is converted into the free amino-group by the addition of excess of triethylamine. This benzyl ester of the 'first' amino-acid residue is now coupled with the N-t-butyloxycarbonyl derivative of the 'second' amino-acid by means of dicyclohexylcarbodi-iiryde. The cycle is then repeated with the N-protected 'third' amino-acid, and so on. When the desired peptide has been synthesised, the ester bond linking it to the resin may be split by dry hydrogen bromide in trifluoro-acetic acid.

3

MECHANISM OF ENZYME ACTION AND CARBONIC ACID

INTRODUCTION

It has been shown that the rate of enzyme-catalysed reactions depends on a number of factors. The pH of the solution has a great effect on enzyme activity, and it has been found that an enzyme behaves efficiently as a catalyst over a narrow range of pH. This optimum pH is characteristic of a particular enzyme and is determined experimentally; it is usually between pH 5 and pH 9. As we have seen extremes of pH denature proteins and so it is reasonable to suppose that the spatial arrangement of the molecular structure plays a part in enzymic activity.

Like all chemical reactions, enzyme-catalysed reactions are affected by changes in temperature, the rate being increased as the temperature rises. However, since enzymes can be denatured by heat too high a temperature destroys the activity of the enzyme. Many enzymes have an optimum temperature between 40° and 50°C, but the range may be higher, particularly for plant enzymes.

The rate of an enzyme-catalysed reaction depends on the concentration of the substrate and that of the enzyme. If the substrate is in excess, the rate is directly proportional to the concentration of the enzyme. On the other hand, if the enzyme concentration is kept constant, then the rate increases rapidly as the substrate concentration increases slowly. However, as the substrate concentration increases further, the rate increases much more slowly and finally reaches a maximum at a high substrate concentration (the rate versus substrate concentration gives a hyperbolic curve). This behaviour has been interpreted as follows. The substrate 'combines' with a particular region on the enzyme surface to form a complex. These regions are the active sites, and the complex is known as the Michaelis complex (1913). An enzyme may have one or more active sites. When all of these sites are occupied, the enzyme is now 'saturated' and consequently no further rate increase is possible. The substrate concentration (of the hyperbolic curve) corresponding to half the maximum rate is called the Michaelis constant, K_m. Its reciprocal ($1/K_m$) is a measure of the affinity of an enzyme for the substrate, *e.g.*, if K_m, is large, $1/K_m$ is small; this indicates that the substrate concentration must be large in order to achieve half the maximum rate.

The general belief is that enzyme-catalysed reactions proceed through a number of steps. If we represent the enzyme (together with its cofactor) as E, the substrate as S, and the products as P, the reaction may be written (in simple terms) as:

$$E + S \rightleftharpoons ES \rightleftharpoons EP \rightleftharpoons E + P$$

The existence of these intermediates has been established by various means *e.g.*, their isolation in some cases, spectroscopic studies, iso topic labelling experiments, etc. The nature of the interactions between enzyme and substrate can be of various types: hydrogen bonds, electrostatic forces, hydro-phobic bonds, and chemical bonds.

Like 'chemical' catalysts, enzymes lower the energy of activation (E) of the reactions which they catalyse but they are far more efficient than the former, *i.e.*, they lower the energy of activation to a much greater extent, *e.g.*, the decomposition of hydrogen peroxide:

$$H_2O_2 \xrightarrow{\text{cat.}} H_2O + \tfrac{1}{2}O_2$$

When platinum is the catalyst, E is ~ 50.2 kJ mol^{-1}, whereas for the enzyme catalase (as catalyst), fis ~2.5 kJ mol^{-1}.

The mechanism whereby enzymes effect these large rate accelerations is still uncertain. It is generally accepted, however, that mechanisms in which enzymes participate involve the usual types of reactions, *i.e.*, nucleophilic, electrophilic, homolytic, rearrangements, etc. Several contributing factors have been suggested to account for the high efficiency of enzyme-catalysed reactions.

(i) **Proximity Effect:** Binding of the reactant molecules (substrate and cofactor) to the enzyme results in an 'increased concentration' of the reactant molecules.

(ii) Binding causes the reactant molecules to be correctly oriented and consequently the transition state is reached more readily.

(iii) Binding produces a straineffect in the reactant molecules and consequently the bonds to be broken are 'deformed', thereby being brought to a state close to those existing in the transition state. Thus. the energy of activation of the reaction is lowered.

It is well established that the catalytic effects of enzymes are due to their three-dimensional structure (see also above). X-ray studies have shown that certain amino-acids, which are not necessarily adjacent in the primary structure, are 'brought together' through folding, thereby producing an active site. Since the mode of folding is dependent on the sequence of the amino-acids (primary structure), the latter must be one factor that contributes to the specificity of enzyme action *i.e.*, there is a steric relationship between the enzyme and the substrate. This was the basis of the 'lock-and-key' theory proposed by Fischer (1894) to explain enzyme specificity. According to this theory, the geometry of the enzyme, the 'lock', is complementary to that of the substrate, the 'key', the result being that the latter fits into the former as a key fits into a lock.

The stereospecificity of an enzyme may be explained on the lock-and key theory as follows. If we assume that an optically active compound can be bound to the enzyme through a minimum of three points (Bergmann *et al*; 1935), then the 'fit' will occur with either the D- or L-enantiomer, but not with both, *e.g.*, if the D-enantiomer fits, the L will not (and *vice versa*). Similarly, reduction of, *e.g.*, pyruvic acid to lactic acid, will occur on one side (enantiotopic or prochiral faces; 2 °7a) to produce one enantiomer of lactic acid. The pyruvic acid molecule fits into the enzyme in one way only and consequently hydrogen transfer must occur to one face only, thereby resulting in the forma-tion of only one enantiomer of lactic acid.

Now let us consider the cofactor NAD^+. This has a pair of enantiotopic (prochiral) faces and when a hydride ion is accepted, NADH is formed and this contains enantiotopic (prochiral) hydrogens at the 4-position:

Experimental work has shown that the NAD^+ -enzyme complex is usually sfereospecific, only one hydrogen (H_a or H_b) reacting exclusively. Which face of NAD+ is attacked and which hydride ion from NADH is transferred depends on the nature of the enzyme.

According to this system, enzymes are divided into six main groups according to the nature of the reaction that is catalysed, and each main group is given a code number. The method of naming enzymes is to add the suffix ase to the name of the substrate, *i.e.*, the substance being acted upon, *e.g.*, esterase acts on esters, amylase on starch (amylum), protease on proteins, urease on urea, etc. Some enzymes, however, have retained their trivial names, *e.g.*, emulsin, pepsin, trypsin, etc. Names are also used for particular enzymes, *e.g.*, urease, amylase, or as general names for groups of enzymes, *e.g.*, esterases, proteases, etc. The main groups are:

1. **Oxidoreductases:** These enzymes catalyse oxidation-reduction reactions, and include oxidases (direct oxidation with molecular oxygen), dehydrogenases (removal of hydrogen from substrates), etc.
2. **Transferases:** This group of enzymes catalyses the transfer of various functional groups, *e.g.*, transaminase.
3. **Hydrolases:** These catalyse hydrolytic reactions, *e.g.*, proteases (proteins), esterases (esters), etc.
4. **Lyases:** There are two types of lyases, one which catalyses addition to double bonds and the other which catalyses removal of groups and leaves double bonds.
5. **Isomerases:** These catalyse various types of isomerisation, *e.g.*, racemases, epimerases, etc.
6. **Ligases:** These enzymes catalyse the formation of a bond between two molecules and is accompanied by the breaking of apyrophosphate bond of ATP or similar triphosphate.

Each of these main groups is divided into subgroups which take the number of their main group followed by another number which specifies the type of group in the substrate that undergoes reaction. The subgroups are also divided into sub-subgroups. These are indicated by a third figure which gives more detailed information on the groups involved in the reaction.

Finally, a fourth figure indicates the serial number of the enzyme in its sub-subgroup. Thus, an enzyme is specified by four numbers (separated by points), *e.g.*, 1.1.1.1 is the oxidoreductase which is involved in hydrogen transfer from a CHOH group to NAD^+ or $NADP^+$ as acceptor. The trivial name of this enzyme is alcohol dehydrogenase.

The systematic names of enzymes consist of two parts, the first part specifying the substrate (or substrates) and the second part, which ends in 'ase', indicates the nature of the reaction that is catalysed. For example, let us consider the reaction: .

L-alanine + 2-oxoglutarate → pyruvate + L-glutamate

This reaction is catalysed by the enzyme transaminase. Since this is a subgroup of the main group of enzymes, the transferases, the common name transaminase has been changed to the more systematic name aminotransferase. Thus, this enzyme is named as L-alanine: 2-oxoglutarate aminotransferase; its Enzyme Commission number is 2.6.1.2. The trivial name of this enzyme is alanine aminotransferase, and was formerly called glutamic-pyruvic transaminase.

Biosynthesis of Amino-acids

As we have seen (Table), man can synthesise some amino-acids but not others; the latter must be provided in the diet. On the other hand, plants and many micro-organisms are capable of synthesising all the amino-acids in proteins.

However, the pathways followed in plants and animals may be different for the non-essential amino-acids.

The common amino-acids are derived from a relatively small number of precursors and, in some cases, an amino-acid may be produced from two different precursors. Histidine is exceptional in that it is produced by a pathway not involved in the biosynthesis of any other common amino-acid.

The glutamate family is derived from 2-oxoglutaric acid (α-ketoglutaric acid) which is synthesised from acetic acid by way of the Krebs cycle (1937), also known as the tricarboxylic acid cycle. One step is the conversion of a hexose molecule into two molecules ofpyruvic acid via phosphoglyceric acid. Pyruvic acid is converted into oxaloacetic acid which then enters into the tricarboxylic acid cycle.

Inspection of the citric acid structural formula shows that the two CH_2CO_2H groups are enantio-topic (prochiral). Potter *et al.* (1949) synthesised, by means of enzymes, citric acid labelled with ^{14}C at one CH_2CO_2H group and showed that the enzymic conversion of this compound gave 2-oxoglutaric acid exclusively labelled at the carboxyl group attached to the carbonyl group.

$$\begin{array}{l} \overset{*}{C}H_2CO_2H \\ |\\ C(OH)CO_2H \\ | \\ CH_2CO_2H \end{array} \xrightarrow{-H_2O} \begin{array}{l} H\overset{*}{C}-CO_2H \\ \| \\ C-CO_2H \\ | \\ CH_2CO_2H \end{array} \xrightarrow{+H_2O} \begin{array}{l} CHOH\overset{*}{C}O_2H \\ | \\ CHCO_2H \\ | \\ CH_2CO_2H \end{array} \xrightarrow{-2H} \begin{array}{l} CO\overset{*}{C}O_2H \\ | \\ CH_2 \\ | \\ CH_2CO_2H \end{array} + CO_2$$

The glutamate family contains four amino-acids: glutamic acid, glutamine, proline, and arginine. Here, we shall deal with the non-essential acids; this excludes arginine (Table). The path-ways are shown in the chart, and starts with 2-oxoglutaric acid (P; is a molecule of phosphoric acid). The various enzymes involved are not given in the charts.

An extremely important feature of the pathway leading to glutamic acid is the utilisation of ammonia in the conversion of the 2-oxo-acid to the amino-acid.

(i) $CH_2{=}C(OP)CO_2H + ADP \rightarrow ATP \rightarrow CH_2COCO_2H \xrightarrow[ATP]{CO_2}$ $HO_2CCH_2COCO_2H + ADP + P_1$

(ii)

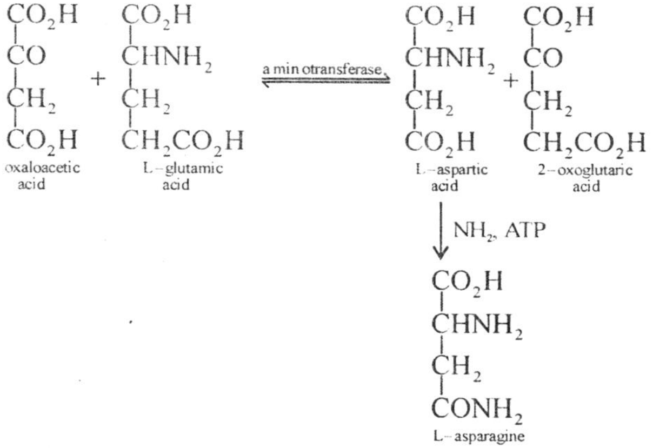

$$\underset{\text{oxaloacetic acid}}{\begin{array}{c}CO_2H\\|\\CO\\|\\CH_2\\|\\CO_2H\end{array}} + \underset{\text{L-glutamic acid}}{\begin{array}{c}CO_2H\\|\\CHNH_2\\|\\CH_2\\|\\CH_2CO_2H\end{array}} \xrightleftharpoons{\text{aminotransferase}} \underset{\text{L-aspartic acid}}{\begin{array}{c}CO_2H\\|\\CHNH_2\\|\\CH_2\\|\\CO_2H\end{array}} + \underset{\text{2-oxoglutaric acid}}{\begin{array}{c}CO_2H\\|\\CO\\|\\CH_2\\|\\CH_2CO_2H\end{array}}$$

$$\downarrow NH_2,\ ATP$$

$$\underset{\text{L-asparagine}}{\begin{array}{c}CO_2H\\|\\CHNH_2\\|\\CH_2\\|\\CONH_2\end{array}}$$

Fig. 1

All experimental work has shown that 2-oxo-acids are produced in the biosynthesis of all the common amino-acids and that these oxo-acids are.aminated by means of L-glutamic acid (see also tryptophan, below). This transamina-tion proceeds under the influence ofaminotransferases. This may be illustrated with a partic acid and asparagine of the aspartate family (which also includes lysine, threonine, and metHionine).

It should also be noted that amino-acid amides are produced by amination of acids with ammonia. The aminotransferases involved in transamination require the presence of a cofactor; this is pyridoxal 5'-phosphate or pyridoxamine 5'-phosphate (see also vitamin B_6).

If we represent the enzyme-cofactor complex as E—CHO and E—CH_2NH_2, we can write the mechanism of the transamination as follows:

$$\text{(i)}\quad E-CHO+H_2N\overset{\overset{CO_2H}{|}}{C}H(CH_2)_2CO_2H \underset{+H_2O}{\overset{-H_2O}{\rightleftharpoons}}$$

$$E-CH=N-\overset{\overset{CO_2H}{|}}{C}H(CH_2)_2CO_2H \rightleftharpoons$$

$$E-CH_2N=\overset{\overset{CO_2H}{|}}{C}H(CH_2)_2CO_2H \underset{-H_2O}{\overset{+H_2O}{\rightleftharpoons}}$$

$$E-CH_2NH_2+O=\overset{\overset{CO_2H}{|}}{C}(CH_2)_2CO_2H$$

$$\text{(ii)}\quad E-CH_2NH_2+O=\overset{\overset{CO_2H}{|}}{C}CH_2CO_2H \underset{+H_2O}{\overset{-H_2O}{\rightleftharpoons}}$$

$$E-CH_2N=\overset{\overset{CO_2H}{|}}{C}CH_2CO_2H \rightleftharpoons$$

$$E-CH=N-\overset{\overset{CO_2H}{|}}{C}HCH_2CO_2H \underset{+H_2O}{\overset{-H_2O}{\rightleftharpoons}}$$

$$E-CHO+H_2N\overset{\overset{CO_2H}{|}}{C}HCH_2CO_2H$$

This mechanism involves the formation of a Schiffbase Which rearranges to the isomeric Schiffbase.

As we have seen, the common amino-acids containing a benzene ring are essential acids (Table). These acids—phenylalanine, tyrosine, and tryptophan—constitute the aromatic family. The aromatic amino-acids are synthesised by micro-organisms by the shikinric acid route, and there is some evidence to show that this route is also followed in higher plants. Two distinct routes to the benzene ring are possible: (i) from acetate; (ii) from carbohydrates. The latter is known as the shikimic acid route.

Both the acetate and the shikimic acid route are followed in the biosynthesis of flavonoids.

The shikimic acid route starts from D-glucose which is converted into phosphoenolpyruvic acid (PEP) and erythrose-4-phosphate. These combine to form 3-deoxy-D-arabino-heptulosonic acid 7-phosphate which is then transformed in a series of steps to shikimic acid which, in turn, is converted into chorismic acid (see chart). Up to this point, the pathway is common to the three aromatic amino-acids (the amino-acids have been written as such and not as the carboxylate form).

SHIKIMIC ACID ROUTE LEADING TO CHORISMIC ACID

Shikimic acid is 3α, 4α, 5β-trihydroxycyclohex-l-ene-l-carboxylic acid, m.p. 190-191°C, $[\alpha]_D$ –157°, and λ_{max} 213 nm (ε 8900). Its structure and stereo-chemistry have been determined by a number of workers, and Hall (1964) has shown, from his NMR spectral studies, that the acid exists predominantly in the half-chair conformation, the boat form contributing only to a very small extent. Davis *et al.* (1951-1958) showed that shikimic acid was an intermediate in the biosynthesis of the three aromatic amino-acids and Gibson *et al.* (1962) established that chorismic acid is another intermediate that was derived from shikimic acid.

These workers also proposed the structure of chorismic acid given (in the chart) and this has been confirmed by NMR studies (Gibson *et al.*, 1963). The stereochemistry of chorismic acid, however, was based on that of (–)-shikimic acid, which had been established by Fischer *et al.* (1937).

The stereochemistry of the elimination of phosphoric acid from shikimic acid 5-phosphate to give chorismic acid has been shown, by means of tritium labelled experiments, to be stereospecifically trans (Hill *et al.*, 1969; Onderka *et al.*, 1969).

Thus, it is H_b which is eliminated, which is the opposite of what would be expected for a concerted reaction.

We now return to the conversion of chorismic acid into the aromatic amino-acids. The pathways are shown in the charts.

SPECIFICITY OF ENZYME ACTION

One of the most characteristic properties of enzymes is their specificity of action. This specificity may be manifested in one of three ways:

(i) An enzyme may catalyse a particular type of reaction, *e.g.*, esterases hydrolyse only esters. Such enzymes are said to be reaction specific. On the other hand, an enzyme may be specific for a particular compound or class of compounds. These enzymes are substrate specific, *e.g.*, urease hydrolyses only urea; phosphatases hydrolyse only phosphate esters.

(ii) Many enzymes exhibit a kinetic specificity, *e.g.*, esterases, although hydrolysing all esters, hydrolyse the various esters at different rates; pepsin hydrolyses the peptide link, but is most active for those links in which, among other things, the amino group belongs to an aromatic amino-acid and the carboxyl group is one of a dicarboxylic amino-acid.

(iii) Many enzymes are stereospecific, *e.g.*, maltase hydrolyses α-glycosides but not β-glycosides, whereas emulsin hydrolyses the latter but not the former.

It should be noted, however, that a given enzyme can exhibit more than one of the specificities, *e.g.*, esterases, while hydrolysing only esters, may also hydrolyse one enantiomer (of an optically active ester) more rapidly than the other.

ENZYME INHIBITORS

Enzyme activity can be reduced or inhibited by the presence of various compounds. There are two major types of inhibition: competitive and irreversible. In competitive inhibition, the inhibitor is a compound whose structure and geometry closely resemble that of the normal substrate. Competition occurs between the two (for the active site), but inhibition can be reversed by increasing the concentration of the normal substrate.

It therefore follows that the enzyme-inhibitor complex readily regenerates the two molecules in competitive inhibition. On the other hand, in irreversible inhibition the inhibitor forms a highly stable enzyme-inhibitor complex (via a covalent bond), and if sufficient inhibitor is present, the catalytic effect of the enzyme towards its normal substrate is completely lost.

The inhibitors discussed above are believed to exhibit inhibition (competitive or irreversible) by combining with the enzyme at its active site. There are however, some compounds which inhibit (or increase) enzyme activity by changing the conformation of the active site of the enzyme.

Enzymes which behave this way are called allosteric enzymes, and are inhibited (or activated) by combination with the allosteric effector at some position of the enzyme which is not the active site. Allosteric effectors are usually small molecules which bear no chemical resemblance to the normal substrate.

PHENYLALANIAE AND TYROSINE

It should be noted that the amino-group is introduced into the 2-position of chorismic acid to give anthranilic acid. Also, the final step appears to be uncertain; one theory is that it occurs as follows:

indole-3-glycerol phosphate → 3-phosphoglyceraldehyde

$$+ \text{ indole} \xrightarrow{\text{serine}} \text{tryptophan}$$

Some other examples of the biosynthesis of amino-acids are :

(a) $$\underset{\text{L-serine}}{HOCH_2CH(NH_2)CO_2H} \xrightarrow{H_2S} \underset{\text{L-cysteine}}{HSCH_2CH(NH_2)CO_2H}$$

$$\xrightarrow{NAD^-} \underset{\text{L-cystine}}{[-SCH_2CH(NH_2)CO_2H]_2} + NADH$$

(b) $$\underset{\text{L-aspanic acid}}{HO_2CCH_2CH(NH_2)CO_2H} \xrightarrow{ATP} \underset{\text{4-phospho-L-aspartic acid}}{PO_2CCH_2CH(NH_2)CO_2H}$$

$$\xrightarrow{NADH} \underset{\text{L-aspartic acid 3-aldehyde}}{OHCCH_2CH(NH_2)CO_2H} \xrightarrow{NADH}$$

$$\underset{\text{L-homoserine}}{HOCH_2CH_2CH(NH_2)CO_2H} \xrightarrow{NADH} \underset{\text{O-phospho-L-homoserine}}{POCH_2CH_2CH(NH_2)CO_2H}$$

$$\xrightarrow[H_2O]{\text{Pyridoxal - P}} \underset{\text{L-threonine}}{CH_3CHOHCH(NH_2)CO_2H}$$

Note the migration of the hydroxyl group in the last step.

A very interesting problem related to the biosynthesis of amino-acids is the work of Miller (1953,1955). This author subjected a mixture of methane, ammonia, hydrogen and water vapour (which possibly made up the atmosphere of the Earth in its early stages) to spark and silent discharges. Analysis of the gases showed that the initial gases were present and, in addition, carbon monoxide, carbon dioxide and nitrogen. The solid product was analysed by means of paper chromatography, and the following aminoacids were identified: glycine, sarcosine (N-methylglycine), D- and L-alanine, β-alanine, D- and L-α-amino-n-butyric acid and a-amino-isobutyric acid. Many other amino-acids (unidentified) were also formed, as well as formic, acetic propionic, glycollic and lactic acids.

Since this work was done, other investigators have carried out experiments in somewhat different ways. Palm *et al.* (1962) irradiated methane in aqueous ammonia with high-energy electrons and observed the formation of glycine, alanine, aspartic acid and other compounds such as hydrocarbons. Oro (1963) subjected a mixture of methane, ammonia, water and ethane to an electrical discharge and obtained amino-acids, amino-acid amides, amines, etc.

Cofactor

Many enzymes require the presence of non-protein compounds in order to perform their catalytic action. These compounds are collectively known as cofactors or activators, and fall into three main groups. Coenzymes are organic molecules which may be separated from the enzyme by, *e.g.*, dialysis.

On the other hand, some cofactors are bound to the enzyme and then referred to as the prosthetic pmf of its enzyme. Finally, cofactors may be inorganic ions. In some cases the metal is tightly bound to the enzyme which is then referred to as a metalloenzyme. In other cases the enzymes are 'metal-activated'. Metal activators are uni- or bi-valent metal cations, *e.g.*, Na^{+}·K^{+},Mg^{2+},Zn^{2+},Ca^{2+}.

The complex, enzyme-cofactor, is known as a holoenzyme, and when the cofactor has been removed the protein that remains is known as an apoenzyme. This has no enzymic activity.

Some enzymes are synthesised in the organism in an inactive form; this is known as a zymogen. Thus, *e.g.*, the enzyme pepsin is synthesised as its zymogen, pepsinogen. This is converted into pepsin in the presence of hydrochloric acid.

Coenzymes and prosthetic groups generally act as carriers of specific functional groups or specific atoms. In order to act in this manner, these cofactors must exist in two forms, one form being converted into the other during a catalysed reaction, and the latter being reconverted into the former by a coupled reaction. These two reactions may, or may not, follow each other. Here, we shall discuss three coenzymes which are nucleotides.

Nicotinamide-adenine Dinucleotide (NAD^{+})

This was formerly known as diphosphopyridine nucleotide (DPN) and has the structure shown. This coenzyme functions as an acceptor of hydrogen atoms and electrons in the presence ofdehydro-genases and is thereby converted into the reduced form NADH. Since only the nicotinamide moiety is involved in this transfer, the reaction may be written as shown (note the hydride ion transfer from the substrate).

NICOTINAMIDE-ADENINE DINUCLEOTIDE PHOSPHATE ($NADP^+$)

This was formerly known as triphospho-pyridine nucleotide (TPN) and has the structure shown (see above). This also behaves as an acceptor of hydrogen atoms and electrons, thereby being converted into the reduced form, NADPH. It appears that NAD^+ and NADH are usually involved in degradative processes, whereas $NADP^+$ and NADPH are usually involved in synthetic processes.

Adenosine triphosphate (ATP) has the structure shown. It is involved in enzyme-catalysed trans-phosphorylation reactions, transferring one phosphate group to the substrate, itself being converted into adenosine diphosphate (ADP). This, in turn, can also transfer a phosphate group and is thereby converted into adenosine monophosphate (AMP). For chemical reactions to proceed, energy must be supplied to overcome the energy barriers. In biosynthetic processes, this energy is supplied by ATP when it is involved in transphosphorylation reactions in the presence of a suitable enzyme, *e.g.*,

$$ROH + ATP \rightarrow R—OPO(OH)_2 + ADP$$

ADP also behaves as a phosphorylating agent, *e.g.*,

$$ROH + ADP \rightarrow R—OPO(OH)_2 + AMP$$

A less usual reaction of ATP is pyrophosphorvlation, *e.g.*,

$$ROH + ATP \rightarrow R—OPO(OH)—O—PO(OH)_2 + AMP$$

Inspection of their structural formulae (see above) shows that the phosphate group in AMP is linked by the normal ester bond. On the other hand, the terminal phosphate groups in ADP and ATP are linked to a phosphate group by an acid anhydride bond.

In hydrolytic reactions, the free energy change (heat of reaction) of an ester bond is ~ –4.0 to –12.5 kJ mol^{-1}, whereas that for the acid anhydride bond is ~ –33.5 kJ mol^{-1}. Hence, in transphosphorylation reactions by ATP or ADP, there is a net free energy change of ~ –29.5 to ~ –21.0 kJ mol^{-1}. It is this energy which is used to 'drive' coupled reactions. These acid anhydride bonds have been referred to as 'energy-rich' bonds, and are sometimes represented by the symbol ~, *e.g.*, ATP has been written as:

adenme-ribose—O—PO(OH) ~ O—PO(OH) ~ O—PO(OH).,

CARBONIC ACID

Orthocarbonic acid, C(OH), is unknown in the free state, but its esters have been prepared. They may be obtained by the reaction between sodium alkoxide and nitrochloroform:

$$4RONa + CCl_3NO_2 \longrightarrow C(OR)_4 + 3NaCl + NaNO_2 \text{ (f.g.)}$$

The alkyl orthocarbonates are ethereal smelling liquids.

Carbonic acid (*metacarbonic acid*), $0 = C(OH)_2$, is unknown in the free state, but its salts and esters have been prepared. Alkyl carbonates may be readily formed by heating alcohols with carbonyl chloride in the presence of pyridine:

They may also be prepared by heating silver carbonate with alkyl iodide:

$$Ag_2CO_3 + 2RI \longrightarrow CO(OR)_2 + 2AgI$$

The alkyl carbonates are ethereal smelling liquids, readily soluble in water. In recent years they have found various uses in organic synthesis, *e.g.*,

(i) They may be used to introduce the carbalkoxy group, CO_2R, into ketones to produce β-ketoesters. The reaction is carried out by heating a ketone with sodium or sodium alkoxide in the presence of a large excess of alkyl carbonate:

$$R^2COCH_3 + (R^1O)_2CO + NaOR^1 \longrightarrow [R^2COCHOR^1]^- Na^+ + 2R^1OH$$

$$\xrightarrow{H^+} R_2COCH_2CO_2R^1$$

(ii) Alkyl carbonates react with primary alkyl cyanides in the presence of sodium alkoxide to form α-cyanoesters:

$$[R^2C(CO_2C_2H_5)_2]^- Na^+ + (R^1O)_2\ CO \longrightarrow R^1R^2C(CO_2C_2H_5)_2 + R^1OCOONa$$

(iii) Alkyl carbonates may be used to alkylate sodiomalonic esters:

$$[R^2C(CO_2C_2H_5)_2]^- Na^+ + (R^1O)_2CO \longrightarrow R^1R^2C(CO_2C_2H_5)_2 + R^1OCOONa$$

The acid chloride of carbonic acid is *carbonyl chloride* (*phosgene*). It may be obtained by the action of chlorine on carbon monoxide under the influence of sunlight, or in the presence of heated charcoal (200°C) as catalyst; the latter is the commercial method.

Carbonyl chloride is a colourless liquid, b.p. 8°C, and has been used as a toxic gas in warfare. It is used in various organic syntheses, behaving as an acid chloride. It is very slowly decomposed by water:

$$COCl_2 + H_2O \longrightarrow CO_2 + 2HCl$$

When treated with slight excess of one equivalent of alcohol in the cold, carbonyl chloride forms chloroformic esters:

$$COCl_2 + ROH \longrightarrow ClCO_2R + HCl$$

When treated with an excess of alcohol in the presence of pyridine, alky carbonates are formed (see above). Carbonyl chloride reacts with ammonia to form urea:

$$COCl_2 + 2NH_3 \longrightarrow CO(NH_2)_2 + 2HCl$$

With primary or secondary amines, substituted ureas are formed, *e.g.*,

$$COCl_2 + 2RNH_2 \longrightarrow CO(NHR)_2 + 2HCl$$

Chloroformic acid (*chlorocarbonic acid*), $ClCO_2H$, is not known in the free state, but its esters have been prepared. These may be readily obtained by treating carbonyl chloride with slight excess of one equivalent of alcohol in the cold (see above). Chloroformic esters are acid-chloride esters, and the chlorine atom reacts with compounds containing active hydrogen; thus ethyl chloroformate, b.p. 94°C, reacts with water, alcohols, ammonia, primary and secondary amines; *e.g.*, are than is formed with ammonia:

$$ClCO_2C_2H_4 + NH_3 \longrightarrow NH_2CO_2C_2H_5 + HCl$$

Ethyl chloroformate is useful for introducing the carbethoxy group on a nitrogen atom; *e.g.*, ethyl *N*-tricarboxylate is formed when ethyl chloroformate is added to a mixture of urethan, ether and sodium, this mixture having been previously heated:

$$NH_2C\ O_2C_2H5 + 2ClCO_2H_2H_5 + 2Na$$
$$\longrightarrow N(CO_2C_2H_5)_3 + 2NaCl + H_2 \quad (51\text{–}57\%)$$

Ethyl chloroformate may also be used to prepare ethyl esters by reaction with Grignard reagents.

Amides of carbonic acid. Since carbonic acid is dibasic, two amides are possible: the mono- and diamide. These are known, respectively, as *carbamic acid*, (I), and *urea*, (II):

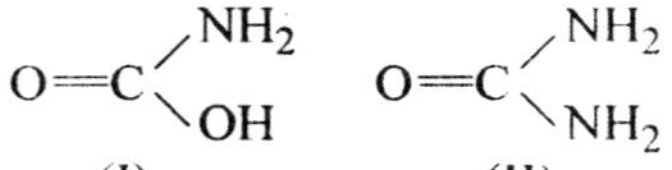

$$O{=}C\begin{matrix}\diagup NH_2\\ \diagdown OH\end{matrix} \qquad O{=}C\begin{matrix}\diagup NH_2\\ \diagdown NH_2\end{matrix}$$

(I) (II)

Carbamic acid is not known in the free state, but its salts and esters have been prepared. Ammonium carbamate is formed when dry ammonia reacts with dry carbon dioxide:

$$2NH_3 + CO_2 \longrightarrow NH_2COONH_4$$

It is a white crystalline solid, very soluble in water. When its aqueous solution is warmed to 60° C, ammonium carbamate is hydrolysed to ammonium carbonate:

$$NH_2COONH_4 + H_2O \longrightarrow CO(ONH_4)_2$$

Esters of carbamic acid are known as **urethans**. These may be prepared:

(i) By treating a chloroformic ester with ammonia, *e.g.*, ethyl chloroformate produces ethyl carbamate:

$$ClCO_2C_2H_5 + NH_3 \longrightarrow NH_2CO_2C_2H_5 + HCl$$

(ii) By refluxing urea in an alcohol, *e.g.*, n-butanol gives n-butyl carbamate when refluxed for 30 hours:

$$CO(NH_2)_2 \longrightarrow NH_3 + HNCO$$

$$HNCO + CH_3CH_2CH_2CHOH \longrightarrow NH_2CO_2CH_2CH_2CH_2CH_3$$

(75–76%)

(iii) N-substituted urethans may be prepared by the Curtius reaction. The acid azide is refluxed in benzene solution and then an alcohol is added:

$$R^1CON_3 \longrightarrow N_2 + R^1NCO \xrightarrow{R^2OH} R^1NHCO_2R^2$$

N-phenylurethans may be prepared by reaction between phenyl isocyanate and alcohol:

$$C_6H_5NCO + ROH \longrightarrow C_6H_5NHCO_2R$$

These are crystalline solids and may be used to characterise the alcohols.

Readily dehydrated alcohols, particularly t-alcohols, do not form urethans; alkenes and diphenyl-urea are produced:

$$R_2C(OH)CH_2CH_3 \longrightarrow R_2C = CHCH_3 + H_2O$$

$$PhNCO + H_2O \longrightarrow CO_2 + PhNH_2 \xrightarrow{PhNCO} (PhNH)_2CO$$

Ethyl carbamate, usually known as *urethan*, is a crystalline solid, m.p. 50°C. It reacts with ammonia to form urea:,

$$NH_2CO_2C_2H_5 + NH \longrightarrow CO(N\ H_2)_2 + C_2H_5OH$$

It is decomposed by aqueous sodium hydroxide on warming:

$$NH_2CO_2C_2H_5 + {}_2NaOH \longrightarrow Na_2CO_3 + NH_3 + C_2H_5OH$$

Ethyl carbamate has hypnotic properties; many urethans are valuable hypnotics, *e.g.*, aponal (t-amyl carbamate, t-amylurethan), $NH_2COOC(CH_3)_2(C_2H_5)$.

Urea (*carbamide*) is very important physiologically. It is the chief nitrogenous product of protein metabolism; adults excrete about 30 g per day in the urine, from which it can be extracted by evaporating the urine to small bulk and adding nitric acid, whereupon the slightly soluble urea nitrate, $CO(NH_2)_2 \cdot HNO_3$, is precipitated.

Urea is historically very important because Wohler (1828) synthesised it by evaporating a solution containing potassium isocyanate and ammonium sulphate, ammonium isocyanate, which is formed first, undergoes molecular rearrangement:

$$NH_4NCO \rightleftharpoons CO(NH_2)_2$$

The reaction is reversible; the solution contains about 5 per cent ammonium isocyanate.

The mechanism of this rearrangement is uncertain. One possibility is as follows (via dissociation):

$$NH_4NCO \rightleftharpoons NH_3 + HNCO$$

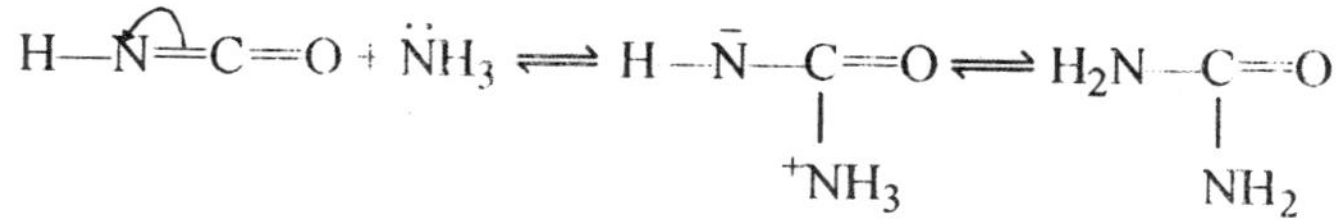

$$H{-}\ddot{N}{=}C{=}O + \ddot{N}H_3 \rightleftharpoons H{-}\bar{N}{-}\underset{\underset{{}^{+}NH_3}{|}}{C}{=}O \rightleftharpoons H_2N{-}\underset{\underset{NH_2}{|}}{C}{=}O$$

Fig. 1

Urea may be prepared in the laboratory by the action of ammonia on carbonyl chloride, alkyl carbonates, chloroformates or urethans, *e.g.*,

$$COCl_2 + 2NH_3 \longrightarrow CO(NH_2)_2 + {}_2HCl \quad (f)$$

$$(C_2H5O)_2CO + {}_2NH_3 \longrightarrow CO(NH_2)_2 + {}_2C_2H_5OH \quad (f)$$

Industrially, urea is prepared by allowing liquid carbon dioxide and liquid ammonia to interact, and heating the ammonium carbamate so formed to 130-150°C under about 35atm pressure:

$$2NH_3 + CO_2 \longrightarrow NH_2COONH_4 \longrightarrow CO(\ NH_2)_2 + H_2O$$

STRUCTURE OF UREA

Although urea appears to be a simple molecule, it is only recently that its structure has been ascertained with any degree of certainty. The diamide structure, $CO(NH_2)_2$, seemed to be indicated by its synthesis from carbonyl chloride and ammonia:

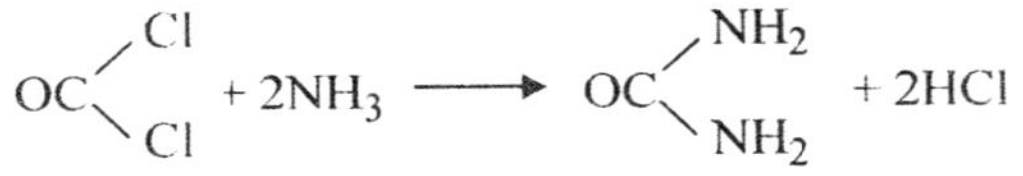

$$OC\begin{smallmatrix} Cl \\ Cl \end{smallmatrix} + 2NH_3 \longrightarrow OC\begin{smallmatrix} NH_2 \\ NH_2 \end{smallmatrix} + 2HCl$$

This structure, however, did not appear to explain all the reactions of urea and so gave rise to a great deal of controversy. Physical methods of structure determination have now shown that the diamide structure is correct, at least in the solid state.

Crystal structure studies have shown that in solid urea both nitrogen atoms are identical. Bond-length measurements in urea give the C—N distance as 1-37Å. In aliphatic amines the C—N bond length is 1-47Å. This indicate; that the C—N bond in urea has some double bond character (about 28 per cent): this can be explained by resonance:

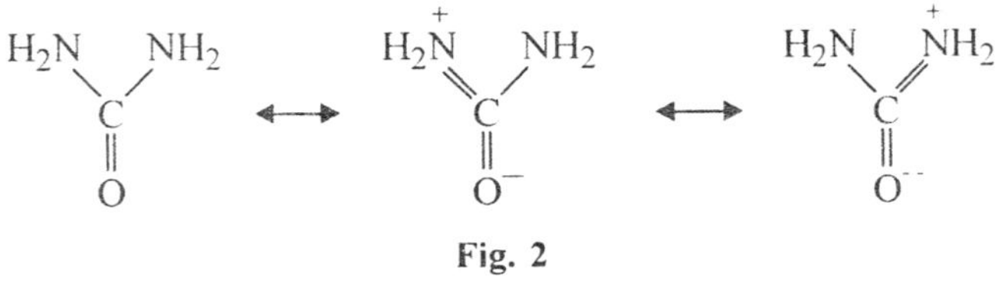

$$H_2N-C(=O)-NH_2 \longleftrightarrow H_2\overset{+}{N}=C(-O^-)-NH_2 \longleftrightarrow H_2N-C(-O^-)=\overset{+}{N}H_2$$

Fig. 2

Both nitrogen atoms are identical in the hybrid molecule. Furthermore, the negatively charge oxygen atom is capable of co-ordinating with one proton (therefore urea will be a 'monoacidic' base), and thus the salt may be formulated as a resonance hybrid:

$$H_2\overset{+}{N}=C(NH_2)-OH \longleftrightarrow H_2N-C(=\overset{+}{N}H_2)-OH \longleftrightarrow H_2N-C(NH_2)=\overset{+}{O}H$$

NMR spectra of crystalline salts of urea (and thiourea) are consistent with protonation at the oxygen atom (or sulphur atom; Redpath *et al.*, 1962).

PROPERTIES AND REACTIONS OF UREA

Urea is a white crystalline solid, m.p. 132°C, soluble in water and ethanol, but insoluble in ether. It is used for preparing formaldehyde-urea plastics, barbiturates, as a fertiliser, etc. A recent use is for the manufacture of hydrazine; urea is treated with alkaline sodium hypochlorite (in effect, the Hofmann preparation of primary amines is applied to urea):

$$NH_2CONH_2 + NaOCl + 2NaOH \longrightarrow N_2H_4 + NaCl + Na_2CO_3 + H_2O$$

(i) Urea behaves as a 'monoacidic' base; the nitrate and oxalate are the most important, since neither is very soluble in water.

When urea nitrate is added to cold concentrated sulphuric acid, nitrourea is formed:

(ii) Urea is hydrolysed by boiling with acids or alkalis:

$$CO(NH_2)_2 + H_2O \longrightarrow CO_2 + 2NH_3$$

The enzyme urease (which occurs in soyabeans) produces the same change.

(iii) When gently heated, urea loses ammonia to form biuret:

$$(NH_2)_2\,CO \rightleftharpoons NH_3 + HNCO \xrightarrow{(NH_2)_2CO} NH_2CONHCONH_2$$

When an aqueous biuret solution is treated with sodium hydroxide solution and a drop of copper sulphate solution, a violet colour is produced. This is known as the *biuret reaction*, which is characteristic of all compounds containing the grouping —CONH—, *e.g.*, proteins.

When heated rapidly, urea evolves ammonia and forms cyanic acid which rapidly polymerises to cyanuric acid. When refluxed with alcohols, urea forms urethans (see above).

(iv) Nitrous acid reacts with urea with the liberation of nitrogen which, however, is not evolved quantitatively:

$$CO(NH_2)_2 + 2HNO_2 \longrightarrow CO_2 + 3H_2O + 2N_2$$

Nitrogen is also evolved, again not quantitatively, when urea is treated with excess of alkaline hypobromite:

(v) Acid chlorides and acid anhydrides react with urea to form weides, *e.g.*, acetyl chloride forms acetylurea:

$$CH_3COCl + NH_2CONH_2 \longrightarrow CH_3CONHCONH_2 + CHl$$

Many of these ureides are useful drugs, particularly when the acid group has a branched chain, *e.g.*, *bromural* (α-bromoisovalerylurea), $(CH_3)_2CHCHBrCONHCONH_2$.

Dicarboxylic acids react with urea in the presence of phosphoryl chloride to form cyclic ureides; *e.g.*, oxalic acid forms parabanic acid (oxalylurea):

$$\begin{matrix} CO_2H \\ | \\ CO_2H \end{matrix} + CO(NH_2)_2 \xrightarrow{POCl_3} + 2H_2O + \begin{matrix} CO—NH \\ | \\ CO—NH \end{matrix}\!\!>CO$$

Cyclic ureides may also be prepared by refluxing a di-ester with urea in ethanolic solution containing sodium ethoxide; *e.g.*, malonic ester forms barbituric acid (malonylurea): Barbituric acid and its 5- or 5,5-derivatives are used in medicine as hypnotics and sedatives, *e.g.*, barbitone (5,5-diethylbarbituric acid), and *phenobarbitone* (5-phenyl-5-ethylbarbituric acid).

Inclusion Complexes

Several types of inclusion complexes are known, and in all of them molecules of one component, the 'guest' molecules, are physically imprisoned in the cavities of the crystalline structure of the second component, the 'host' molecules. Two important types of inclusion complexes are the *channel* (*canal*) *complexes* and the *clathrate* (*cage*) *complexes* (for a discussion of the latter). Channel complexes are those cases where the host crystallises in a form with parallel, approximately cylindrical channels in which molecules of the guest are enclosed lengthwise. Urea normally forms a crystal structure which is closely packed, but in the presence of various *straight-chain* molecules, *e.g.*, n-alkanes, n-alcohols, n-acids, n-esters, etc., urea crystallises in a more open structure which contains long channels enclosing the guest molecules.

These channels contain a number of guest molecules and, in general, the number is inversely proportional to the length of the guest molecule. Branched-chain and cyclic structures cannot 'fit' into these channels, and so this property affords a means of separating straight-chain from branched-chain compounds (but see below). The complexes are decomposed by melting or by dissolving away the urea with water.

The formula of the channel complexes is usually A_nB (A is urea) where n, usually not a whole number, has values of 4 or more and increases as the length of B increases.

Furthermore, since channel complexes are characterised by the fact that guest molecules bear some structural resemblance to each other, it is not possible to separate homologues by channel complexes. For each homologous series there is a certain minimum chain-length for channel complex formation, *e.g.*, six carbons for alkanes, seven for alcohols, five for acids, etc.

Channel complex formation has been used to resolve racemic modifications, and also to separate geometrical isomers. It is also interesting to note that if a molecule is sufficiently long, then 'small' side-chains will not prevent such a molecule behaving as a guest, *e.g.*, Schiessler *et al.* (1957) have shown that 2-methyldecane forms a channel complex with urea.

X-ray analysis of these channel complexes has shown that the diameter of the channel is about 5 Å (Smith, 1952; see also thiourea).

Substituted ureas may be prepared by a reaction similar to Wohler's synthesis of urea; the hydrochloride or sulphate of a primary or secondary amine is heated with potassium isocyanate, *e.g.*, methylamine hydrochloride (I molecule) forms methylurea:

$$CH_3NH_2{\cdot}HCl + KNCO \longrightarrow KCl + CH_3NH_2{\cdot}HNCO \longrightarrow CH_3NHCONH_2$$

If excess of amine salt is used, N, N'-disubstituted ureas are obtained, *e.g.*, excess of aniline hydrochloride forms N, N'-diphenylurea::

$$C_6H_5NH_2{\cdot}HCl + KNCO \longrightarrow KCl + C_6H_5NH_2{\cdot}HNCO \longrightarrow$$

$$C_6H_5NHCONH_2 \xrightarrow{C_6H_5NH_2{\cdot}HCl} C_6H_5NHCONHC_6H_5 + NH_4Cl$$

Alternatively, N,N'-diphenylurea may be obtained by refluxing an aqueous solution of aniline hydrochloride with urea; the urea produces ammonia and isocyanic acid, and the reaction proceeds as shown above. Both phenylurea (38-40 per cent) and N,N′-diphenylurea (52-55 per cent) may be isolated.

N,N′-Substituted ureas may be obtained by the action of carbonyl chloride on a primary or secondary amine:

$$COCl_2 + 2RNH_2 \longrightarrow CO(NHR)_2 + 2HCl$$

The reaction between phenyl isocyanate and a primary or secondary amine also produces an 5-disubstituted urea, *e.g.*, with ethylamine, N-ethyl-N'-phenylurea is formed:

$$C_6H_5NCO + C_2H_5NH_2 \longrightarrow C_6H_5NHCONHC_2H_5$$

This reaction is used to characterise amines.

A very convenient method of preparing alkylureas is to evaporate an aqueous or ethanolic solution of an amine and nitrourea (see above); the latter decomposes into isocyanic acid and nitramide, NH_2NO_2, which readily decomposes into nitrous oxide:

$$NH_2CONHNO_2 \longrightarrow NHCO + NH_2NO_2$$

$$HNCO + R_2NH \longrightarrow R_2NCONH_2$$

$$NH_2NO_2 \longrightarrow N_2O + H_2O$$

When urea is treated with methyl sulphate in faintly alkaline solution, methylisourea is obtained:

$$NH_2CONH_2 \xrightarrow{Me_2SO_4} NH = C(OMe)NH_2$$

Compounds Related to Urea

Semicarbazide (*aminourea*), m.p. 96°C, may be prepared by treating hydrazine sulphate with potassium cyanate:

$$H_2NNH_2 \cdot HNCO \longrightarrow NH_2CONHNH_2$$

It is manufactured by heating urea with hydrazine hydrate.

$$NH_2CONH_2 + N_2H_4 \cdot H_2O \longrightarrow NH_2CONHNH_2 + NH_3 + H_2O$$

A more recent method is the electrolytic reduction of nitrourea in sulphuric acid solution using a lead anode:

$$NH_2CONHNO_2 \xrightarrow{e;H^+} NH_2CONHNH_2 + 2H_2O \text{ (61–69\%)}$$

Semicarbazide is an important reagent for the identification of carbonyl compounds, with which it forms *semicarbazones*; it is also used in the Wolff-Kishner reduction.

Guanidine (*ammomethanamidine*), m.p. 50° C, is found in beet juice, and is one of the degradation products of the purines. It may be prepared by heating cyanamide with ammonium chloride:

$$NH_2CN + NH_4Cl \longrightarrow (NH_2)_2C = NH \cdot HCl$$

Guanidine is a *strong 'monoacid'* base, even forming a carbonate. Its strength as a base may be explained by resonance. There are unlike charges in the neutral molecule, whereas the ion has only one positive charge which, is spread, and at the same time all the resonating structures are equivalent (cf. urea).

Neutral molecule

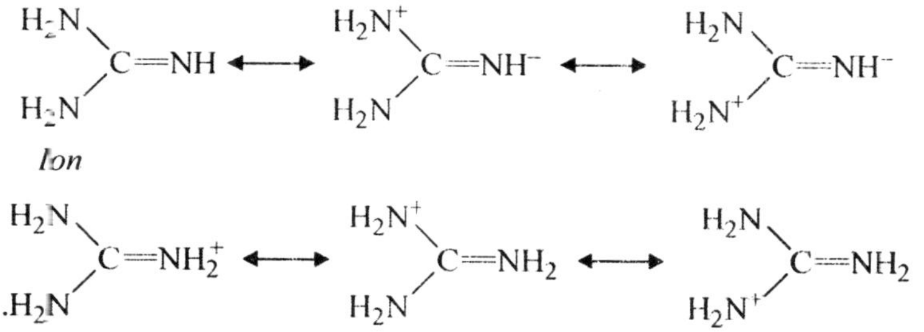

$$(H_2N)_2C{=}NH \longleftrightarrow (H_2N^+)(H_2N)C{=}NH^- \longleftrightarrow (H_2N)(H_2N^+)C{=}NH^-$$

Ion

$$(H_2N)_2C{=}NH_2^+ \longleftrightarrow (H_2N^+)(H_2N)C{=}NH_2 \longleftrightarrow (H_2N)(H_2N^+)C{=}NH_2$$

Fig. 3

X-ray analysis of guanidinium iodide shows that the three nitrogen atoms are symmetrically placed round the carbon atom. Furthermore, the C—N distance has been found to be 1.38Å (Theilacker, 1935; cf. urea). These facts are in keeping with the assumption that the guanidinium ion is a resonance hybrid. Careful hydrolysis with barium hydroxide solution converts guaridine into urea:

$$(NH_2)_2C = NH + H_2O \longrightarrow (NH_2)_2C = O + NH_3$$

Guanidine nitrate may be prepared by heating dicyandiamide with ammonium nitrate:

$$(NH_2)_2C = NCN + 2NH_4NO_3 \longrightarrow 2(NH_2)_2C = NH \cdot HNO_3 \ (85\%)$$

When treated with concentrated sulphuric acid, guanidine nitrate is converted into nitroguanidine, $(NH_2)_2C = NNO_2$ (cf. urea nitrate). Nitroguanidine is used for making flashless powders.

Thiourea (*thiocarbamide*), m.p. 180°C, may be prepared by heating ammonium thiocyanate at 170°C for some time:

The structure of thiourea is similar to that of urea; it behaves as a 'monoacidic' base and, like urea, forms channel complexes. However, in this case, the diameter of the channel is about 7Å, and so can accommodate larger guest molecules (than can urea).

Thus, thiourea readily encloses branched-chain Carbohydrates and Proteins and even cycloalkanes. Apparently the cross-section of straight-chain alkanes is too small to support the channels in thiourea and consequently these compounds do not form channel complexes. When heated with alkalis, thiourea is hydrolysed:

$$S = C(NH_2)_2 + 2H_2O \xrightarrow{NaOH} CO_2 + H_2S + 2NH_3$$

Oxidation with *alkaline* permanganate or alkaline hydrogen peroxide converts thiourea into urea:

$$S = C(NH_2)_2 + [O] \longrightarrow O = C\ (NH_2)_2 + S$$

Oxides of lead, silver or mercury remove a molecule of hydrogen sulphide from thiourea into room temperature to form cyanamide:

$$(NH_2)_2CS + HGO \longrightarrow NH_2CN + HgS + H_2O$$

When treated with alkyl halide, thiourea forms S-*alkyl*-ψ-thiouronium salts (S-*alkyliso- thiouronium salts*):

$$CH_3I + S{=}C(NH_2)_2 \longrightarrow \left.CH_3SC{\begin{matrix} /\!\!/ NH_2 \\ \backslash NH_2 \end{matrix}}\right\} I^-$$

These compounds are used to characterise sulphonic acids, with which they form insoluble salts. They may also be used to prepare thioalcohols, and also to prepare sulphonyl chlorides by oxidation with chlorine-water.

$$RSC{\begin{matrix} /\!\!/ \overset{+}{N}H_2Cl^- \\ \backslash NH_2 \end{matrix}} \quad \begin{matrix} Cl_2 \\ H_2O \end{matrix} \quad RSO_2Cl$$

s-*Diphenyllhiourea* or *diphenylthiocarbanilide*, $(C_6H_5NH)_2CS$, which is used as a rubber accelerator, may be prepared by heating aniline with carbon disulphide in an ethanolic solution containing potassium hydroxide.

Dithiocarbamic acid, S = C(NH_2)SH, although unstable in the free state, gives rise to stable salts. These may be prepared by heating a primary or secondary mine with carbon disulphide. This reaction is complicated and the mechanism is uncertain. A possibility is via the formation of the unstable N-alkyldithioca basic acid which forms the stable dithiocarbamic acid salt with another molecule of amine:

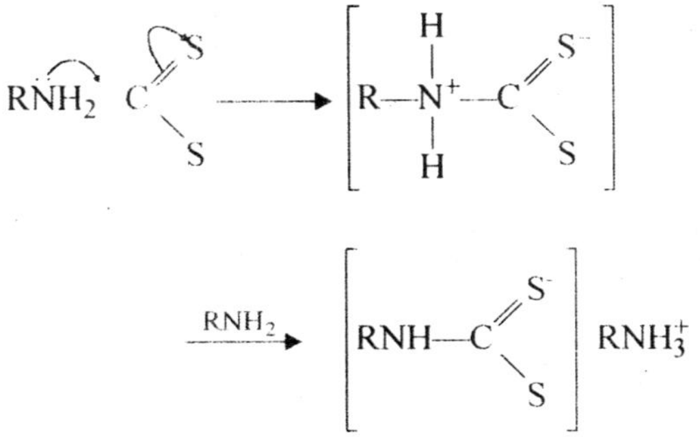

$$R\ddot{N}H_2 \quad S{=}C{=}S \longrightarrow \left[R{-}\overset{H}{\underset{H}{N^+}}{-}C\begin{matrix} {=}S^- \\ {-}S \end{matrix} \right]$$

$$\xrightarrow{RNH_2} \left[RNH{-}C\begin{matrix} {=}S^- \\ {-}S \end{matrix} \right] RNH_3^+$$

Fig. 4

The Hofmam mustard oil reaction is believed to take place via the formation of a dithio-carbamic acid salt, which is then decomposed by the mercuric chloride, possibly as follows:

MONOCARBOXYLIC ACIDS AND THEIR DERIVATIVES

The saturated monocarboxylic acids are also referred to as the fatty acids, a name which arose from the fact that some of the higher members, particularly palmitic and stearic acids, occur in natural fats.

Their general formula is $C_nH_{2n}O_2$ but, because their functional group is the carboxyl group, —CO_2H, they are more conveniently expressed as $C_nH_{2n+1}CO_2H$, since these show the nature of the functional group.

The structure of the carboxyl group is written as (or —COOH), but, as we shall later, it does not represent accurately the behaviour of this group. Also, since only the carboxyl group hydrogen atom is replaceable by a metal, the monocarboxylic acids are *monobasic*.

Nomenclature

The monocarboxylic acids are commonly known by the trivial names, which have been derived from the source of the particular acid; *e.g.*, formic acid, HCO_2H, was so named because it was first obtained by the distillation of ants; the Latin word for ant is *formica*. Acetic acid, CH_3CO_2H, is the chief constituent of vinegar, the Latin word for which is *acetum*; etc.

Another system of nomenclature considers the acids, except formic acid, as derivatives of acetic acid, *e.g.*,

$CH_3CH_2CO_2H$ methylacetic acid

$(CH_3)_3CCO_2H$ trimethylacetic acid

$(CH_3)_2CHCH_2CO_2H$ isopropylacetic acid

In the above two systems the positions of substituents in the chain are indicated by Greek letters, the α-carbon atom being the one joined to the caboxyl group, *e.g.*,

$CH_3CH(OH)CH_2CH_2CO_2H$ γ-hydroxyvaleric acid

$CH_3CHClCH(CH_3)CO_2H$ β-chloro α-methylbutyric acid

According to the I.U.P.A.C. system of nomenclature, the suffix of the monocarboxylic acids is *-oic*, which is added to the name of the alkane corresponding to the longest carbon chain containing the carboxyl group, *e.g.*,

HCO_2H methanoic acid

$CH_3CH_2CO_2H$ propanoic acid

The positions of side-chains (or substituents) are indicated by numbers, *the carboxyl group always being given number* 1:

$CH_3CH(CH_3)CH(CH_3)CH_2CO_2H$ 3,4-dimethylpentanoic acid

Alternatively, the carboxyl group is regarded as a substituent, and is denoted by the suffix carboxylic acid, *e.g.*,

$\overset{4}{C}H_3\overset{3}{C}H_2\overset{2}{C}H(CH_3)\overset{1}{C}H_2CO_2H$ 2-methylbutane-1-carboxylic acid

General Methods of Preparation

1. Oxidation of alcohols, aldehydes or ketones with acid dichromate yields acids:

$$RCH_2OH \xrightarrow{[O]} RCHO \xrightarrow{[O]} RCO_2H \ (g.\text{–}v.g.)$$

$$R_2CHOH \xrightarrow{[O]} R_2CO \xrightarrow{[O]} R^1CO^2H + R^2CO_2H \ (g.)$$

Many *primary* alcohols, on oxidation with acid dichromate, form esters in addition to acids, *e.g.*, n-butanol gives n-butyric acid and n-butyl n-butyrate (41–47 per cent). Several mechanisms have been proposed, but the generally accepted one involves hemiacetal formation.

$$RCH_2OH \xrightarrow{[O]} RCHO \xrightarrow{RCH_2OH} RCH(OH)OCH_2R \xrightarrow{[O]} RCO_2CH_2R$$

2. A very good synthetic method is the hydrolysis of cyanides with or alkali:

The amide, $RCONH_2$, may be isolated if suitable precautions are taken.

3. By the reaction between a Grignard reagent and carbon dioxide:

$$RMgX + CO_2 \longrightarrow RCO_2MgX \xrightarrow{H_2O} RCO_2H$$

4. Many monocarboxylic acids may be conveniently synthesised from alkyl halides and ethyl malonate or acetoacetic ester. These syntheses will be discussed in detail later.

5. A number of higher acids may be obtained by the hydrolysis of natural fats or may be prehalf-ester of a dicarboxylic acid (Linstead *et al.*, 1950; *cf.*).

$$RCO_2H + HO_2C(CH_2)_nCO_2CH_3 \longrightarrow H_2 + 2CO_2 + R(CH_2)_nCO_2CH_3$$

The chain of a monocarboxylic acid can be extended by five or six carbon atoms in a synthesis involving enamines.

6. Monocarboxylic acids may be obtained by heating a sodium alkoxide with carbon monoxide under pressure.

$$R\overset{-}{O}\overset{+}{Na} + CO \longrightarrow RCO_2^-\overset{+}{Na}$$

Acids may also be obtained by heating the alcohol with carbon monoxide under pressure in the presence of a catalyst.

The sodium salts of acetic and propionic acids are formed by the interaction of methylsodium or ethylsodium, respectively, and carbon dioxide, *e.g.*,

$$C_2\overline{H}_5\overset{+}{Na} + CO_2 \longrightarrow C_2H_5CO_2Na$$

Monocarboxylic acids may be prepared by heating an alkene with carbon monoxide and steam under pressure at 300 – 400°C in the presence of a catalyst, *e.g.*, phosphoric acid:

$$CH_2{=}CH_2 + CO + H_2O \longrightarrow CH_3CH_2CO_2H$$

$$CH_3CH{=}CH_2 + CO + H_2O \longrightarrow (CH_2)_2CHCO_2H$$

GENERAL PROPERTIES OF THE MONOCARBOXYLIC ACIDS

The first three acids are colourless, pungent-smelling liquids; the acids from butyric, $C_4H_8O_2$, to nonanoic, $C_9H_{18}O_2$, are oils which smell like goats' butter; and those higher than decanoic acid, $C_{10}H_{20}O_2$, are odourless solids. The lower members are far less volatile than is to be expected from their molecular weight. This can be explained by hydrogen bonding, and electron diffraction studies (Pauling, 1934) have shown that an eight-membered ring is present, *i.e.*, the acids exist as cyclic dimers. The lower members exist as dimers in the vapour phase and in aqueous solution, but in the liquid phase they exist as 'polymers'. The melting point of the n-monocarboyxlic acids show alternation or oscillation from one member to the next, the melting point of an 'even' acid being higher than that of the 'odd' acid immediately below and above it in the series (The physical constants of the individuals). A number of homologous series follow this oscillation or 'saw-tooth' rule.

The first four members are very soluble in water, and the solubility decreases as the molecular weight increases. This solubility is due to the acids being capable of forming hydrogen bonds with water.

The infra-red absorption region of the carboxyl group contains a number of bands. The carbonyl group (stretch) absorbs in the same region as that of acyclic ketones. 1725–1700 cm^{-1}(s), but the (free) hydroxyl group (stretch) absorbs at 3560–3500 cm^{-1}(m), which is in the lower frequency region than for a alcohols (3670–3580 cm^{-1}). The infra-red spectrum of stearic acid (as a 1 per cent solution in carbon tetrachloride). It should be noted that the band for the bonded O—H (str.) is hidden by the strong C—H (str.)) 1706 cm^{-1}: C=O (str.) in (saturated ketones and) saturated aliphatic acids. 2942 and 2857 cm^{-1}: C—H (str.) in Me and/or CH_2. 1460cm^{-1}: C—H (def.) in Me and/or CH_2.

On the other hand is the spectrum of stearic acid as a Nujol mull. In this case, the stearic acid is in the solid (dispersed) state, and it can be seen that there is an extremely large difference in the region 1300–1180 cm^{-1}. In this region there are eight maxima, and this pattern is characteristic of the spectra of long-chain n-alkyl compounds (in the solid state), the number of peaks depending on the chain length Nujol (a petroleum fraction), since it is a hydrocarbon mixture, has the bands characteristic of alkanes. The O—H (str.) is hidden. These two spectra illustrate the influence of the physical state of a compound on its infra-red spectrum. 2940, 1460 and 1370 cm^{-1}: Bands for Nujol. 1709cm^{-1}: C=O (str.) 1312, 1295, 1279, 1258, 1235, 1220, 1203 and 1188 cm^{-1}: These bands are characteristic of long-chain n-alkyl compounds (in the solid state).

Because of hydrogen bonding (intermolecular) and exchange reactions, the proton peak of the carboxyl group can be found in many different positions in the NMR spectrum. This applies to hydroxyl protons in general. However, it has been found that in non-polar solvents, the τ-value of the carboxyl proton in aliphatic acids is around – 1.0 p.p.m. (*i.e.*, there is a great shift to lower field due to strong hydrogen bonding).

Mass spectra. Acids, esters, and amides undergo similar fragmentation patterns, *e.g.*, α-cleavage for lower members (Z=OH, OR, NH_2): Also, R and Z may carry the positive charge. Hence, peaks are obtained at *M* – 17 and *m/e* 45 (Z = OH), *M* – 31, *M*–45, and *m/e* 59, 73 (Z = OMe and OEt, respectively), and *M* – 16 and *m/e* 44 (Z = NH_2). Since they are more volatile than their corresponding acids, esters are used preferentially. All can undergo the McLafferty rearrangement if they contain a γ-hydrogen atom: When the alkyl group in O of esters is higher than methyl, esters can also undergo the McLafferty rearrangement to give alkene and acid, *e.g.*,

General Reactions of the Monocarboxylic Acids

1. The acids are acted upon by the strongly electropositive metals with the liberation of hydrogen and formation of a salt:

$$RCO_2H + Na \longrightarrow RCO_2^-Na^+ + \frac{1}{2}H_2$$

Salts are also formed by the reaction between an acid and an alkali:

$$RCO_2H + NaOH \longrightarrow RCO_2^-Na^+ + H_2O$$

2. Monocarboxylic acids react with alcohols to form *esters*:

$$R^1CO_2H + R^2OH \rightleftharpoons R^1CO_2R^2 + H_2O$$

3. Phosphorus trichloride, phosphorus pentachloride or thionyl chloride act upon the monocarboxylic acids to from *acid chlorides*:

$$3RCO_2H + PCl_3 \longrightarrow 3RCOCl + H_3PO_3$$
$$RCO_2H + PCl_5 \longrightarrow RCOCl + HCl + POCl_3$$
$$RCO_2H + SOCl_2 \longrightarrow RCOCl + HCl + SO_3$$

The acid series may be 'stepped up' via the acid chloride, but since the process involves many steps, the yield of the higher acid homologue is usually only *f.-f.g.*:

$$(a)\ RCO_2H \xrightarrow{PCl_5} RCOCl \xrightarrow[\text{catalyst}]{H_2} RCHO \xrightarrow[\text{catalyst}]{H_2}$$

$$RCH_2OH \xrightarrow{PBr_3} RCH_2Br \xrightarrow{KCN} RCH_2CN \xrightarrow{H^+} RCH_2CO_2H$$

$$(b)\ RCO_2H \xrightarrow{PCl_5} RCOCl \xrightarrow{KCN} RCOCN$$

$$\xrightarrow{H^+} RCOCO_2H \xrightarrow[\text{reduction}]{\text{Clemmensen}} RCH_2CO_2H$$

Although method (*b*) involves fewer steps than (*a*), the final yield of acid by route (*a*) is higher than that by (*b*), since the yields in each step of (*b*) are generally low. Furthermore, the yield from (*a*) can be improved by direct reduction of acid to alcohol by means of various metallic hydrides.

4. When the ammonium salts of the monocarboxylic acids are strongly heated, the *acid amide* is formed by the elimination of water:

$$RCO_2NH_4 \longrightarrow RCONH_2 + H_2O$$

5. When the anhydrous sodium salt of a monocarboxylic acid is heated with soda-lime, an alkane and other products are formed.

The mechanism of this decarboxylation is uncertain, but there is much evidence to show that *salts* decompose by an S_E1 reaction : Monocarboxylic acids may be 'stepped down' by heating the calcium salt with calcium acetate, and the methyl ketone produced is oxidised with acid dichromate:

$$RCH_2CO_2Ca/_2 + CH_3CO_2Ca/_2 \longrightarrow RCH_2COCH_3$$
$$\xrightarrow{[O]} RCO_2H + CH_3CO_2H$$

It should be noted that the second step is based on the general rule that when an unsymmetrical ketone is oxidised, the carbonyl group remains chiefly with the smaller alkyl group. Since monocarboxylic acids containing an odd number of carbon atoms occur only in small amounts, this method of descending the series affords a useful means of obtaining 'odd' acids from 'even'.

When a mixture of the calcium salt (or any of the others mentioned above) and calcium formate is heated, an aldehyde is obtained. The pyrolysis of iron salts gives very good yields of symmetrical ketones. The pyrolysis of iron slats gives very good yields of symmetrical ketones.

Metallic salts, particularly the silver slats, are converted by bromine into the alkyl bromide. This reaction may also be used to 'step down' the acid series.

6. When a concentrated aqueous solution of the sodium or potassium salt of a monocarboxylic acid is electrolysed, an alkane is obtained.

$$2RCO_2K + 2H_2O \longrightarrow R{-}R + 2CO_2 + 2KOH + H_2$$

7. Monocarboxylic acids react slowly with chlorine or bromine in the cold, but at higher temperatures and in the presence of a small amount of phosphorus, reaction proceeds smoothly to give a-halogeno-acids; *e.g.*,

$$RCH_2CO_2H \xrightarrow{P/Br_2} RCHBrCO_2H \xrightarrow{P/Br_2} RCBr_2CO_2H$$

8. All the acids, except formic acids, are extremely resistant to oxidation, but prolonged heating with oxidising agents ultimately produces carbon dioxide and water.

9. The reduction products of monocarboxylic acids depend on the nature of the reducing agent. Thus, heating with hydrogen iodide-red phosphorous under pressure, or with hydrogen under pressure at elevated temperature in the presence of a nickel catalyst, produces an alkane.

$$RCO_2H + 3H_2 \xrightarrow{Ni} RCH_3 + 2H_2O$$

If the catalyst is Ru—C, the alcohol is formed. Alcohols are also produced by reduction with various metallic hydrides.

10. The acids undergo the *Schmidt reaction* to form a primary amine:

$$RCO_2H + HN_3 \xrightarrow{H_2SO_4} RNH_2 + CO_2 + N_2$$

Schmidt's reaction with acids is a modification of the Curtius reaction. The following mechanism has been proposed : The rearrangement (1,2-shift) has been shown to be intramolecular, *e.g.*, Keynon *et al.* (1939)

showed retention of optical activity when α-phenylpropionic acid underwent the Schmidt reaction:

$$C_6H_5CH(CH_3)CO_2H \longrightarrow C_6H_5CH(CH_3)NH_2$$

Formic acid (*methanoic acid*) is prepared industrially by heating sodium hydroxide with carbon monoxide at 210°C, and at a pressure of 6–10 atmospheres:

$$NaOH + CO \longrightarrow HCO_2Na$$

Anhydrous formic acid is obtained from the aqueous solution (70–77 per cent) by the addition of butyl formate followed by distillation. The first fraction is an azeotrope of ester and water, and then the excess of ester is removed from the formic acid by fractionation.

The most convenient laboratory preparation of formic acid is to heat glycerol with oxalic acid at 100–100 C. Glyceryl monoxalate is produced, and decomposes into glyceryl monoformate (monoformin) and carbon dioxide. When the evolution of carbon dioxide ceases, more oxalic acid is added, whereupon formic acid is produced:

The distillate contains formic acid and water. The aqueous formic acid solution cannot be fractionated to give anhydrous formic acid because the boiling point of the acid is 100.5°C. The procedure adopted is to neutralise the aqueous acid solution with lead carbonate, and concentrate the solution until lead formate crystallises out. The precipitate is then recrystallised, dried, and heated at 100°C in a current of hydrogen sulphide:

$$(HCO_2)_2Pb + H_2S \longrightarrow 2HCO_2H + PbS$$

The anhydrous formic acid which distils over contains a small amount of hydrogen sulphide, and may be freed from the latter by adding some dry lead formate and redistilling. This procedure for obtaining the anhydrous acid from its aqueous solution can only be used for volatile acids.

Formic acid is a pungent corrosive liquid, m.p. 8.4°C, b.p. 100.5°C, miscible in all proportions with water, ethanol and ether. It forms salts which, except for the lead and silver salts, are readily soluble in water. Formic acid is a stronger acid than any of its homologues.

Formic acid is dehydrated to carbon monoxide by concentrated sulphuric acid and when heated under pressure at 160°C, is decomposed into carbon dioxide and hydrogen:

$$CO + H_2O \xleftarrow{H_2SO_4} HCO_2H \xrightarrow{heat} CO_2 + H_2$$

The same decomposition takes place at room temperature in the presence of a catalyst such as iridium, rhodium, etc.

When metallic formates are heated with an alkali, hydrogen is evolve.

$$HCO_2Na + NaOH \longrightarrow H_2 + Na_2CO_3$$

When calcium or zinc formate is strongly heated, formaldehyde is produced:

$$(HCO_2)_2Ca \longrightarrow HCHO + CaCO_3$$

When sodium or potassium formate is rapidly heated to 360°C, hydrogen is evolved and the oxalate is formed.

$$2HCO_2Na \longrightarrow (COONa)_2 + H_2$$

Formic acid forms esters, but since it is a relatively strong acid it is not necessary to use a catalyst: refluxing 90 per cent formic acid with the alcohol is usually sufficient.

Formic acid differs from the rest of the members of the monocarboxylic acid series in being a powerful reducing agent; it reduces ammoniacal silver nitrate and the salts of many of the heavy metals, *e.g.*, it converts mercuric chloride into mercurous chloride.

Acetic acid (*ethanoic acid*) is prepared industrially by various methods, *e.g.*, (*i*) the air oxidation of acetaldehyde in acetic acid in the presence of manganous ion as catalyst: (*ii*) the air oxidation of n-butane in the liquid phase under pressure and at 130–230°C in the presence of a suitable catalyst, *e.g.*, manganous ion; (*iii*) by reaction between carbon monoxide and methanol under pressure in the presence of cobalt octacarbonyl at about 210°C:

$$CO + CH_3OH \longrightarrow CH_3CO_2H.$$

One of the earliest methods for preparing acetic acid was by the destructive distillation of wood to give pyroligneous acid. This contains about 10 per cent acetic acid, and was originally treated by neutralising with lime and then distilling off the volatile compounds (these are mainly methanol and acetone). On distillation with dilute sulphuric acid, the residue gives dilute acetic acid. More recently, the acetic acid is extracted by means of solvents, *e.g.*, isopropyl ether.

Vinegar, which is 6–10 per cent aqueous solution of acetic acid, may be made in several ways. Malt vinegar is prepared by the oxidation of wort by means of the bacteria *Mycoderma aceti*:

$$CH_3CH_2OH + O_2 \longrightarrow CH_3CO_2H + H_2O$$

In the 'quick' vinegar process', beech shavings, contained in barrels, are moistened with strong vinegar containing the bacteria. A 10 per cent aqueous solution of ethanol containing phosphates and inorganic salts (which are necessary for the fermentation) is then poured through the barrels, and the ethanol is thereby oxidised to acetic acid. A plentiful supply of air is necessary, otherwise the oxidation is incomplete and acetaldehyde is produced.

Acetic acid is a pungent corrosive liquid, m.p. 16.6°C, b.p. 118°C, miscible in all proportions with water, ethanol ad ether It is stable towards oxidising agents, and so is a useful solvent for chromium trioxide oxidations. Acetic acid is commonly used as a solvent, and in the preparation of acetates, acetone, acetic anhydride, etc.

STRENGTHS OF THE MONOCARBOXYLIC ACIDS

If the structure of the carboxyl group were simply (I), then, because of the – I effect of the carbonyl group (as shown in (II), proton release will be facilitated as compared with alcohols. Thus carboxylic acids will be stronger acids than alcohols. However, the inductive effect alone cannot account for the very large difference in acid strength. One explanation is that the carboxyl group is a resonance hybrid and because of the positive charge on the oxygen atom of the OH group, proton release is facilitated; this positive charge is absent in alcohols since they are not resonance hybrids. This argument can be extended as follows. When the acid ionises, *i.e.*, donates its proton to water (which is the solvent), the conjugate base of the acid. *i.e.*, the carboxylate ion, is a resonance hybrid of two *equivalent* resonating structures. At the same time, since the negative charge is spread, the resonance energy of this anion will be greater than that of the un-ionised acid, which is a resonance hybrid of two different resonating structures, one with separation of *unlike* charges. Therefore the internal energy of the anion is lower than that of the un-ionised acid. In the case of alcohols, thee is no resonance stabilisation in the alkoxide ion. It therefore follows that $\Delta G^{\ominus}$ for the ionisation of monocarboyxlic acids will be more negative than that of alcohols. Hence, the equilibrium constant of the former will be greater than that of the latter, *i.e.*, the pK_a values of the acids will be lower than those of the alcohols. Thus, alcohols are weaker acids than the monocaboxylic acids.

X-ray and electron-diffraction measurements of the carboxylate ion have shown that the two C—O bonds are of equal length. This is in keeping with the ion being a resonance hybrid.

The treatment of the carboxyl group and the carboxylate ion from the M.O. point of view also leads to the same results as those obtained by the V.B. method, the increased stability now being due to *delocalisation*. Let us first consider the *undissociated* carboxylic acid. The carbon atom is linked to the hydroxyl group by a σ-bond, and to the oxygen of the carbonyl group by one σ- and one π-bond. If the oxygen atom of the hydroxyl group has sp^2 hybridisation, this would leave a p orbital perpendicular to the plane of the carboxyl group, and it would contain a pair of electrons. This orbital could now overlap with both the p orbital of the carbon atom and that of the oxygen atom of the carbonyl group. The combination of these three orbitals would give rise to M.O.s embracing *three* nuclei. Since *four* electrons are involved, they must be accommodated, in the ground state, in the lowest two M.O.s. The M.O. with the lowest energy level has *one* node and the next energy level has *two* nodes.

Table 1

Acid	pK_a	*Acid*	pK_a
Formic	3.75	Chloroacetic	2.85
Acetic	4.76	Dichloroacetic	1.25
Propionic	4.87	Trichloroacetic	0.66
n-Butyric	4.82	Trifluoroacetic	0.23
Isobutyric	4.86	Bromoacetic	2.9
n-Valeric	4.86	Iodoacetic	3.16
Isovaleric	4.78	α-Chloropropionic	2.80
Trimethylacetic	5.05	β-Chloropropionic	4.08
Hexanoic	4.88	Nitroacetic	1.68
Heptanoic	4.89	Hydroxyacetic	3.83
Octanoic	4.90	Hydrochloric	-- 7.00

The ground state of the undissociated carboxyl group. Because of delocalisation, this group has become more stable. Since a hydroxyl group is more electron-attracting than an oxygen atom, the delocalisation of the pair of electrons in the p orbital is not very large, *i.e.*, although this lone pair embraces three nuclei, this pair is far more likely to be found in the region of the donor atom than anywhere else. Since the oxygen atom of the hydroxyl group has lost 'full-control' of this lone pair, it will acquire a small positive charge, and since the oxygen atom of the CO group has acquired a small share of this lone pair, this oxygen atom acquires a small negative charge (actually this oxygen atom already has a small negative charge due to the inductive effect).

The greater electronegativity of oxygen in a hydroxyl group over that of an oxygen atom alone is due to the hydrogen atom sharing one pair of electrons in forming the bond and thereby decreasing, to some extent, the electron density on that oxygen atom.

The oxygen atom of the original hydroxyl group retains the σ-electrons when the hydrogen atom is removed as a proton. As in the case of the *carboxy group*, one lone pair enters into conjugation with the π-bond of the C=O group. In the latter case, however, delocalisation is complete, *i.e.*, the lone pair is now likely to be found equally well or either oxygen atom (each atom will have a charge of – 1/2). Since delocalisation is complete, the resonance (delocalisation) energy is much greater than in the case of the undissoicated carboxyl group. Because the carbon atom is sp^2 hybridised in both the carboxyl group and carboxylate ion, the O—C—O angle in both would be expected to be about 120°. This has been observed in practice.

The orbitals drawn in thicker type indicate that the atoms concerned contain *two* electrons in these orbitals. It should also be noted that lobes and not spheres (with + and – signs) have been used in these diagrams, and that overlapping orbitals have been drawn in an alternative way.

Swern *et al.* (1962) have prepared per-acids (85–97 per cent yield) by treating a solution or slurry of an aliphatic (or aromatic) acid in methanesulphonic acid with 90–95 per cent hydrogen peroxide. This appears to be the best method so far.

Peracetic acid is an unpleasant-smelling liquid, f.p. + 0.1°C, soluble in water, ethanol and ether. It explodes violently when heated above 110°C. It is a powerful oxidising agent; it oxidises the olefinic bond to the oxide.

It also oxidises primary aromatic amines to nitroso-compounds, *e.g.*, aniline is converted into nitrosobenzene:

$$C_6H_5NH_2 + 2CH_3COO_2H \longrightarrow C_6H_5NO + 2CH_2CO_2H + H_2O$$

Infra-red absorption spectra measurements of per-acids show that these acids exist in solution very largely in the monomeric, intra-molecularly hydrogen-bonded form (*inter alia*, Minkoff, 1954).

Propionic acid (*propanoic acid*) is prepared industrially by the oxidation of n-propanol:

$$CH_3CH_2CH_2OH \xrightarrow{[O]} CH_3CH_2CO_2H$$

It is a colourless liquid with an acrid odour, m.p. – 22°C, b.p. 141°C, miscible with water, ethanol and ether in all proportions.

Butyric acids. There are two isomers possible, and both are known.

n-Butyric acid occurs as the glyceryl ester in butter, and as the free acid in perspiration. It is prepared industrially by the oxidation of n-butanol.

n-Butyric acid is a viscous unpleasant-smelling liquid, m.p. – 4.7°C, b.p. 162°C, miscible with water, ethanol and ether. It is the liberation of free n-butyric acid that gives stale butter its rancid odour.

Isobutyric acid occurs in the free state and as its esters in many plants. It is prepared industrially by the oxidation of isobutanol and may be prepared synthetically as follows:

$$(CH_3)_2CHOH \xrightarrow{PBr_3} (CH_3)_2CHBr \xrightarrow{KCN}$$

$$(CH_3)_2CHCN \xrightarrow[acid]{H_2O} (CH_3)_2CHO_2H$$

Isobutyric acid is a liquid, m.p. – 47°C, b.p. 154°C. Its calcium salt is more soluble in hot water than in cold, whereas calcium butyrate is more soluble in cold water than in hot.

Valeric acids. There are four isomers possible, and all are known:

n-Valeric acid, $CH_3CH_2CH_2CH_2CO_2H$, m.p. – 34.5°C, b.p. 187°C.

Isovaleric acid, $(CH_3)_2CHCH_2CO_2H$, m.p. – 51°, b.p. 175°C.

Ethylmethylacetic acid or *active valeric acid*, CH_3CH_2CH-$(CH_3)CO_2H$, p.b. 175°C.

Trimethylacetic acid or *pivalic acid*, $(CH_3)_3CCO_2H$, m.p. 35.5°, p.b. 164°C.

The higher monocarboxylic acids which occur in nature usually have straight chains, and usually contain an *even* number of carbon atoms. **Caproic** (*hexanoic*), $C_6H_{12}O_2$ (m.p. – 9.5°, b.p. 205°C), **caprylic** (*octanoic*), $C_8H_{16}O_2$ (m.p. 16°, p.b. 237°C) and *capric* (*decanoic*) acid, $C_{10}H_{20}O_2$ (m.p. 31.5° p.b. 270°C), are present as glyceryl esters in goats' butter. *Lauric* (*dodecanoic*), $C_{12}H_{24}O_2$ (m.p. 44°C) and **myristic** (*tetradecanoic*) acid, $C_{14}H_{28}O_2$ (m.p. 58°C), occur as their glyceryl esters in certain vegetable oils. The most important higher acids are *palmitic* (*hexadecanoic*), $C_{16}H_{32}O_2$ (m.p. 64°C), and **stearic** (*octadecanoic*), $C_{18}H_{36}O_2$ (m.p. 72°C), which are very widely distributed as their glyceryl esters (together with oleic acid) in most animal ad vegetable oils and fats. The sodium and potassium salts of palmitic and stearic acids are the constituents of ordinary soaps.

Some still higher acids are found in waxes: **arachidic** (*eicosanoic*), $C_{20}H_{40}O_2$ (m.p. 77°C), **behenic** (*docosanoic*), $C_{22}H_{44}O_2$ (m.p. 82°C), **lignoceric** (*tetracosanoic*), $C_{24}H_{48}O_2$ (m.p. 83.5°C), **cerotic** (*hexacosanoic*), $C_{26}H_{52}O_2$ (m.p. 87.7°C), and **melissic** acid (*triacontanoic*), $C_{30}H_{60}O_2$ (m.p. 90°C). The odd fatty acids may be obtained by degrading an even acid. Two odd acids which may be prepared readily are **n-heptanoic** or **oenanthylic** acid, $C_7H_{14}O_2$ (m.p. – 10°, p.b. 223.5°C), and **nonanoic** or **pelargonic** acid $C_9H_{18}O_2$ (m.p. 12°, p.b. 254°C). n-Heptanoic acid is prepared by the oxidation of n-heptaldehyde, which is obtained by destructively distilling castor oil which contains ricinoleic acid (*q.v.*):

$$CH_3(CH_2)_5CHO + [O] \xrightarrow[H_2SO_4]{KMnO_4} CH_3(CH_2)_5CO_2H \ (68\text{–}70\%)$$

Nonanoic acid is obtained by the oxidation of oleic acid (*q.v.*). *Margaric acid* (*heptadecanoic acid*), $C_{16}H_{33}CO_2H$, m.p. 61°C, is prepared by heating a mixture of the calcium salts of stearic and acetic acids, and then oxidising the heptadecyl methyl ketone so produced.

$$C_{17}H_{35}CO_2ca + CH_3CO_2ca \longrightarrow CaCO_3 + C_{17}H_{35}COCH_3 \xrightarrow{[O]} C_{16}H_{33}CO_2H$$

Odd-numbered monocarboxylic acids occur naturally in very small amounts up to $n = 23$, and their main source is butter fat. Branched-chain acids have also been isolated from butter fat.

General Reactions of the Monocarboxylic Acids and Their Derivatives

Acids and their derivatives may be represented as follows, where $Z = OH, OR, X, NH_2$, etc.

Thus, all are resonance hybrids and the behaviour of the 'carbonyl' group depends on the nature of Z. Z may have a – I and a + R effect and consequently the actual weights of the canonical forms will depend on the relative contributions of these two electronic effects, *e.g.*, for Z = NH_2 (acid amide), + R > – I, whereas for Z = Cl (acid chloride), – I > + R. Hence, the rates of reaction of the carbonyl group in these compounds can be expected to be different.

These reactions are typical of the behaviour of the carbonyl group (in aldehydes and ketones), and the greater the positive charge on the carbonyl atom, the more easily is this carbon atom attacked by a nucleophile (in the r/d step). Since the + R effect tends to neutralise the positive charge, the greater the + R effect of Z, the slower will be the reaction. Experimental work has shown the rates of hydrolysis are in the order $RCONH_2 < RCO_2R < RCOCl$, and this is the reverse order of the + R effects of the groups, *i.e.*, $NH_2 > OR > Cl$.

Esters

Esters are compounds which are formed when the hydroxylic hydrogen atom in oxygen acids is replaced by an alkyl group; the acid may be organic or inorganic. The most important esters are derived from the caboxylic acids. The general formula of the carboxylic esters is $C_2H_{2n}O_2$, which is the same as that of the carboxylic acids, and they are named as the alkyl salts of the acid. *e.g.*,

$CH_3COOC_2H_5$ ethyl acetate

$(CH_3)_2CHOOCH(CH_3)_2$ isopropyl isobutyrate

Carboxylic esters are formed by the action of the acid on an alcohol: acid – alcohol:

acid + alcohol $\rightleftharpoons$ ester + water

The reaction is reversible, the *forward* reaction being known as *esterification*, and the *backward* reaction as *hydrolysis* (for mechanisms).

GENERAL METHODS OF PREPARATION OF THE CARBOXYLIC ESTERS

1. The usual method is *esterification*. The reaction is always slow, but is speeded up by the presence of small amounts of inorganic acids as catalysts, *e.g.*, when the acid is refluxed with the alcohol in the presence of 5–10 per cent concentrated sulphuric acid:

$$R^1COOH + R^2OH \xrightarrow{H_2SO_4} R^1COOR^2 + H_2O \ (v.g.)$$

Alternatively, hydrogen chloride is passed into the mixture of alcohol and acid until there is a 3 per cent increase in weight, and the mixture is refluxed (the yields are very good). This is known as the *Fische–Speier method* (1895), and is more satisfactory for secondary and tertiary alcohols than the sulphuric acid method, which tends to dehydrate the alcohol to alkene.

Esterification without the use of catalysts, and starting with one mole of acid and one mole of alcohol, gives rise to about 2/3 mole of ester. The yield of ester may be increased by using excess of acid or alcohol, the cheaper usually being the one in excess. Increased yields may also be effected by dehydrating agents, *e.g.*, concentrated sulphuric acid behaves both as a catalyst and a dehydrating agent. The same effect may be obtained by removing the water or ester from the reaction mixture by distillation, which is particularly useful for high-boiling acids and alcohols. On the other hand, the water may be removed from the reaction mixture by the addition of benzene or carbon tetrachloride, each of which forms a binary mixture with water (and may form a ternary mixture with water and the alcohol), the azeotropic mixtures boiling at a lower temperature than any of the components.

2. Acid chlorides or anhydrides react rapidly with alochols to form esters:

$$R^1COCl + R_2OH \longrightarrow R^1CO_2R^2 + HCl \ (v.g.\text{–}ex.)$$

$$(R^1CO)_2O + R^2OH \longrightarrow R^1CO_2R^2 + R^1CO_2H \ (v.g.\text{–}ex.)$$

The reaction with tertiary alcohols is very show and is often accompanied by side reactions (alkene or alkyl halide formation), but by using the appropriate conditions, esters are produced. Esters of tertiary alochols may also be conveniently prepared by means of a Grignard reagent.

3. Esters may be prepared by refluxing the silver salt of an acid with an alkyl halide in ethanolic solution:

$$R^1CO_2Ag + R^2Br \longrightarrow R^1CO_2R^2 + AgBr \ (v.g. \text{–}ex.)$$

This method is useful where direct esterification is difficult.

4. *Methyl* esters are verv conveniently prepared by treating an acid with an ethereal solution of diazomethane (*q.v.*):

$$RCO_2H + CH_2N_2 \longrightarrow RCO_2CH_3 + N_2 \ (ex.)$$

According to Morrison *et al.* (1961), the BF_3—MeOH reagent is better than diazomethane for the methylation of carboxylic acids.

5. The interaction between an ether and carbon monoxide at 125–180°C, under 500 atmospheres pressure, in the presence of boron trifluoride plus a little water, produces esters:

$$R_2O + CO \longrightarrow RCO_2R$$

Esters may also be prepared by the Tischenko reaction (with aldehydes); ethyl acetate is made commercially this way.

GENERAL PROPERTIES OF ESTERS

The carboxylic esters are pleasant-smelling liquids or solids. The boiling points of the straight chain isomers are higher than those of the branched-chain isomers. The boiling points of the methyl and ethyl esters are lower than those of the corresponding acid, and this is due to the fact that the esters are not associated since they cannot form intermolecular hydrogen bonds. The esters of low molecular weight are fairly soluble in water–hydrogen bonding between ester and water is possible–and the solubility decreases as the series is ascended; all esters are soluble in most organic solvents. The structure of the esters, like their parent acids, is best represented as a resonance hybrid: The infra-red absorption region of the carbonyl group (stretch) in saturated aliphatic esters is 1750–1735 cm^{-1}(s). This is a higher frequency than that of the carbonyl group in ketones and acids, and so esters may usually be identified.

Characteristic bands are also shown by the C—O (stretch) in formates, 1200–1180 cm^{-1}(s); acetates, 1250–1230 cm^{-1}(s); propionates, etc., 1200–1170 cm^{-1}(s).

The infra-red (film) ad NMR (in $CDCL_3$ at 60 MHz) spectra of ethyl acetate and respectively.

General Reactions of Esters

1. Esters are hydrolysed by acids or alkalis:

$$R^1CO_2H + R^2OH \xleftarrow{H_3O^+} R^1CO_2R^2 \xrightarrow{OH^-} R^1CO_2^- + R^2OH$$

When hydrolysis is carried out with alkali, the salt of the acid is obtained, and since the alkali salts of the higher acids are soaps, alkaline hydrolysis is known as *saponification* (derived from Latin word meaning soap); saponification is far more rapid than acid hydrolysis.

Mechanisms of carboxyl esterification and hydrolysis of carboxylic esters. Mechanistic studies have shown that in alkaline or neutral hydrolysis, it is the neutral ester molecule which undergoes reaction, whereas in acid-catalysed hydrolysis, it is the conjugate acid of the ester which undergoes reaction. The protonated carbonyl oxygen form of the cA is preferred, and this is based on the work of Fraenkel (1961), who examined the NMR spectrum of methyl formate in, *e.g.*, 100 per cent sulphuric acid and concluded from his results that the ester is protonated chiefly on the carbonyl oxygen. This formulation is also used for *acids*, partly from analogy with esters and partly from the fact that of the two possible conjugate acids, (I) and (II) , (I) may be written as a resonance hybrid to which two identical canonical forms contribute, whereas for (II) the two contributing structures are different and one has fewer bonds and carries unlike charges. The hydrolysis of carboxylic esters may be formulated in two ways:

Both modes of fission have been demonstrated experimentally, and kinetic studies have shown that two types of mechanism may operate for each mode, *unimolecular* and *bimolecular*. Hence, *eight* mechanisms are possible for ester hydrolysis but, so far, *six* have been observed Table. Also, since in principle, the *acid-catalysed* reactions (esterification and hydrolysis) are reversible, but not *alkaline* hydrolysis, then *four* mechanisms are possible for esterification.

Table 2

`Type of mechanism	*Hydrolysis*	*Esterification*
$B_{AC}1$	—	—
$B_{AC}2$	Very common	—
$A_{AC}1$	Special cases	Special cases
$A_{AC}2$	Very common	Very common
$B_{AL}1$	Special cases	—
$B_{AL}2$	Rare	—
$A_{AL}1$	Very common for t-alcohols	Very common for t-alcohols
$A_{AL}2$	—	—

So far, *three* have been observed (Table). The eight possible mechanisms are indicated by the symbols shown in Table, where *A* represents the cA of the ester or acid (reaction in *acid* solution), *B* represents the unprotonated ester (reaction in *basic* or *neutral* solution), subscripts AC and AL, respectively, denote acyl- and alkyl-heterolysis, and numbers 1 and 2 represent the molecularity of the rate-determining step (Ingold, 1941):

Most of the experimental work has been carried out on hydrolysis because this process can be studied in acidic, basic, or neutral media, whereas esterification can be studied only in acidic media. It might appear at first sight that esterification could also be studied in neutral media, but since carboxylic acids are acids, reaction in the absence of added acid (as catalyst) is still effectively in an acid medium. Because of this, mechanisms of esterification are partly based on the principle of microscopic reversibility.

Bimolecular basic hydrolysis with acyl-oxygen heterolysis (*B_{AC}* 2 mechanism). It has long been known that the alkaline hydrolysis of esters is a second-order reaction, *i.e.*, rate = *k* [ester] [OH⁻]. Evidence for acyl-oxygen heterolysis has been obtained in several ways, *e.g.*, Polanyl *et al.*, (1934) showed that the alkaline hydrolysis of n-pentyl acetate in water enriched with ^{18}O gave n-pentyl alcohol containing *no* ^{18}O. Therefore acyl-oxygen heterolysis must have occurred:

$$CH5CO—OC5H11 + \dot{O}H^- \longrightarrow CH_3CO\dot{O}H + C_5H_{11}O^- \longrightarrow$$

$$CH_3CO\dot{O}^- + C_5H_{11}OH$$

Another method uses optical activity as a diagnostic test for demonstrating acyl-oxygen heterolysis. If the alcohol group is optically active, and if the $B_{AC}2$ mechanism operates, then the ion RO^- will be liberated and will retain its optical activity since the bond R—O is never broken. Various examples of retention are known when the reaction is bimolecular.

A mechanism consistent with the above facts (and with the work of Bender (1951), who used esters labelled with ^{18}O in the carbonyl group) is: Any R^1 group with – I effect should accelerate hydrolysis by creating a positive charge on the attached carbon atom, thereby facilitating attack by the hydroxide ion. On the other hand, because the r/d step is bimolecular, the larger R^1 is the larger can be expected to be the steric retardation, since the hybridisation of the carbon atom of the carbonyl group changes from sp^2 to sp^3 (with consequent increased crowding). These anticipated results have been observed in practice, *e.g.*,

– I effect

Ester	$MeCO_2Me$	$ClCH_2CO_2Me$	$CHCl_2CO_2Me$
Relative rates	1	761	16000

Steric effect

Ester	$MeCO_2Et$	$EtCO_2Et$	$isoPrCO_2Et$
Relative rates	1.00	0.47	0.10

Bimolecular acid-catalysed hydrolysis and esterification with acyl-oxygen heterolysis ($A_{AC}2$ mechanism). Acid-catalysed hydrolysis and acid-catalysed esterification are the reverse of each other, and so according to the principle of microscopic reversibility the mechanism of acid-catalysed hydrolysis will be that of acid-catalysed esterification, but in reverse.

The rate law for acid-catalysed hydrolysis has been shown to be second-order, first-order in both ester and hydrogen ion concentration, *i.e.*, rate = k [ester] $[H_3O^+]$.

Also, evidence for acyl=oxygen heterolysis has been obtained in several ways, *e.g.*, Ingold *et al.* (1939) showed that the acid-catalysed hydrolysis of methyl hydrogen succinate in water enriched with ^{18}O gave methanol with no extra ^{18}O. Thus, acyl-oxygen heterolysis must have occurred:

Furthermore, Robers *et al.* (1938) esterified benzoic acid with methanol enriched with 18O and obtained water *not* enriched with 18O. Thus, acyl-oxygen heterolysis must have occurred:

A mechanism that fits these facts is:It should be noted that the r/ d step for esterification is the addition of alcohol, whereas that for hydrolysis is the addition of water.

In both cases, the r/d step involves change of hybridisation of the carbon atom of the carbonyl group from sp_2 to sp_3, and consequently steric retardation can be expected to increase as R1 increase as size (*cf.* the BAC_2 mechanism), *e.g.* (esterification with methanol):

Acid	$MeCO_2H$	$n\text{-}PrCO_2H$	$MeCCO_2H$	Et_3CCO_2H
Relative rates	1.0	0.51	0.037	0.00016

Unimolecular acid-catalysed hydrolysis and esterification with acyl-cxygen heterolysis ($A_{AC}1$ mechanism). Let us consider the following *observed* results (*M* represents the mesityl group, 1,3,5-trimethylphenyl, in *mesitoic acid*, $Me_3C_6H_2CO_2H$):.

$$PhCO_2H + MeOH \xrightarrow{HCl} PhCO_2Me \quad \textit{occurs readily}$$

$$PhCO_2ME + OH^- \longrightarrow PhCO_2^- + MeOH \quad \textit{occurs readily}$$

$$MCO_2H + MeOH \xrightarrow{HCl} \textit{No reaction}$$

$$MCO_2Me + OH^- \longrightarrow MCO_2^- + MeOH \quad \textit{very slow}$$

Hammett *et al.* (1937) showed that when methyl benzoate is dissolved in corcentrated sulphuric acid and the solution then poured into ice/water, almost all of the ester is recovered. On the other hand, methyl mesticate, on the same treatment, gave a quantitative yield of mestioic acid. Then Newman (1941) showed that when a solution of mesitoic acid in concentrated sulphuric acid is poured into a cold methanol, methyl mesitcate is formed.

This different behaviour of mesitoyl derivatives from benzoyl derivatives suggests that different mechanisms are operating in esterification and hydrolysis for these two sets of compounds. Now, Hammett (1937) had shown that when mesitoic acid is dissolved in concentrated sulphuric acid, the van't Hoff factor *i* is 4 (measured by depression of freezing point). Under the same conditions, benzoic acid gives *i* = 2. Hammett therefore proposed that the mesitoyl reactions proceed through an intermediate *acylium* (*oxocarbonium*) ion (*mesitoylium* ion in this case) formed by acyl-oxygen heterolysis:

Unimolecular basic hydrolysis with alkyl-oxygen heterolysis ($B_{AL}1$ mechanism). This mechanism has been shown to occur when the alkyl group of the alcohol is capable of forming a relatively stable carbonium ion, the solvent has high ionising power (thereby encouraging alkyl-cxygen heterolysis), and the hydroxide ion concentration is low (thereby suppressing the bimolecular mechanism). An example of the $B_{AL}1$ mechanism is the hydrolysis of optically active α-phenylethyl hydrogen phthalate in *faintly* alkaline solution. Kenyon *et al.* (1936) showed that the product was recemissed α-phenylethanol (phenylmethylmethanol); this is in keeping with the formation of an intermediate carbonium ion (sp^2 hybridised)

Bimolecular basic hydrolysis with alkyl-oxygen heterolysis ($B_{AL}2$ mechanism). This mechanism is rare. One example is the reaction between methyl benzoate and sodium methoxide in methanol (Bunnett *et al.*, 1950). The products were dimethyl ether and sodium benzoate; these results are readily explained on the basis of the $B_{AL}2$ mechanism:

Unimolecular acid-catalysed hydrolysis and esterification with alkyl-oxygen heterolysis ($A_{AL}1$ mechanism). Definite evidence for this mechanism was obtained by Hughes, Ingold *et al.* (1939), who showed that the esterification of acetic acid with optically active octan-2-ol in the presence of sulphuric acid gave a large amount of racemised ester. Racemisation is to be expected if an intermediate carbonium ion is formed, and so the reaction may be formulated as $A_{AL}1$ (R = n-C_6H_{11}CHMe—):

Since all steps are reversible, the $A_{AL}1$ mechanism should, in principle, be possible for ester hydrolysis, Evidence for its occurrence has been obtained by, *e.g.*, Bunton *et al.* (1951), who carried out the acid-catalysed hydrolysis of t-butyl acetate in water enriched with ^{18}O and obtained t-butanol containing ^{18}O.

The $A_{AL}1$ mechanism is common for t-alcohols; these from relatively stable carbonium ions (*cf.* the $B_{AL}1$ mechanism).

2. Esters are converted into alcohols by the Bouveault-Blanc reduction, catalytic hydrogenation, and by metallic hydrides.

3. Esters react with ammonia to form amides. This reaction is example of *ammonolysis* (which means, literally, splitting by ammonia):

$$R^1COOR^2 + NH_3 \longrightarrow R^1CONH_2 + R^2OH \ (g.)$$

With hydrazine, esters form acid hydrazides:

$$R^1COOR^2 + H_2NNH_2 \longrightarrow R^1CONHNH_2 + R^2OH$$

Esters react with phosphorus pentachloride or thionyl chloride to form acyl chlorides, *e.g.*,

$$R^1CO_2R^2 + PCl_5 \longrightarrow R^1COCl + POCl_3 + R^2Cl$$

4. By means of *alcoholysis* (splitting by alcohol), an alcohol residue in an ester can be replaced by another alcohol residue. Alcoholysis is carried out by refluxing the ester with a large excess of alcohol, preferably in the presence of a small amount of acid or sodium alkoxide as catalyst.

$$CH_3CO_2C_2H_9 + C_2H_5OH \longrightarrow CH_3CO_2C_2H_5 + C_4H_9OH$$

Alcoholysis of esters is also known as *transesterification*.

The Mechanism of alcoholysis is uncertain; a possibility is (in basic media):

In *acidolysis*, the acid residue is displaced from its ester by another acid residue, *e.g.*,

$$CH_3CO_2C_2H_5 + C_5H_{11}CO_2H \rightleftharpoons C_5H_{11}CO_2C_2H_5 + CH_3CO_2H$$

Acidolysis is a useful reaction for converting the natural ester of a dibasic acid into its acid ester.

According to Parker *et al.* (1968), who used ^{18}O-labelled toluic acids and their esters, the mechanism of acidolysis is:

5. When an ester – preferably the methyl or ethyl ester – is treated with sodium in an inert solvent, *e.g.*, ether, benzene or toluene, and subsequently with acid, an **acyloin** is formed (50–70 per cent yield). It is important that the reaction be carried out *in the absence of any free alcohol*. Acyloins are α,β-keto-alcohols, and according to Kharasch and his co-workers (1939), their formation takes place via a free-redical mechanism, *e.g.*, propionin from ethyl propionate:

Acyloins are reduced by the Clemmensen method to alkanes.

6. Carboxylic esters which contain α-hydrogen atoms react with sodamide in liquid ammonia solution to form the acid amide and a condensation product involving two molecules of the ester, *e.g.*, ethyl acetate gives acetamide and acetoacetic ester :

$$CH_3CO_2C_2H_5 + NaNH_2 \longrightarrow CH_3CONH_2 + C_2H_5ONa$$

$$CH_3CO_2C_2H_5 + NaNH_2 \longrightarrow NH_3 + NH_3 + [\ddot{C}H_2CO_2C_2H_5]^- Na^+$$

$$\xrightarrow{CH_3CO_2C_2H_5} CH_3COCH_2CO_2C_2H_5 + C2H_5OH$$

Of particular interest is the *carbonation* of esters, *i.e.*, the introduction of the carboxyl group.

$$R^1CH_2CO_2R^2 + NaNH_2 \xrightarrow{\text{liquid } NH_3}$$

$$NH_3 + [R^1CHCO_2R^2]^- Na^+ \xrightarrow[\text{ether}]{CO_2}$$

The yields of these *malonic acid derivatives* are 54–60 per cent with acetates, and becomes progressively lower as the molecular weight of he acid ncreases.

7. The pyrolysis of esters, particularly acetates, gives alkenes.

Esters are used as solvents for cellulose, oils, gums, resins, etc., and as plasticisers. They are also used for making artificial flavours and essences, *e.g.*, isoamyl acetate – banana oil; amylbutyrate–apricot; isoamyl isovalerate–apple; methyl butyrate–banana oil; amyl butyrate–apicot; isoamyl isovalerate–apple; methyl butyrate–pineapple; etc.

Ortho-esters are compounds of the type $R^1C(OR^2)_3$. They are derived from the ortho-acids, $RC(OH)_3$, which have not yet been isolated, but which are possibly present in aqueous solution: The most important ortho-esters are the orthoformic esters, particularly ethyl orthoformate. This may be prepared by running ethanol and chloroform into sodium covered with ether:

$$2CHCl_3 + 6C_2H_5OH + 6Na \longrightarrow 2HC(OC_2H_5)_3 + 6NaCl + 3H_3\ (70\%)$$

A possible mechanism is via the formation of dichloromethylene:

$$CHCl_3 \xrightarrow{EtO^-} :CCl_2 \xrightarrow{EtOH} EtOCHCl_2 \xrightarrow{2EtO^-}$$
$$HC(OEt)_3 + 2Cl^-$$

Ethyl orthoformate may be used for preparing ketals and for preparing aldehydes by means of a Grignard reagent.

Esters of the Inorganic Acids

Alkyl sulphates, $(RO)_2SO_2$. Only methanol and ethanol give a good yield of the alkyl sulphate by reaction with concentrated sulphuric acid; the higher alochols give mainly alkenes and ethers.

On the other hand, all the alcohols give a fair yield of alkyl hydrogen sulphate when a mixture of alcohol and concentrated sulphuric acid is heated on a steam bath.

Methyl sulphate is prepared industrially:

(*i*) By heating methanol with concentrated sulphuric acid, and then distilling the methyl hydrogen sulphate under reduced pressure:

$$CH_3OH + H_2SO_4 \longrightarrow H_2O + CH_3OSO_3H \xrightarrow{2} (CH_3O)_2SO_2 + H_2SO_4$$

(*ii*) By treating methanol with sulphur trioxide at low temperatures:

$$2SO_3 + 2CH_3OH \longrightarrow (CH_3)_2SO_4 + H_2SO_4$$

Ethyl sulphate may be prepared by the same methods as methyl sulphate, but in addition, there is the industrial preparation by passing ethylene in excess into cold concentrated sulphuric acid:

$$2C_2H_4 + H_2SO_4 \longrightarrow (C_2H_5)_2SO_4$$

Sulphates of the higher alcohols are conveniently prepared by the oxidation of sulphites as follows:

$$2ROH_4 + SOCl_2 \longrightarrow 2HCl + (RO)_2SO \xrightarrow{KMnO_4} (RO)_2SO_2$$

Methyl sulphate, b.p. 188°C, and ethyl sulphate, b.p. 208°C, are heavy poisonous liquids. They are largely used as alkylating agents since the alkyl group will replace the hydrogen atom of the groups —OH, —NH— or —SH. The alkylation may be carried out by treating the compound with the alkyl sulphate in sodium hydroxide solution. Usually only one of the alkyl groups takes part in the reaction, *e.g.*, methylation of a primary amine:

$$RNH_2 + (CH_3)_2SO_4 + NaOH \longrightarrow RNHCH_3 + CH_3NaSO_4 + H_2O$$

The methyl and ethyl esters of the carboxylic acids maybe conveniently prepared by treating the sodium salt of the acid with respectively methyl or ethyl sulphate:

$$RCO_2Na + (C_2H_5)_2SO_4 \longrightarrow RCO_2C_2H_5 + C_2H_5NaSO_4$$

The sodium alkyl sulphates of the higher alcohols are used as detergents, *e.g.*, sodium lauryl sulphate:

Alkyl nitrates, $RONO_2$. The only important alkyl nitrate is ethyl nitrate, $C_2H_5ONO_2$. This may be prepared by heating ethyl iodide with silver nitrate in ethanolic solution:

$$C2H_5I + AgNO3 \longrightarrow C_2H_5ONO_2 + AgI$$

When concentrated nitric acid is added to ethanol, the reaction is usually violent; part of the ethanol is oxidised, and some of the nitric acid is reduced to nitrous acid. Apparently it is the presence of the nitrous acid which produces the violent oxidation of the alcohol by the nitric acid.

This danger may be avoided by first boiling the nitric acid with urea, which destroys any nitrous acid present, and then adding this mixture to cool ethanol, any nitrous acid produced being immediately destroyed by the urea:

$$CO(NH_2)_2 + 2HNO_2 \longrightarrow CO_2 + 2N_2 + 3H_2O$$

Ethyl nitrate is a pleasant-smelling liquid, b.p. 87.5°C. When reduced with tin and hydrochloric acid, it forms hydroxylamine and ethanol:

$$C_2H_2ONO_2 \xrightarrow{Sn/HCl} C_2H_5OH + NH_2OH + H_2O$$

Alkyl nitrites, RONO. Alkyl nitrites are isomeric with the nitroalkanes. The only important alkyl nitrites are the ethyl and amyl nitrites; the latter is actually mainly isoamyl nitrite, since the amyl alcohol used is the isopentanol from fusel oil. Ethyl and amyl nitrites are prepared by adding concentrated hydrochloric acid or sulphuric acid to aqueous sodium nitrite and the alcohol: Ethyl nitrite, p.b. 17°C, and amyl nitrite, p.b. 99°C, are pleasant-smelling liquids, and are used as a means of preparing nitrous acid in anhydrous media; *e.g.*, an ethanolic solution of nitrous acid may be prepared by passing dry hydrogen chloride into amyl nitrite dissolved in ethanol.

ACID OR ACYL CHLORIDES

The general formula of the *acyl* groups Acid chlorides, which may be prepared by the replacement of the hydroxyl in the carboxyl group by chlorine, are also known as *acyl chlorides* because they contain the acyl group.

Nomenclature

According to the I.U.P.A.C. system the class suffix of the acyl chlorides is *-oyl*; the common names are formed by changing the suffix *-ic* of the *trivial* name of the acid into *-yl*, *e.g.*,

If the carboxyl group is considered as a substituent, then according to the I.U.P.AC. system, the nomenclature of all substances containing acyl groups is in all cases based on the name 'carbonyl' for the CO group, *e.g.*,

$CH_3CH_2CH_2COCl$ propane-1-carbonyl chloride

When naming amides, the 'yl' is elided before the suffix amide, *e.g.*,

$CH_3CH_2CH_2CONH_2$ propane-1-carbonamide

General Methods of Preparation

1. The acid is heated with phosphorus trichloride or pentachloride, *e.g.*,

$$3RCOOH + PCl_3 \longrightarrow 3RCOCl + H_3PO_3 \ (g.)$$

$$RCOOH + PCl_5 \longrightarrow RCOCl + HCl + POCl_3 \; (v.g.)$$

The reaction with phosphorus trichloride is accompanied by the formation of small amounts of volatile phosphorus compounds, *e.g.*,

$$RCO_2H + PCl_3 \longrightarrow RCO_2PCl_2 + HCl$$

Thionyl chloride may be used instead of the phosphorus chlorides:

$$ROC_2H + SOCl_2 \longrightarrow RCOCl + SO_2 + HCl \; (v.g.)$$

The inorganic chloride is chosen according to the boiling point of the acyl chloride formed. Phosphorous acid decomposes at 200°C; the boiling point of phosphoryl chloride is 107°C, and that of thionyl chloride is 76°C. Since acetyl chloride boils at 52°C, any of the three inorganic halides may be used, but it is difficult to separate acetyl chloride from thionyl chloride (which is generally used in excess) by fractionation. n-Butyryl chloride boils at 102°C, and so phosphorus pentachoride cannot be used. Usually thionyl chloride is the most convenient.

The mechanism of the formation of acid chlorides is uncertain; it may be similar to that when an alcohol is the substrate.

Acyl halides undergo the following exchange reaction with acids (probably via the aci[illegible] ydride):

$$R^1COCl + \text{[illegible]} \rightleftharpoons [R^1COOOCR^2 + HCl] \rightleftharpoons R^1CO_2H + R^2COCl$$

This therefore offers a means of preparing acyl halides, but the success depends on displacement of the equilibrium to the right. This is readily achieved by using oxalyl chloride (or bromide) as the acyl halide.

The half-acid chloride of oxalic acid produced is decomposed to carbon monoxide and carbon dioxide under the conditions of the experiment.

2. By distilling the salt of the acid with either phosphorus trichloride, phosphoryl chloride or sulphuryl chloride, *e.g.*,

$$3CH_3CO_2Na + PCl_3 \longrightarrow 3CH_3COCl + Na_3PO_3$$

$$2CH_3CO_2Na + POCl_3 \longrightarrow 2CH_3COCl + NaCl + NaPO_3$$

$$(CH_3CO_2)_2Ca + SO_2Cl_2 \longrightarrow 2CH_3COCl + CaSO_4$$

This method is used industrially since the salts are cheaper than the acid.

Acyl bromides may be prepared by the action of phosphorus tribromide or pentabromide (red phosphorus and bromine) on the acid, or by the action of excess of hydrogen bromide on the acyl chloride:

$$RCOCl + HBr \longrightarrow RCOBr + HCl$$

Acyl iodides may be prepared by the action of phosphorus tri-iodide on acid anhydrides. *Acyl fluorides* are also known, and may be prepared by the action of hydrogen fluoride on acid anhydrides:

$$(RCO)_2O + HF \longrightarrow RCOF + RCO_2H.$$

4

GENERAL PROPERTIES AND REACTIONS OF THE ACYL CHLORIDES

INTRODUCTION

The lower acyl chlorides are colourless liquids with irritating odours; the higher members are colourless solids. The chlorine atom is very reactive and so the acid chlorides are important reagents.

The infra-red absorption region of the carbonyl group (stretch) in acyl chlorides is 1815–1785 cm^{-1}(s).

1. The acyl chlorides are readily hydrolysed by water, the lower members reacting vigorously:

$$RCOCl + H_2O \longrightarrow RCO_2H + HCl$$

Acyl chlorides usually react rapidly with compounds containing 'active' hydrogen atoms, *i.e.* hydrogen attached to oxygen, nitrogen or sulphur; *e.g.*, esters are formed with alcohols:

$$R^1COCl + R^2OH \longrightarrow R^1COOR^2 + HCl$$

Amides are formed with ammonia, and *N*-substituted amides with primary and secondary amines:

$$RCOCl + 2NH_3 \longrightarrow RCONH_2 + NH_4Cl$$

$$R^1COCl + R^2NH_2 \longrightarrow R^1CONHR^2 + HCl$$

Hydrazides are formed with hydrazine, and hydroxamic acids with hydroxylamine:

$$RCOCl + H_2NNH_2 \longrightarrow RCONHNH_2 + HCl$$

$$RCOCl + NH_2OH \longrightarrow RCONHOH + HCl$$

The mechanisms of these reactions are uncertain; a possibility is:

2. Acyl chlorides may be reduced catalytically to aldehydes or to alcohols:

$$RCOCl \xrightarrow[Pd]{H_2} RCHO \xrightarrow{H_2} RCH_2OH$$

Acid chlorides are reduced to esters of an enediol by sodium amalgam in an inert solvent, *e.g.*, ether : Under suitable conditions, small yields of α-diketones have been obtained. Acid chlorides are reduced to alcohols by lithium aluminium hydride or sodium borohydride, and to aldehydes by lithium hydride or lithium tri-t-butoxyaluminium hydride.

3. Acyl chlorides react with salts of the acids to form *acid anhydrides* (*q.v.*):

$$R^1COCl + R^2COONa \longrightarrow R^1CO{\cdot}O{\cdot}OCR^2 + NaCl$$

4. Acyl chlorides may be used in the Friedel–Crafts reaction to produce an aromatic ketone: *e.g.*, acetyl chloride reacts with benzene in the presence of anhydrous aluminium chloride to form acetophenone:

$$C_6H_6 + CH_3COCl \xrightarrow{AlCl_3} C_6H_5COCH_3 + HCl$$

5. Acyl chlorides react with Grignard reagents to produce ketones or tertiary alcohols, according to the conditions; *e.g.*, acetyl chloride forms butanone with ethylmagnesium iodide:

$$CH_3COCl + C_2H_5MgI \longrightarrow CH_3COC_2H_5 + MgClI$$

6. Acyl chlorides are readily halogenated in the α-position.

7. Acyl chlorides from esters (and other products) when heated with an ether in the presence of anhydrous zinc chloride as catalyst:

$$R^1COCl + R_2{}^2O \xrightarrow{ZnCl_2} R^1CO_2R^2 + R^2Cl$$

8. Acyl chlorides add on to the double bond of an alkene in the presence of a catalyst, *e.g.*, zinc chloride or aluminium chloride, to form a chloroketone which, on heating, eliminates a molecule of hydrogen chloride to form an unsaturated ketone:

$$(CH_3)_2C{=}CH_2 + CH_3COCl \xrightarrow{ZnCl_2} (CH_3)_2CClCH_2COCH_3$$

$$\xrightarrow{heat} (CH_3)_2C{=}CHCOCH_2 + HCl$$

Formyl fluoride, HCOF, has been isolated by Olah *et al.* (1961). Formyl chloride, HCOCl, has been obtained in solution by the action of hydrogen chloride on a solution of 1-formylimidazole in chloroform at – 10°C (Staab *et al.*, 1964).

A mixture of carbon monoxide and hydrogen chloride behaves as if it were formyl chloride in the Gattermann–Koch aldehyde synthesis.

Acetyl chloride is the most important acyl halide. It is a colourless fuming liquid, b.p. 52°C, and is soluble in ether and chloroform. It is largely used as an *acetylating* agent, *i.e.*, as a means of introducing the acetyl group, into compounds containing 'active' hydrogen atoms.

It is also used to detect the presence of hydroxyl groups in organic compounds, and to estimate their number.

The accepted abbreviation for the acetyl group is Ac; *e.g.*, AcOH is acetic acid; Ac_2O, acetic anhydride; AcCl, acetyl chloride; C_6H_5NHAc, acetanilide; etc.

ACID ANHYDRIDES

The acid anhydrides may be regarded theoretically as being derived from an acid by the removal of one molecule of water form two molecules of the acid: 1818 and 1748 cm^{-1}: C=O (str.) in acyclic anhydrides. 1035^{-1}:C—O (str.) in —C—C— (acyclic anhydrides). 2985 and 2890 cm^{-1}: C—H (str.) in Me and/or CH_2. 1460 and 1408 cm^{-1}: C—H (def.) in Me and/or CH_2. 1379 cm^{-1}: C—H (def.) in Me.

It s possible in practice to prepare many acid anhydrides by direct dehydration, but this method is usually confined to the anhydrides of the higher members of the acid series (see below).

The infra-red absorption regions of the carbonyl group (stretch) in acyclic anhydrides are 1840–1800 (s) and 1780–1740 cm^{-1} (s). There is also a band in the region 11801030 cm^{-1} (s) due to the C—O (stretch) in the grouping —C—O—C—.

The infra-red spectrum of n-butyric anhydride (as a film).

Nomenclature

The acid anhydrides are named as the anhydride of the acid groups present. the *trivial* name of the acid being used, *e.g.*,

$(CH_3CO)_2O$	acetic anhydride
$(CH_3CH_2CO)_2O$	propionic anhydride

Acetic anhydride may be conveniently prepared by distilling a mixture of anhydrous sodium acetate and acetyl chloride: Aceti anhydride is prepared industrially in a number of ways, *e.g.*, (*i*) By passing acetylene into glacial acetic acid in the presence of mercuric ions as catalyst, and distilling the resulting ethylidene diacetate:

$$C_2H_2 + 2CH_3CO_2H \xrightarrow{Hg^{2+}} CH_3CH(OCOCH_3)_2 \xrightarrow{heat}$$
$$(CH_3CO)_2O + CH_3CHO$$

(*ii*) By passing keten into glacial acetic acid:

$$CH_2{=}C{=}O + CH_3CO_2H \longrightarrow (CH_3CO)_2O$$

Anhydrides of the higher acids may be prepared by heating, and then fractionating, a mixture of the acid and acetic anhydride:

$$2RCO_2H + (CH_3CO)_2O \rightleftharpoons (RCO)_2O + 2CH_3CO_2H$$

This method is only satisfactory for anhydrides which have higher boiling points than acetic acid. A much better method of preparation is to the acid chloride with pyridine and benzene, and the acid, and then heat:

$$RCOCl + RCO_2H + C_5H_5N \longrightarrow (RCO)_2O + C_5H_5\overset{+}{N}HCl^- \ (g.\text{–}ex.)$$

Suschitzky *et al.* (1964) have shown that acid anhydrides which are unaffected by cold water may be readily prepared in high yield and purity by shaking an aqueous solution of a metal salt of the carboxylic acid with an acyl halide at room temperature in the presence of a t-amine, *e.g.*, pyridine; *e.g.*, heptanoic anhydride:

$$CH_3(CH_2)_5CO_2Na + CH_3(CH_2)_5COCl \xrightarrow{pyridine} [CH_3(CH_2)_5CO]_2O$$

Properties of Acetic Anhydride

Acetic anhydride is a colourless liquid, b.p. 139.5°C, with an irritating smell. It is neutral when pure, and is slightly soluble in water, but readily soluble in ether and benzene. It is hydrolysed slowly in water, but rapidly by alkali:

$$(CH_3CO)_2O + H_2O \longrightarrow 2CH_3CO_2H$$

In undergoes reactions similar to those of acetyl chloride, but with less vigour; only half of the acetic anhydride molecule is used in acetylation, the other half being converted into acetic acid:

$$(CH_3CO)_2O + C_2H_5OH \longrightarrow CH_3CO_2C_2H_5 + CH_3CO_2H$$
$$(CH_3CO)_2O + NH_3 \longrightarrow CH_3CONH_2 + CH_3CO_2H$$

Acetylation with aceti anhydride is usually best carried out in pyridine solution or in the presence of a small amount of sodium acetate or concentrated sulphuric acid as catalyst. Acetic anhydride reacts with dry hydrogen chloride to form acetyl chloride:

$$(CH_3CO)_2O + HCl \longrightarrow CH_3COCl + CH_3CO_2H$$

It is readily halogenated, and may be used in the Friedel–Crafts reaction. It reacts with aldehydes to form alkylidene diacetates; *e.g.*, with acetaldehyde it forms ethylidene diacetate:

$$(CH_3CO)_2O + CH_3CHO \longrightarrow CH_3CH(OCOCH_3)2$$

Acetic anhydride is reduced to ethanol by some metallic hydrides.

Acetic anhydride reacts with nitrogen pentoxide to form **acetyl nitrate**, b.p. 221°C/70 mm, which explodes violently if heated suddenly. It is hydrolysed by water to acetic and nitric acid:

$$CH_3COONO_2 + H_2O \longrightarrow CH_3CO_2H + HNO_3$$

Acetyl nitrate is very useful for preparing certain aromatic nitro-compounds, but is dangerous to handle.

Acylation of a hydroxy-compound to give an ester is generally a better procedure than direct esterification of the acid. It has already been pointed out that sterically hindered acids are difficult to esterify, but Parish *et al.* (1965) have shown that such acids may be readily converted into their esters by dissolving the aid in trifluoroacetic anhydride and adding the hydroxy-compound (yield: up to 97 per cent). According to the authors, the active agent is the mixed anhydride, $RCOOCOCF_3$, that is formed.

Acetyl peroxide, $CH_3COOOCOCH_3$, may be prepared by the action of barium peroxide on acetic anhydride. It is colourless, pungent-smelling liquid, p.b. 63°C/21 mm, and tends to explode on warming. It is powerful oxidising agent.

Propionic anhydride, p.b. 169°C, is manufactured, *e.g.*,

$$CH_3CH_2CO_2H + C_2H_4 + CO \xrightarrow[\text{pressure}]{\text{heat;}} (CH_3CH_2CO)_2O$$

It is mainly used for the preparation of cellulose esters.

Formic anhydride may be prepared (about 90 per cent purity) by the distillation of a mixture of diethoxymethyl formate and n-butyric formic anhydride (Schijf *et al.*, 1965).

$HCO_2CH(OEt)_2$ + PrCOOOCH ⟶ $(HCO)_2O$ + $PrCO_2CH(OEt)_2$

Mixed anhydrides containing the formic acid group have been known for some time and may be prepared as follows, *e.g.*,

CH_3COCl + HCO_2Na ⟶ $CH_3COOOCH$ + NaCl

This behaves as a *formylating* reagent.

ACID AMIDES

Acid amides are compounds in which the hydroxyl of the carboxyl group has been replaced by the amino-group, NH_2, to form the *amido-group*, —$CONH_2$.

There are three classes of amides: primary, $RCONH_2$; secondary, $(RCO)_2NH$; and tertiary, $(RCO)_2N$. Only the primary amides are important, and it an be seen from their formulae that all three classes may be regarded as the acyl derivatives of ammonia.

Nomenclature

According to the I.U.P.A.C. system of nomenclature, the suffix *-oic* of the parent acid is replaced by *amide,* but they are commonly named by replacing the suffix *-ic* of the trivial name of the parent acid by *amide*.

$HCONH_2$ methanamide; formamide

CH3$CONH_2$ ethanamide *or* methylcarbonamide; acetamide

GENERAL METHODS OF PREPARATION OF THE AMIDES

1. By heating the ammonium salt of the acid:

$RCOONH_4$ ⟶ $RCONH_2$ + H_2O (*g.–v.g.*)

Since the ammonium salts tend to dissociate on heating, the reaction is best carried out in the presence of some free acid, which represses the hydrolysis and the dissociation of the ammonium salt.

Amides may also be prepared by heating an acid with urea (Cherbuliez *et al.*, 1946):

RCO_2H + $CO(NH_2)_2$ ⟶ $RCONH_2$ + CO_2 + NH_3 (*g.–v.g.*)

2. By *ammonolysis*, *i.e.*, the action of concentrated ammonia solution on acid chlorides, acid anhydrides or esters:

RCOCl + $2NH_3$ ⟶ $RCONH_2$ + NH_4Cl (*v.g.*)

$$(RCO)_2O + 2NH_3 \longrightarrow RCONH_2 + RCO_2NH_4 \qquad (v.g.)$$

$$R^1COOR_2 + NH_3 \longrightarrow R^1CONH_2 + R_2OH \quad (g.)$$

The mechanisms of these reactions are uncertain, but a possibility is: *N*-substituted amides may be prepared by using primary or secondary amines instead of ammonia, *e.g.*,

$$R^1COCl + R^2NH_2 \longrightarrow R^1CONHR^2 + HCl$$

3. By the *graded* hydrolysis of alkyl cyanides. This may be carried out satisfactorily by dissolving the alkyl cyanide in concentrated sulphuric acid and then pouring the solution into cold water, or by shaking the alkyl cyanide with cold concentrated hydrochloric acid:

$$RC{\equiv}N + H_2O \xrightarrow{HCl} RCONH_2 \qquad (g.)$$

Synder *et al.* (1954) have shown that the hydrolysis of cyanides with polyphosphoric acid gives very good yields of amide.

A possible mechanism for this hydrolysis is:

Sometimes the conversion of an alkyl cyanide into the amide may be effected by means of alkaline hydrogen peroxide. A possible mechanism is (Wiberg 1953): Other methods for preparing amides are the Ritter reaction, the action of ammonia on diazoketones, the Willgerodt reaction, and the Beckmann rearrangement.

GENERAL PROPERTIES AND REACTION OF THE AMIDES

Except for formamide, which is a liquid, all the amides are colourless crystalline solids, and those of low molecular weight are soluble in water (with which they can from hydrogen bonds). The lower amides have much higher melting points and boiling points than are to be expected from their molecular weights; this indicates association (through hydrogen bonding):

This association has been demonstrated from infra-red spectroscopic studies on amides in solution and in the solid state (Richards *et al.*, 1947).

Infra-red spectroscopic evidence has also indicated that amides are resonance hybrids, and this is supported by N.M.R. studies, *e.g.*, the two methyl groups in dimethylformaide give two separate signals at room temperature. This can be explained on the basis that, because of the partial double bond character of the C—N bond, the two methyl groups have *different* environments, one methyl group being *cis* and the other *trans* to the hydrogen atom. At room temperature the rate of rotation about the C—N bond is slow enough for the two methyl groups to be non-equivalent. As the temperature is raised, the signals broaden and finally collapse to a single line. At these higher temperatures, the molecule has absorbed sufficient energy to overcome the energy barrier to rotation to make the average environment of the two methyl groups the same *i.e.*, the two methyl groups are *now* equivalent.

The infra-red absorption regions of the carbonyl group (stretch) in primary amides are near 1690 (s) and 1650 cm^{-1}(s), and those of the free NH group (stretch) are near 3500 (s) and 3400 cm^{-1} (s). For a hydrogen-bonded NH group, the regions are near 3350 (m) and 3200 cm^{-1} (m).

Dimethylformamide, p.b. 152°C, is a very good solvent for polar and non-polar compounds, and is used as s formylating agent in conjunction with phosphoryl chloride.

It is manufactured by reaction between methyl formate and dimethylamine:

$$HCO_2Me + Me_2NH \longrightarrow HCONMe_2 + MeOH$$

1. Amides are hydrolysed slowly by water, rapidly by acids, and far more rapidly by alkalis:

$$RCONH_2 + H_2O \longrightarrow RCO_2H + NH_3$$

Possible mechanisms are:

2. Amides are very feebly basic and from unstable salts with strong inorganic acids, *e.g.*, $RCONH_2{\cdot}HCl$. The structure of these salts may be (I) or (II): Kutzennigg *et al.* (1962), from their proton magnetic resonance studies, have provided evidence that the salts of both acetamide and the *N,N*-disubstituted acetamides are protonated on the oxygen atom. This would be in agreement with the fact that amides are resonance hybrids, the O carrying a negative charge and the N a positive charge.

3. Amides are also feebly acidic; *e.g.*, they dissolve mercuric oxide to form covalent mercury compounds:

$$2RCONH_2 + HgO \longrightarrow (RCONH)_2Hg + H_2O$$

When amides are treated with sodium or sodamide in ethereal solution, the sodium *salt* is formed, the structure of which is probably a resonance hybrid (ambident cation): Treatment of the *sodium* salt with iodide gives the *N*-alkyl derivative, whereas with the silver salt, the *O*-alkyl derivative (*imido ester*) is obtained. Imido esters are obtained by alkylation of amides with alkyl sulphates or oxonium salts, *e.g.*, $R_3O^+BF_4^-$

4. Amides are reduced by sodium and ethanol, catalytically, etc.

$$RCONH_2 \xrightarrow{Na/C_2H_5OH} RCH_2NH_2 + H_2O$$

5. When heated with phosphorus pentoxide, or better, with PCl_5, $POCl_3$, or $SOCl_2$, amides are dehydrated to alkyl cyanides:

$$RCONH_2 \xrightarrow{-H_2O} RCN$$

The mechanisms of these dehydrations are not certain, but according to Kirsanov (1954) and Lapidot *et al.* (1962), the mechanism with phosphorus pentachloride is: Alkyl cyanides are also formed when the amides of the higher acids are heated to a high temperature:

$$2RCONH_2 \longrightarrow RCN + RCO_2H + NH_3$$

If this reaction is carried out in the presence of excess of ammonia, then all the amide is converted into alkyl cyanide:

$$RCO_2H + NH_3 \longrightarrow RCO_2NH_4 \xrightarrow{-H_2O} RCONH_2 \xrightarrow{-H_2O} RCN$$

6. When amides are treated with nitrous acid, nitrogen is evolved and the acid is formed:

$$RCONH_2 + HNO_2 \longrightarrow RCO_2H + N_2 + H_2O$$

This reaction is usually slow; a better method is use butyl nitrite:

$$RCONH_2 + BuONO \longrightarrow RCO_2H + N_2 + BuOH$$

7. **Hofmann preparation of primary amines** (1881). This is the conversion of an amide into a primary amine with one carbon atom less by means of bromine (or chlorine) and alkali. The overall equation for the reaction may be written:

$$RCONH_2 + Br_2 + 4KOH \longrightarrow RNH_2 + 2KBr + K_2CO_3 + 2H_2O$$

The mechanism of this reaction has been the subject of a great deal of work. A number of intermediates have been isolated: *N*-bromamides, RCONHBR; salts of these bromamides, $[RCONBr]^-K^+$; isocyanates, RNCO.

On this basis, a possible mechanism is the following, which is an example of the 1,2-shift: The uncertainty of this mechanism lies mainly in the formation of the acyl nitrene. According to Wright (1968), who used ^{14}C- and ^{15}N-lableed compounds, the Hofmann, Curtius and Lossen rearrangements may not involve nitrenes as intermediates; he has proposed the following *concerted* mechanism:

Hauser *et al.* (1937) have shown that electron-releasing groups in the *para*-position of a migrating aryl group accelerate the reaction and electron-withdrawing groups retard the reaction. This supports the formation of a bridged ion (a phenonium ion), (I*a*), where the migrating group is aromatic, and by inference, a bridged ion (I*b*) when the migrating group is

alkyl. This is further supported by the fact that the rearrangement is intramolecular, *e.g.*, Wallis *et al.* (1931), using the amide of optically active α-benzylpropionic acid, found that there was retention of configuration in the product, α-bonzylethylamine; thus the migrating group is never free:

$$C_6H_5CH_2CHMeCONH_2 \longrightarrow C_6H_5CH_2CHMeNH_3$$

The Hofmann amine preparation can be carried out on all monocarboxylic acids, and the yields are 70–90 per cent for amides containing up to seven carbon atoms; amides with more than seven carbon atoms in the chain give mainly alkyl cyanide:

$$RCH_2CONH_2 \xrightarrow{Br_2KOH} RCN$$

The alkyl cyanide, however, may readily be reduced to the primary amine RCH_2NH_2.

It is possible to obtain the amine in good yield from the long-chain amides by modifying the procedure as follows. Bromine is added rapidly to a methanolic solution of the amide containing sodium methoxide:

$$RCOHN_2 + 2CH_3ONa + Br_2 \longrightarrow RNHCO_2CH_3 + 2NaBr + CH_3OH$$

The *N*-alkylurethan, on hydrolysis with alkali, gives the amine:

$$RNHCO_2CH_3 + 2NaOH \longrightarrow RNH_2 + Na_2CO_3 + CH_3OH$$

The Hofmann amine preparation offers an excellent method of preparing primary amines free from secondary and tertiary amines; it also affords a means of descending a series.

If the amide contains an electron-attracting group, the product is a bromide, *e.g.*, heptafluorobutyramide gives bromoheptafluoropropane (Haszeldine *et al.*, 1956):

$$C_3F_7CONH_2 \xrightarrow[NaOH]{Br_2} C_3F_7Br\ (85\%)$$

Another method for converting primary amides into amines is as follows (Beckwith *et al.*, 1965). On treatment with lead tetra-acetate, the amide gives an acylamine (40–81 per cent); this is readily hydrolysed to the amine:

$$RCONH_2 \xrightarrow[\text{benzene, reflux}]{Pb(OAc)_4} RNHAc$$

ACID HYDRAZIDES AND ACID AZIDES

Acid hydrazides may be prepared by the action of hydrazine on esters or acyl chlorides:

$$RCOOC_2H_5 + N_2H_4 \longrightarrow RCONHNH_2 + C_2H_5OH$$

$$RCOCl + N_2H_4 \longrightarrow RCONHNH_2 + HCl$$

Girard's reagent 'T' is trimethylaminoacetohydrazide, which may be prepared by interaction between trimethylamine, ethyl chloroacetate and hydrazine:

$$(CH_3)_3N + ClCH_2CO_2C_2H_5 + N_2H_4 \longrightarrow [(CH_3)_3NCH_2CONHNH_2]^+ Cl^- + C_2H_5OH$$

Girard's reagent 'P' is formed in a similar manner, except that pyridine is used instead of trimethylamine.

The acid hydrazides resemble the acid amides, but differ in the following ways:

(*i*) They are less readily hydrolysed than the amides:

$$RCONHNH_2 + H_2O \longrightarrow RCO_2H + N_2H_4$$

(*ii*) They are reducing agents–hydrazine is a powerful reducing agent.

(*iii*) They from acid azides when treated with nitrous acid; no nitrogen is evolved:

$$RCONHNH_2 + HNO_2 \longrightarrow RCON_3 + 2H_2O$$

Acid azides may also be prepared by the reaction between an acyl chloride and sodium azide:

$$RCOCl + NaN_3 \longrightarrow RCON_3 + NACl$$

The structure of the acid azides is best represented as a resonance hybrid:

When boiled with an alcohol, acid azides undergo rearrangement to form *N*-alkyl-substituted urethans:

$$R^1CON_3 + R^2OH \longrightarrow RNHCO_2R^2 + N_2 \ (v.g.\text{-}ex.)$$

The mechanism of this rearrangement – the **Curtius rearrangement** (1894)– is similar to that of the Hofmann and Lossen rearrangements, and so may be formulated:

This mechanism is supported by the fact that when the terminal nitrogen atom is ^{15}N in 3,5-dinitrobenzazide, all of this tracer is found in the nitrogen eliminated in the reaction (Bothner–By *et al.*, 1951).

The Curtius reaction may be used to step down a series (*cf.* the Hofmann amine preparation):

$$RCO_2H \xrightarrow{C_2H_5OH} RCO_2C_2H_5 \xrightarrow{N_2H_4} RCONHNH_2 \xrightarrow{HNO_2}$$

$$RCON_3 \xrightarrow{CH_3OH} RNHCO_2CH_3 \xrightarrow{NaOH} RNH_3$$

The Curtius reactions offers a very good method for preparing primary amines free from secondary and tertiary amines, and may also be used for preparing isocyanates and urethans; *e.g.*, when heated in benzene or chloroform solution, an acid azide rearranges to the alkyl isocyanate which can be isolated. If the isocyanate is warmed with an alcohol, an *N*-substituted urethan is formed:

$$R^1CON_3 \xrightarrow[\text{solution}]{\text{benzene}} R^1NCO \xrightarrow{R^2OH} R^1NHCO_2R^2$$

HALOGEN DERIVATIVES OF THE MONOCARBOXYLIC ACIDS

The halogen derivatives of the monocarboxylic acids are compounds in which one or more hydrogen atoms in the *carbon chain* have been replaced by halogen.

Nomenclature

The following examples illustrate the methods:

CH_2ClCO_2H	monochloroacetic acid
$CH_3CHClCO_2H$	α-chloropropionic acid or 2-chloropropanoic acid
$(CH_3)_2CClCHBrCO_2H$	α-bromo-β-chloroisovaleric acid or 2-bromo-3-chloro-3-methylbutanoic acid

α-Halogeno-acids. Although the monocarboxylic acids themselves are not readily halogenated, their acid chlorides and anhydrides may be halogenated very easily. Bromination takes place only at the α-position, but chlorination, although it occurs mainly at the α-position, may also take place further in the chain; *e.g.*, chlorination of propionic acid results in the formation of the α- and β-chloro-derivatives (see also below):

$$CH_3CH_2CO_2H \xrightarrow{Cl_2} CH_3CHClCO_2H + ClCH_2CH_2CO_2H$$

Fluorination of the monocarboxylic acids has not yet been studied in great detail, but the work done so far seems to indicate that fluorine enters the chain somewhat indiscriminately; *e.g.*, fluorination of butyric acid in carbon tetrachloride solution gives the β-and γ-fluoro-derivatives.

Iodo-derivatives are prepared from the chloro- or bromo-compound by the action of potassium iodide in methanolic or acetaone solution.

The usual method for preparing α-chloro- or bromo-acids is by the **Hell-Volhard-Zelinsky reaction (H.V.Z. reaction),** which is carried out by treating the acid with chlorine or bromine in the presence of a small amount of red phosphorus.

The mechanism of the H.V.Z. reaction is uncertain, but there is some evidence to show that it may proceed through an enol intermediate (Zimmerman *et al.*, 1959):

Enolisation of the acid bromide is far more rapid than that of the acid.

The second α-hydrogen atom may be replaced by chlorine or bromine by using excess of the halogen, but whereas bromination ceases when both α-hydrogen atoms have been replaced, chlorination proceeds further in the chain (see above).

Since the H.V.Z. reaction with bromine is specific for α-hydrogen atoms, it can be used to detect the presence of α-hydrogen in an acid; *e.g.*, trimethylacetic acid does not undergo the H.V.Z. reaction with bromine.

The H.V.Z. reaction with bromine is applicable to di- and polycarboxylic acids, all the α-positions being substituted if sufficient bromine is used.

Another convenient method of preparing a-bromo-acids is by brominating an alkylmalonic acid and heating the bromo-acid, whereupon it is decarboxylated to the monobasic acid:

$$RCH(CO_2H)_2 \xrightarrow{Br_2} RCBr(CO_2H)_2 \xrightarrow{heat} RCHBrCO_2H + CO_2$$

Sulphuryl chloride, in the presence of small amount of iodine, chlorinates monocarboxylic acids in the α-position, but in the presence of organic peroxides (to give a free-radical reaction), the α-, β-, γ-, etc. positions are substituted. On the other hand, Louw (1966) has shown that cupric chloride (plus LiCl) chlorinates acid chlorides, anhydrides, and carboxylic acid (containing the acid chloride or anhydride as 'catalyst') at the α-position only (yields > 95 per cent). The reaction is carried out in a polar inert solvent, *e.g.*, sulpholane, and only the α-monochloro derivative is obtained regardless of the number of α-hydrogen atoms in the compound.

β-Halogeno-acids may be prepared by treating in α,β-unsaturated acid with halogen acid; *e.g.*, acrylic acid forms β-bromopropionic acid when treated with hydrogen bromide in acetic acid solution:

$$CH_2{=}CHCO_2H + HBr \longrightarrow BrCH_2CH_2CO_2H$$

β-Halogeno-acids may also be prepared by treating an α,β-unsaturated aldehyde with halogen acid, and oxidising the β-chloroaldehyde produced, *e.g.*, acraldehyde gives b-chloropropionic acid when treated with concentrated hydrochloric acid, and the product then oxidised with concentrated nitric acid:

$$CH_2{=}CHCHO + HCl \longrightarrow ClCH_2CH_2CHO \xrightarrow{[O]} ClCH_2CH_2CO_2H \ (60\text{-}65\%)$$

The addition of all the halogen acids to α,β-unsaturated carbonyl compounds takes place very readily, and in a direction contrary to Markownikoff's rule.

When an alkene cyanohydrin is refluxed with 40 per cent hydrobromic acid, the β-bromoacid is obtained; *e.g.*, ethylene cyanohydrin gives β-bromopropionic acid:

$$CH_2(OH)CH_2CN + 2HBr + H_2O \longrightarrow BrCH_2CH_2CO_2H + NH_4Br \ (82\text{-}83\%)$$

Alternatively, when an alkene halohydrin is oxidised with concentrated nitric acid, the halogeno-acid is produced; *e.g.*, trimethylene chlorohydrin gives β-chloropropionic acid:

$$ClCH_2CH_2CH_2OH \longrightarrow ClCH_2CH_2CO_2H \ (78\text{-}79\%)$$

γ-Halogeno-acids may be prepared by the addition of halogen acid to a β,γ-unsaturated acid; the addition occurs contrary to Markownikoff's rule (*cf.* above); *e.g.*, pent-3-enoic acid gives γ-bromo-n-valeric acid when treated with hydrobromic acid:

$$CH_3CH{=}CHCH_2CO_2H + HBr \longrightarrow CH_3CHBrCH_2CH_2CO_2H$$

γ-Halogeno-acids may also be prepared by heating a γ-hydroxyacid with oncentrated halogen acid solution *e.g.*, γ-hydroxybutyric acid gives γ-chlorobutyric acid when treated with concentrated hydrochloric acid:

$$HOCH_3CH_2CH_2CO_2H + HCl \longrightarrow ClCH_3CH_2CH_2CO_2H + H_2O$$

The methods described above apply to particular types of halogeno-acids. However, the following method may be used for preparing any halogeno-acid; *e.g.*, a δ-halogeno-acid:

$$MeO_2C(CH_2)_4CO_2H \xrightarrow[KOH]{AgNO_3} MeO_2(CH_2)_4CO_2Ag \xrightarrow[CCl_4]{Br_2}$$
$$MeO_2C(CH_2)_3CH_2Br\ (65\text{-}68\%)$$

Properties of Halogeno Acids

The α-halogeno-acids undergo most of the reactions of the alkyl halides, but the halogen atom in the acid is far more reactive than that in the alkyl halide; the enhanced reactivity is due to the – I effect of the adjacent carbonyl group. The reactions of the carboxyl group are uncharged. β-, γ- and δ-halogeno-acids undergo some of the reactions of the alkyl halides, but tend to eliminate a molecule of hydrogen halide to form an unsaturated acid or a lactone. They do not form Grignard reagents, but the halogeno-acids esters react with Grignard reagents to form halogen-substituted tertiary alcohols. The halogeno-acids are reduced to the corresponding acids by sodium amalgam, but the reduction of acids of the type $R^1R^2CClCO_2H$ with lithium aluminium hydride gives chlorohydrins (I), alcohols (II), aldehydes (III) and glycols (IV).

$$\underset{(I)}{R^1R^2CClCH_2OH} \qquad \underset{(II)}{R^1R^2CHCH_2OH} \qquad \underset{(III)}{R^1R^2CHCHO}$$

$$\underset{(IV)}{R^1R^2C(OH)CH_2OH}$$

The amount of each depends on the nature of R^1 and R^2, *e.g.*, very little glycol is obtained unless at least one of these groups is phenyl (Eliel *et al.*, 1956).

The behaviour of an halogeno-acid with alkali depends on the position of the halogen atom relative to the carboxyl group.

(*a*) α-Halogeno-acids are converted into the corresponding α-hydroxyacid:

$$RCHXCO_2H + H_2O \xrightarrow[\text{(ii) } H^+]{\text{(i) NaOH}} RCH(OH)CO_2H$$

On the other hand, if the α-halogeno-acid ester is heated with a tertiary amine, the α,β-unsaturated acid ester is formed by the elimination of a molecule of hydrogen halide; *e.g.*, ethyl α-bromobutyrate gives ethyl crotonate when heated with dimethylaniline, $C_6H_5N(CH_3)_2$:

$$CH_3CH_2CHBrCO_2C_2H_5 \xrightarrow{-HBr} CH_3CH{=}CHCO_2C_2H_5$$

(*b*) β-Halogeno-acids are converted into the corresponding β-hydroxyacid, which, on continued reflux with alkali, eliminates a molecule of water to form the α,β-unsaturated acid:

$$RCHBrCH_2CO_2H \xrightarrow{NaOH} RCH(OH)CH_2CO_2^- \xrightarrow[-H_2O]{NaOH} RCH{=}CHCO_2^-$$

(*c*) γ- and δ-Halogeno-acids are converted into lactones; *e.g.*, γ-chlorobutyric acid gives γ-butyrolactone:

This is an example of neighbouring group participation.

(*d*) ε-Halogeno-acids, etc., give the corresponding hydroxyacid; *e.g.*; ε-bromocaproic acid gives ε-hydroxycaproic acid:

$$BrCH_2CH_2CH_2CH_2CH_2CO_2H \xrightarrow[\text{(ii) } H^+]{\text{(i) NaOH}} HOCH_2CH_2CH_2CH_2CH_2CO_2H$$

The most characteristic reaction of the α-halogeno-acid esters is the *Reformatsky reaction.*

The halogeno-acids are all stronger acids than the parent acid, and for a group of isomeric acids, the further the halogen is removed from the carboxyl group the weaker is the acid.

The increase in acid strength may be explained by high electronegativity of the halogen atom which exerts a strong inductive effects, thereby facilitating the release of proton from the carboxyl group.

The larger the number of halogen atoms on the α-carbon atom, the stronger is the inductive effect, and consequently the stronger is the acid;

also, the further removed the halogen atom is from the carboxyl group, the weaker is the inductive effect at the carboxyl group, and consequently the weaker is the acid. In the same way, since an alkyl group is electron-repelling, increasing their number on the α-carbon atom should *decrease* the strength of the acid since release of the proton will be hindered. Since – I groups in the α-position, *e.g.*, X, NO_2, OH, increase the strength of the acid, and + I groups, *e.g.*, Me, weaken the acid, the measurement of the pK_a's of various substituted acetic acids affords a means of measuring the *relative* inductive effects of different groups.

Relative inductive effects (or electronegativities) may also be measured by NMR spectroscopy. The results, however, must be accepted with reserve, since the applied field itself may affect the electron distribution in the molecule.

To study the strengths of the different members of the carboxylic acids quantitatively, we must use the equations:

(*i*) HA (solvated) $\rightleftharpoons$ H^+ (solvated) + A^- (solvated)

(*ii*) $-RT \ln K_a = \Delta G^\theta$ (ionisation) = ΔH^θ (ionisation – $T\Delta S^\theta$) (ionisation)

$$\Delta H_1 = \Delta H_2 + \Delta H_3 + \Delta H_4 + \Delta H_5 + \Delta H_6 + \Delta H_7$$

It then follows that:

$$\Delta S_1 = \Delta S_2 + \Delta S_3 + \Delta S_4 + \Delta S_5 + \Delta S_6 + \Delta S_7$$

Unfortunately, very few of these values are known for the various organic acids. However, it is possible to evaluate ΔH_1 from experimental values of K_a over a range of temperatures, and since ΔG is also, obtained, ΔS_1 can thus be calculated. Some values of ΔH_1 (kJ) and ΔS_1 (JK^{-1}), respectively, are in water: formic aicd, – 0·08368, – 73·64; acetic, – 0·3766, – 92·47; propionic, – 0·6694, – 95·40; chloroacetic, – 4·853, – 71·13

It can be seen that ΔH_1 is quite small, but ΔS_1 can vary a great deal. Thus the difference in acid strengths must be due to entropy effects. An interesting point to note is the smaller entropy loss for chloroacetic acid makes this a stronger acid than acetic. ΔS_1 for trichloroacetic acid is – 8·368 JK^{-1}, and this acid is a very strong acid. These entropy changes may be explained as follows. Acetic acid is far less polar than the acetate ion, and so the latter is far more solvated than the former. Since solvation produces some sort of arrangement of the solvent molecules around the anion, this loss of randomness of solvent molecules means a decrease in entropy (*i.e.*, ΔS is negative). Because trichloroacetic acid is highly polar, it is highly solvated both in the un-ionised form and in the anion, and so the loss in entropy in this case is very much smaller. It should also be noted that when ionisation occurs, the liberated proton is also solvated.

This solvation causes the 'freezing' of four molecules of water (to form $H^+(H_2O)_4$), and so this contributes to the decrease in entropy when the acid ionises. However, since the solvated proton is the same for all acids, only the change in entropy of the anion need be considered.

Chloroacetic acid may be prepared by the H.V.Z. reaction; the reaction can be carried out in direct sunlight, and in the absence of phosphorus. Chloroacetic acid is prepared industrially:

(*i*) by heating trichloroethylene with 74 per cent sulphuric acid:

$$CHCl{=}CCl_2 + 2H_2O \xrightarrow[140°C]{H_2SO_4} ClCH_2CO_2H + 2HCl$$

(ii) By the chlorination of acetic acid in the presence of acetic anhydride-sulphuric acid mixture.

Chloroacetic acid is a deliquescent solid, m.p. 61°C, soluble in water and ethanol. It finds many uses in organic syntheses, and is used in the industrial preparation of indigotin (*q.v.*). It is readily esterified with ethanol in the presence of sulphuric acid.

Ethyl chloroacetate is converted into chloroacetamide when shaken with aqueous ammonia:

$$ClCH_2COOC_2H_5 + NH_3 \longrightarrow ClCH_2CONH_2 + C_2H_5OH \ (78\text{-}84\%)$$

Chloroacetamide is dehydrated to chloroacetonitrile (p.b. 123°C) by phosphorus pentoxide:

$$ClCH_2CONH_2 \xrightarrow[-H_2O]{P205} ClCH_2CN \ (61\text{-}70\%)$$

Dichloroacetic acid may be prepared in the laboratory and industrially by adding calcium carbonate to a warm aqueous solution of chloral hydrate, then adding an aqueous solution of sodium cyanide, and finally heating the mixture:

$$2CCl_3CH(OH)_2 + 2CaCO_3 \xrightarrow{NaCN} (CHCl_2CO_2)_2Ca + 2CO_2 +$$

$$CACl_2 + 2H_2O \xrightarrow{HCl} CHCl_2CO_2H \ (88\text{-}92\%)$$

The action of the sodium cyanide is not understood.

Dichloroacetic acid, p.b. 194°C, gives glyoxylic acid (*q.v.*) on hydrolysis:

$$CHCl_2CO_2H + H_2O \xrightarrow[(ii) H^+]{(i) NaOH} OCHCO_2H + 2HCl$$

Vigorous hydrolysis with concentrated alkali gives oxalate and glycollate, due to the glyoxylic acid undergoing he Cannizzaro reaction:

$$2OCHCO_2H + 3NaOH \longrightarrow (CO_2Na)_2 + CH_2(OH)CO_2Na + 2H_2O$$

Trichloroacetic acid is best prepared by oxidising chloral hydrate with concentrated nitric acid:

$$CCL_3CH(OH)_2 + [O] \xrightarrow{HNO_3} CCl_3CO_2H + H_2O$$

Trichloroacetic acid is a deliquescent solid, m.p. 58°C, and is one of the strongest organic acids.

The presence of three chlorine atoms on a carbon atom adjacent to a carbonyl group causes the C—C bond to break very easily. Thus, when trichloroacetic acid is boiled with dilute sodium hydroxide, or even water, chloroform is obtained:

$$CCL_3CO_2H \longrightarrow CHCl_3 + CO_2$$

The ready fission of the C—C bond may be attributed to the strong inductive effect of the chlorine atoms, and a possible mechanism is:

With *concentrated* alkali, formates are produced, and these probably result via the formation of dichloromethylene.

Trifluoroacetic acid is conveniently prepared by the oxidation of *p*-trifluoromethyltoluidine with chromic acid:

Trifluoroacetic acid is a liquid fumes in the air, and is one of the strongest organic acids known. It does not form fluoroform when heated with alkali, and is reduced to trifluroacetaldehyde by lithium a aluminium hydride.

Peroxytrifluoroacetic acid, CF_3CO_3H, is a useful oxidising agent (Emmons, 1954; *cf.* peroxyacetic acid).

Fluoroacetic acid may be prepared by heating a mixture of carbon monoxide, hydrogen fluoride and formaldehyde under pressure.

$$CO + HF + HCHO \longrightarrow FCH_2CO_2H$$

The simplest chloro-acid would be chloroformic acid, $ClCO_2H$, but it is unknown; chlorination of formic acid results in the formation of hydrogen chloride and carbon dioxide:

$$HCO_2H \xrightarrow{Cl_2} 2HCl + CO_2$$

On the other hand, esters of chloroformic acid are known.

MONOCARBOXYLIC ACIDS

The general formula of the saturated dicarboxylic acids is $C_nH_{2n}(CO_2H)_2$ (n = 0 for oxalic acid), and the best-known examples are those which have the two carboxyl groups at the opposite ends of the carbon chain. The dicarboxylic acids are commonly known by names which indicate their source, *e.g.*, HO_2CCO_2H oxalic acid; this occurs in plants of the *oxalis* group (for further examples, see the individual acids).

In this trivial system of nomenclature, the positions of side-chains or substituents are indicated by Greek letters, *e.g.*,

$HO_2CCH(CH_3)CH_2CH_2CHClCO_2H$ α-chloro-α′-methyladipic acid

According to the I.U.P.A.C. system of nomenclature, the class suffix is *-dioic*, *e.g.*,

HO_2CCH,CO,H propanedioic acid

$HO_2CCH(CH_3)CH(CH_3)CH_2CO_2H$ 2,3-dimethylpentanedioic acid

When this method leads to cumbrous names, the alternative' scheme is. to regard the carboxyl group as a substituent, and the name of the acid is then obtained by adding the suffix carboxylic acid, *e.g.*,

$HO_2CCH_2CH_2CO_2H$ 1,2-ethanedicarboxylic acid

or

ethane-1,2-dicarboxylic acid

GENERAL METHODS OF PREPARATION

1. The *cyanide synthesis* of dicarboxylic acids is a very useful method: the starting material may be either a halogen-acid or a polymethylene dibromide:

(i) $ClCH_2CO_2H \xrightarrow{KCN} CH_2(CN)CO_2H \xrightarrow{H_3O^+}$

$CH_2(CO_2H)_2$

(ii) $Br(CH_2)_2Br \xrightarrow{KCN} NC(CH_2)_2CN \xrightarrow{H_3O^+}$
$HO_2C(CH_2)_2CO_2H$

2. The **Crum-Brown** and **Walker electrolytic method** (1891, 1893). This is the electrolysis of an aqueous solution of the potassium alkyl esters of the dicarboxylic acids (cf. Kolbe's method):

$$\begin{matrix} K\text{—}O_2C\text{—}(CH_2)_n\text{—}CO_2C_2H_5 \\ K\text{—}O_2C\text{—}(CH_2)_n\text{—}CO_2C_2H_5 \end{matrix} + 2H_2O \longrightarrow$$

$$\begin{matrix} (CH_2)_nCO_2C_2H_5 \\ | \\ (CH_2)_nCO_2C_2H_5 \end{matrix} + 2CO_2 + 2KOH + H_2 \quad (f.g.\text{-}g.)$$

It is obvious that this method can be used to prepare only *even* homologues.

Dicarboxylic acids may also be prepared by the electrochemical hydrodimerisation of suitable α, β-unsaturated compounds.

3. Dicarboxylic acids may be prepared by the acetoacetic ester synthesis and better, by the malonic ester synthesis.

4. Cyclic ketones may be oxidised to dicarboxylic acids, *e.g.*, cyclohexanone gives adipic acid: Alternatively, lactones can be converted into dicarboxylic acids.

5. The preparation of higher homologues from lower homologues may be carried out in several ways; which method is used depends on the homologue desired. Many *even* higher homologues can be prepared by the Crum-Brown and Walker method (method 2). Any dicarboxylic acid can be stepped up by two carbon atoms as follows:

$$(CH_2)_n(CO_2H)_2 \xrightarrow{LAH} (CH_2)_n(CH_2OH)_2 \xrightarrow{HBr}$$
$$(CH_2)_n(CH_2Br)_2 \xrightarrow{KCN} (CH_2)_n(CH_2CN)_2$$
$$\xrightarrow{H_3O^+} (CH_2)_n(CH_2CO_2H)_2$$

By using malonic ester instead of potassium cyanide, the acid may be stepped up by four carbon atoms:

$$(CH_2)_n(CH_2Br)_n + 2[CH(CO_2C_2H_2)_2]^- Na^+ \longrightarrow$$

$$2NaBr + (CH_2)_n [CH_2CH(CO_2C_2H_5)_2]_2 \xrightarrow[(ii) HCl]{(i) KOH}$$

$$(CH_2)_n[CH_2CH(CO_2H)_2]_2 \xrightarrow{150-200°C} (CH_2)_n(CH_2CH_2CO_2H)_2 + 2CO_2$$

Dicarboxylic acids can be stepped up by ten carbon atoms, by starting with a di-acyl chloride and an examine.

Dicarboxylic acids may also be prepared from oc-bromo-esters as follows.

General Properties

All the dicarboxylic acids are crystalline solids, the lower members being soluble in water, the solubility decreasing with increase in molecular weight: the odd acids are more soluble than the even. None is steam volatile, and except for oxalic acid, the dicarboxylic acids are stable towards oxidising agents. Their melting points follow the oscillation rule, the even acids having higher melting points than the odd. This alternation is due to the relative positions of the carboxyl groups in the crystal structures. These acids dissociate in two steps, the dissociation constant of the first being much greater than that of the second. Furthermore, the acid strength of the dicarboxylic acids decreases as the series is ascended.

Table 1

Acid	pK_a^1	pK_a^2	K_a^1/K_a^2
Oxalic	1.271	4.266	989
Malonic	2.86	5.70	692
Succinic	4.21	5.64	26.9
Glutaric	4.34	5.27	8.51
Adipic	4.41	5.28	7.41

Since these acids possess two carboxyl groups, ionisation pair occur in two steps:

$$HO_2C(CH_2)_nCO_2H + H_2O \rightleftharpoons HO_2C(CH_2)_n \quad ...(1)$$

$$HO_2C(CH_2)_nCO_2^- + H_2O \rightleftharpoons {}^-O_2C(CH_2)_nCO_2^- + H_3O^+ \quad ...(2)$$

In the-first-stage ionisation (1), the acid can lose a proton from two positions, but the 'half-ion' (conjugate base, cB_1) can add a proton to only one position. Hence, if no other factors operated, it would be anticipated that the acid in equilibrium (1) is twice as strong as a monocarboxylic acid, RCO_2H, *i.e.*, $K_a^1 = 2 \times K_a$, of RCO_2H. In the second-stage ionisation (2), the half-ion can lose a proton from only one position, but the dianion (cB_2) can add a proton to two positions. Therefore for equilibrium (2) $K_a^2 = 1/2\ K_a$, of RCO_2H. Thus, equilibrium (1) is statistically four times as favoured as equilibrium (2). Hence, on the assumption above, *i.e.*, that no other factors operate, the value of K_a^1 would be expected to be four times K_a^2. In practice, the two constants differ from this factor of 4 Table.

An explanation for the experimental results may be as follows. Bjerrum (1923) assumed that the electrostatic effect of the negatively charged carboxylate ion, cB_1, would require additional energy to separate the proton of the carboxyl group, *i.e.*, there is a field effect through the solvent and consequently increases the ratio K_a^1/K_a^2. At the same time, the carboxylate anion exerts a + I effect with resultant weakening of the acid. It would therefore follow that as the chain increases in length, *i.e.*, as n increases, both the field and + I effects would decrease, and so the ratio would approach 4. This is borne out in practice, Table : see also conformational effects.

The reactions of the dicarboxylic acids depend, to a large extent, on the length of the carbon chain. The dicarboxylic acids, therefore, will be described individually.

Oxalic acid (*ethanedioic acid*) is one of the most important dicarboxylic acids. It occurs in rhubarb, in sorrel and other plants of the *oxalis* group (hence its name). Oxalic acid is one of the final products of oxidation of many organic compounds, *e.g.*, sugars, starch, etc., give oxalic acid when oxidised with concentrated nitric acid.

Preparation

Oxalic acid is prepared industrially by heating sodium formate rapidly to 360°C.

$$2HOC_2Na \longrightarrow (CO_2Na)_2 + H_2$$

The free acid is obtained from its sodium salt by adding calcium hydroxide solution, then treating the precipitated calcium oxalate with the calculated amount of dilute sulphuric acid, removing the precipitated calcium sulphate, and finally evaporating the filtrate to crystallisation.

The usual laboratory method for preparing oxalic acid is to oxidise sucrose or molasses with concentrated nitric acid in the presence of vanadium pentoxide as catalyst:

$$C1_2H_{22}O_{11} + 18\ [O] \xrightarrow[V_2O_5]{HNO_3} 6(CO_2H)_2 + 5H_2O\ (25\%)$$

Properties and Reactions

Oxalic acid crystallises from water as colourless crystals with two molecules of water of crystallisation; the melting point of the hydrate is 101.5°C; that of the anhydrous acid is 189.5°C. Oxalic acid is poisonous, soluble in water and ethanol but almost insoluble in ether. The dihydrate loses water when heated at 100-105°C, and when heated at about 200°C, oxalic acid decomposes into carbon dioxide, carbon monoxide, formic acid and water:

$$(CO_2H)_2 \longrightarrow CO_2 + HCO_2H \longrightarrow CO + H_2O$$

The anhydrous acid is conveniently obtained by heating the hydrate with carbon tetrachloride.

When heated with concentrated sulphuric acid at 90°C, oxalic acid is decomposed:

$$(CO_2H)_2 \xrightarrow{H_2SO_4} CO + CO_2 + H_2O$$

It is oxidised by permanganate to carbon dioxide:

$$(CO_2H)_2 + [O] \longrightarrow 2CO_2 + H_2O$$

It is only very slowly oxidised by concentrated nitric acid. When fused with potassium hydroxide, it evolves hydrogen:

$$(CO_2H)_2 + 2C_2H_5OH \longrightarrow (CO_2C_2H_5)_2 + 2H_2O\ (80\text{–}90\%)$$

The anhydride of oxalic acid is unknown. When anhydrous oxalic acid is refluxed with ethanol, ethyl oxalate is formed:

$$(CO_2)_2 + 2C_2H_5OH \longrightarrow (CO_2C_2H_5)_2 + 2H_2O\ (80\text{–}90\%)$$

A more general method for preparing di-esters is to heat a mixture of the dicarboxylic acid, alcohol, toluene and concentrated sulphuric acid. Sulphuric acid is the catalyst, and the toluene removes the water by forming a ternary azeotrope of alcohol, water and toluene; the yield of di-ester is 94-98 per cent from oxalic to sebacic acid.

The esterification may be carried out in the absence of toluene: a larger amount of sulphuric acid is required and the yields are lower (70-90 per cent).

The ethyl and propyl esters of oxalic acid are liquid; the methyl ester is solid, and hence has been used to prepare pure methanol by hydrolysis with sodium hydroxide solution. Ethyl oxalate undergoes the Claisen condensation with esters containing two α-hydrogen atoms to form keto-esters, *e.g.*, with ethyl acetate, oxalacetic ester is formed:

$$C_2H_5O_2CCO_2C_2H_5 + CH_3CO_2C_2H_5 \xrightarrow[\text{ethanol}]{C_2H_5ONa\text{ in}}$$

$$\begin{matrix} CHCO_2C_2H_5 \\ \| \\ C(ONa)CO_2C_2H_5 \end{matrix} \xrightarrow{CH_3CO_2H} \begin{matrix} CH_2CO_2C_2H_5 \\ | \\ COCO_2C_2H_5 \end{matrix} \quad (60\text{–}70\%)$$

When oxalic acid is heated with ethylene glycol, the cyclic compound, ethylene oxalate, is obtained:

$$HOCH_2CH_2OH + HO_2CCO_2H \longrightarrow O\begin{matrix} \diagup CH_2\text{—}CH_2 \diagdown \\ \diagdown CO\text{—}CO \diagup \end{matrix}O + 2H_2O$$

This reaction is characteristic of oxalic acid; the other dicarboxylic acids usually react with glycol to form polyesters. When oxalic acid is heated with glycerol, formic acid or allyl alcohol is obtained according to the conditions.

Oxalic acid forms the diamide, *oxamide*. This may be prepared by shaking ethyl oxalate with concentrated ammonium hydroxide solution:

$$C_2H_5O_2CCO_2C_2H_5 + 2NH_3 \longrightarrow NH_2COCONH_2 + 2C_2H_5OH$$

It may also be prepared by passing cyanogen into *cold* concentrated hydrochloric acid:

$$(CN)_2 + 2H_2O \xrightarrow{HCl} (CO_2NH_4)_2$$

Oxamide (decomp. 350°C). in aqueous solution slowly changes into ammonium oxalate, a change which is brought about rapidly by hydrochloric acid (on warming):

$$(CONH_2)_2 \xrightarrow{H_3O^+} (CO_2NH_4)_2$$

When heated with phosphorus pentoxide, oxamide is dehydrated to cyanogen:

$$(CONH_2)_2 \xrightarrow{P_2O_5} (CN)_2 + 2H_2O$$

The monoamide of oxalic acid is also known. It is called oxamic acid, the suffix of the names of all the monoamides of the dicarboxylic acids is -amide, Oxamic acid, m.p. 210°C, may be prepared by the action of concentrated ammonium hydroxide solution on ethyl hydrogen oxalate, or by heating ammonium hydrogen oxalate.

When reduced with zinc and sulphuric acid, oxalic acid forms *glycollic acid*, $HOCH_2CO_2H$. Electrolytic reduction with a lead cathode gives glycollic and glyoxylic acids, the latter being obtained in better yield by reducing oxalic acid with magnesium and sulphuric acid.

When oxalic acid is treated with *excess* of phosphorus pentachloride, *oxalyl chloride*, b.p. 64°C, is formed:

$$(CO_2H)_2 \xrightarrow{PCl_5} (COCl)_2 \qquad (f.\text{–}g)$$

If an excess of phosphorus pentachloride is not used, oxalic acid is decomposed, possibly via the intermediate formation of the half-acid chloride.

$$(CO_2H)_2 \xrightarrow{PCl_5} [HO_2CCOCl] \longrightarrow CO_2 + CO + HCl$$

Alkanes can be carboxylated by oxalyl chloride under the influence of light (Kharasch and Brown, 1942):

$$RH + (COCl)_2 \longrightarrow RCOCl + CO + HCl$$

Oxalic acid is used for the manufacture of ink and for bleaching straw. Its antimony salts are used as mordants in printing and dyeing.

Malonic acid (*propanedioic acid*), m.p. 136°C, was first obtained by the oxidation of malic acid (hence its name):

$$HO_2CCHOHCH_2CO_2H + [O]$$

$$\xrightarrow{K_2Cr_2O_7/H_2SO_4} \underset{\text{oxalacetic acid}}{HO_2CCOCH_2CO_2H}$$

$$\xrightarrow{[O]} HO_2CCH_2CO_2H + CO_2$$

Malonic acid may be prepared by heating potassium chloroacetate with aqueous potassium cyanide and hydrolysing the product, potassium cyanoacetate, with hydrochloric acid:

$$ClCH_2CO_2K \xrightarrow{KCN} CH_2(CN)CO_2K \xrightarrow{H_3O^+} CH_2(CO_2H)_2\ (84\%)$$

When heated to 140-150°C, or when refluxed in sulphuric acid solution, it eliminates carbon dioxide.

$$HO_2CCH_2CO_2H \longrightarrow C_3CO_2 + CO_2$$

All dicarboxylic acids which have both carboxyl groups attached to the *same* carbon atom are decomposed in a similar manner. When malonic acid is heated with phosphorus pentoxide, a small amount of *carbon suboxide* is obtained:

$$HO_2CCH_2CO_2H \xrightarrow{P_2O_5} O = C = C = C = O + 2H_2O$$

Carbon suboxide, b.p. 7°C, may be regarded as a diketen; it combines with water to form malonic acid. Malonic acid does not form a cyclic anhydride, but dimethylmalonic acid does.

When malonic acid is treated with nitrous acid and the product hydrolysed, **mesoxalic acid** *(keto-malonic acid)* is obtained:

$$(HO_2C)_2CH_2 \xrightarrow{HNO_2} (HO_2C)_2C = NOH \xrightarrow{H_3O^+} HO_2CCOCO_2H$$

Mesoxalic acid crystallises from water with one molecule of water which is held firmly, and hence is believed to be water of constitution: $(HO)_2C(CO_2H)_2$ (cf. glyoxylic acid). Ethyl malonate may be oxidised directly to ethyl mesoxalate by selenium dioxide (yield: 23 per cent), or with dinitrogen trioxide (HNO_3 + AS_2O_3; yield: 74-76 per cent).

Ethyl malonate is far more important than the acid because of its synthetic uses; the acid contains an active methylene group, the reactivity of which is more pronounced in the ester.

Both the acid and ester may be readily brominated, *e.g.*, monobromomalonic acid, CHBr(CO,H), is formed when a suspension of malonic acid in ether is treated with bromine; owing to the high reactivity of the methylene group, no red phosphorus is required as catalyst.

Ethyl malonate does not form the diamide, *malonamide,* when shaken with concentrated ammonium hydroxide solution; the dimethyl ester, however, gives a good yield of malonamide. Ethyl malonate forms *barbituric acid* when heated with urea in the presence of sodium ethoxide.

Succinic acid *(butanedioic acid)*, m.p. 185°C, was originally obtained by the distillation of amber (Latin: *succinum*, amber). It may be synthesised by the following methods:

(i) From ethylene dibromide:

$$\begin{matrix} CH_2Br \\ | \\ CH_2Br \end{matrix} \xrightarrow{KCN} \begin{matrix} CH_2CN \\ | \\ CH_2CN \end{matrix} \xrightarrow{H_3O^+} \begin{matrix} CH_2CO_2H \\ | \\ CH_2CO_2H \end{matrix} \quad (80\%)$$

(ii) By the reaction between malonic ester (I molecule) and ethyl chloroacetate, or between malonic ester (2 molecules) and iodine.

Alkyl-substituted succinic acids may be prepared by using monoalkylmalonic ester and α-halogeno-acid esters or iodine.

Alternatively, they may be prepared by the addition of hydrogen cyanide to α, β-unsaturated esters and hydrolysing the Bcyanacomplex produced.

(iii) By heating malic acid, in a sealed tube, with constant boiling hydriodic acid and red phosphorus:

$$\begin{matrix} CHOHCO_2H \\ | \\ CH_2CO_2H \end{matrix} + 2HI \longrightarrow \begin{matrix} CH_2CO_2H \\ | \\ CH_2CO_2H \end{matrix} + I_2 + H_2O \quad (60\%)$$

Succinic acid is prepared industrially by the catalytic (or by the electrolytic) reduction of maleic acid:

$$\begin{matrix} CHCO_2H \\ \| \\ CHCO_2H \end{matrix} + H_2 \xrightarrow{Ni} \begin{matrix} CH_2CO_2H \\ | \\ CH_2CO_2H \end{matrix}$$

When heated, a large amount sublimes, the rest being converted into the *cyclic* anhydride, succinic anhydride:

$$\begin{matrix} CH_2CO_2H \\ | \\ CH_2CO_2H \end{matrix} \longrightarrow \begin{matrix} CH_2CO_2H \diagdown \\ | \qquad\quad O \\ CH_2CO_2H \diagup \end{matrix} + H_2O$$

When heated with excess of glycol, succinic acid forms high-polymer esters (*polyesters*, which belong to the group known as the *alkyl resins*). These esters are acidic, the end groups being succinic acid residues:

$$HO_2C(CH_2)_2CO_2H + HOCH_2CH_2OH + HO_2C(CH_2)_2CO_2H + HOCH_2CH_2OH + \ldots \longrightarrow HO_2C(CH_2)_2CO{-}[{-}O(CH_2)_2OCO(CH_2)_2CO{-}]{-}OH$$

Dicarboxylic acids from malonic to adipic acid form these polyesters with ethylene glycol.

These esters are linear polymers, but if glycerol is used instead of glycol, three dimensional polymeric esters are obtained.

Succinic anhydride (*butanedioic anhydride*) is obtained in excellent yield by distilling succinic acid with acetic anhydride, acetyl chloride, or phosphoryl chloride.

The anhydride, and not the acid chloride, is obtained when succinic acid is heated with thinly chloride.

Succinic anhydride, m.p. 119°C, when boiled with water or alkalis, is converted into succinic acid:

$$\begin{array}{l} CH_2CO \\ | \quad\quad\quad\ \ \backslash \\ | \quad\quad\quad\quad O \\ | \quad\quad\quad\ \ / \\ CH_2CO \end{array} + H_2O \longrightarrow \begin{array}{l} CH_2CO_2H \\ | \\ CH_2CO_2H \end{array}$$

When reduced with sodium and ethanol, succinic anhydride is converted first into γ-butyro- lactone, and finally into tetramethylene glycol:

$$\begin{array}{l} CH_2CO \\ | \quad\quad\quad\ \ \backslash \\ | \quad\quad\quad\quad O \\ | \quad\quad\quad\ \ / \\ CH_2CO \end{array} \xrightarrow{e;\ H^+} \begin{array}{l} CH_2CH_2 \\ | \\ CH_2CO \end{array} \xrightarrow{e;\ H^+} \begin{array}{l} CH_2CH_2OH \\ | \\ CH_2CH_2OH \end{array}$$

If the reduction is carried out with sodium amalgam in *acid* solution, the lactone is obtained in good yield; some butyric acid is also formed.

Succinic anhydride is very useful for preparing the 'half-derivatives' of succinic acid, *e.g.*, with alcohols it forms the acid-ester:

$$\begin{array}{l} CH_2CO \\ | \quad\quad\quad\ \ \backslash \\ | \quad\quad\quad\quad O \\ | \quad\quad\quad\ \ / \\ CH_2CO \end{array} + ROH \longrightarrow \begin{array}{l} CH_2CO_2R \\ | \\ CH_2CO_2H \end{array}$$

With Grignard reagents, succinic anhydride forms γ-keto-acids:

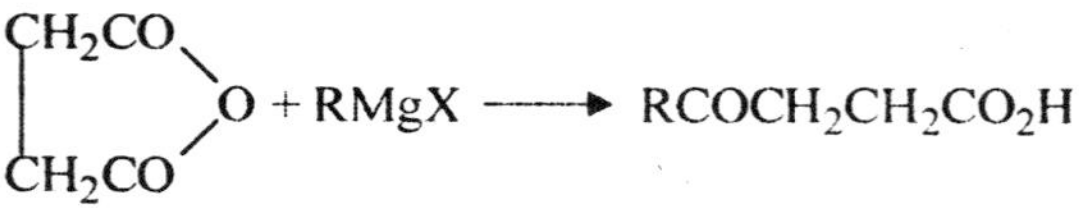

$$\begin{matrix} CH_2CO \\ | \\ CH_2CO \end{matrix}\!\!>O + RMgX \longrightarrow RCOCH_2CH_2CO_2H$$

The carbonyl groups (C = O stretch) in *five*-membered ring anhydrides absorb in the regions 1870-1830 (s) and 1800-1760 cm^{-1}(s). Thus these anhydrides can be distinguished from acyclic anhydrides. The molecular formula should also help here for all cyclic anhydrides.

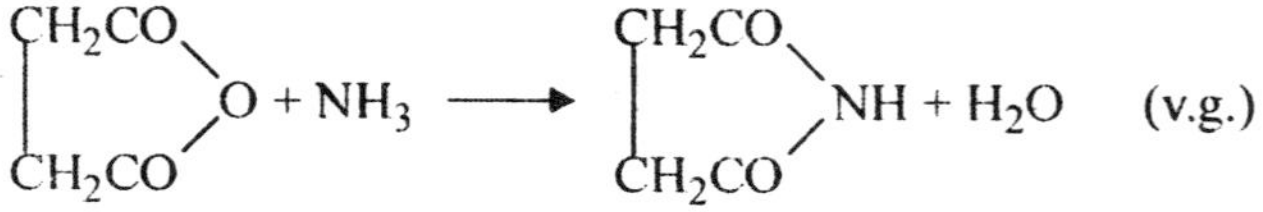

$$\begin{matrix} CH_2CO \\ | \\ CH_2CO \end{matrix}\!\!>O + NH_3 \longrightarrow \begin{matrix} CH_2CO \\ | \\ CH_2CO \end{matrix}\!\!>NH + H_2O \quad (v.g.)$$

The infrared spectrum of succinic anhydride (as a Nujol mull).

2940, 1460 and 1370cm^{-1} : Bands for Nujol. 1859 and 1786cm^{-1}: C = O (str.) in 5-ring anhydrides

Succinimide (*butanimide*) is formed when succinic anhydride is heated in a current of dry ammonia:

Succinamic acid (m.p. 157°C), (I), and succinamide (m.p. 243°C), (II), are both readily converted into succinimide when heated:

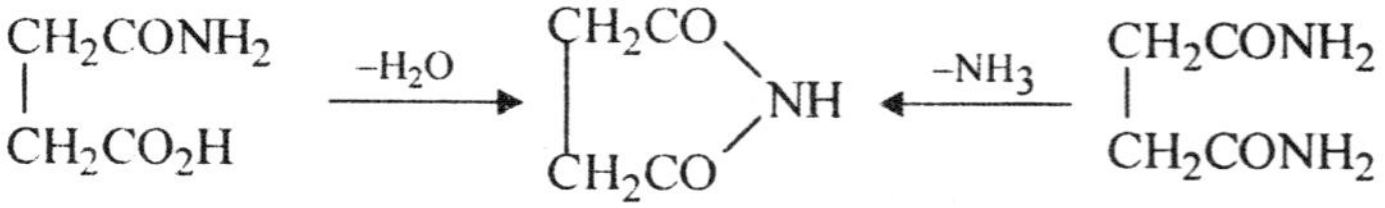

$$\begin{matrix} CH_2CONH_2 \\ | \\ CH_2CO_2H \end{matrix} \xrightarrow{-H_2O} \begin{matrix} CH_2CO \\ | \\ CH_2CO \end{matrix}\!\!>NH \xleftarrow{-NH_3} \begin{matrix} CH_2CONH_2 \\ | \\ CH_2CONH_2 \end{matrix}$$

Succinimide, m.p. 125°C, when boiled with water or alkalis is converted into succinic acid:

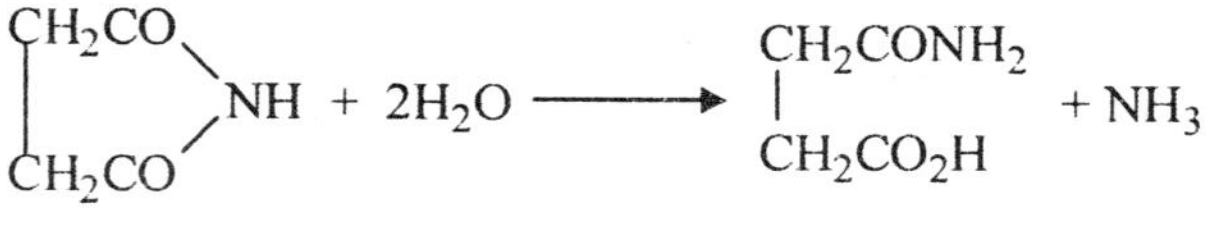

$$\begin{matrix} CH_2CO \\ | \\ CH_2CO \end{matrix}\!\!>NH + 2H_2O \longrightarrow \begin{matrix} CH_2CONH_2 \\ | \\ CH_2CO_2H \end{matrix} + NH_3$$

When succinimide is distilled with zinc dust, *pyrrole*, (III), is obtained. When reduced will sodium and ethanol, succinimide forms *pyrrolidine*, (IV), and when reduced electrolytical it forms *pyrrolidone*, (V):

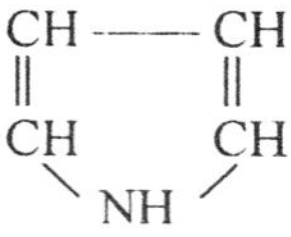

$$\underset{\text{(III)}}{\begin{matrix} CH & — & CH \\ \| & & \| \\ CH & & CH \\ & \diagdown NH \diagup & \end{matrix}} \qquad \underset{\text{(IV)}}{\begin{matrix} CH_2 & — & CH_2 \\ | & & | \\ CH_2 & & CH_2 \\ & \diagdown NH \diagup & \end{matrix}} \qquad \underset{\text{(V)}}{\begin{matrix} CH_2 & — & CH_2 \\ | & & | \\ CH_2 & & CO \\ & \diagdown NH \diagup & \end{matrix}}$$

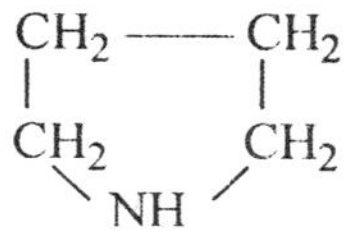

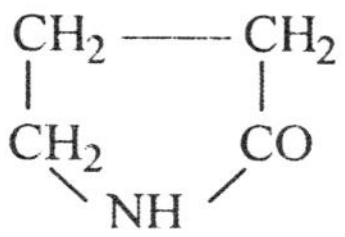

Succinimide is acidic, *e.g.*, it reacts with potassium hydroxide to form potassium succinimide:

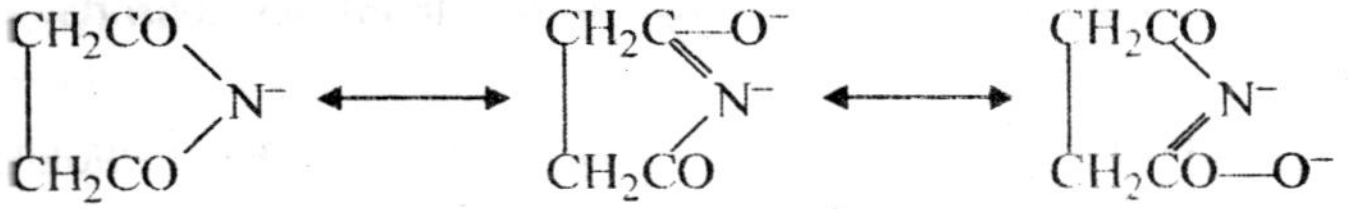

$$\begin{matrix} CH_2CO \\ | \\ CH_2CO \end{matrix} \!\!> N^- \longleftrightarrow \begin{matrix} CH_2C{=}O^- \\ | \\ CH_2CO \end{matrix} \!\!> N^- \longleftrightarrow \begin{matrix} CH_2CO \\ | \\ CH_2CO{-}O^- \end{matrix} \!\!> N^-$$

Succinimide, however, is only very weakly acidic; the potassium salt is decomposed by carbon dioxide, the imide being regenerated.

A very important derivative of succinimide is N-*bromosuccinimide*, which may be prepared by the action of bromine on succinimide at 0°C in the presence of sodium hydroxide:

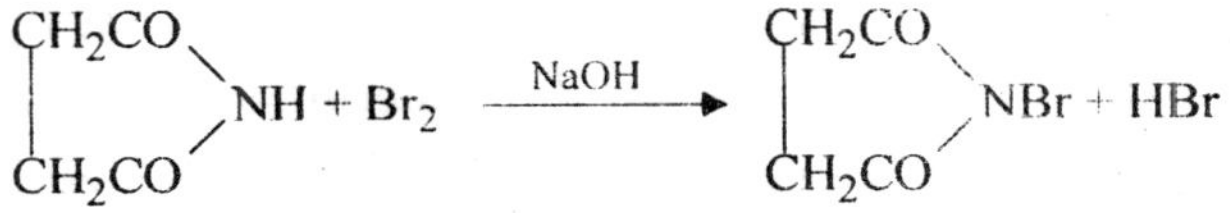

$$\begin{matrix} CH_2CO \\ | \\ CH_2CO \end{matrix} \!\!> NH + Br_2 \xrightarrow{NaOH} \begin{matrix} CH_2CO \\ | \\ CH_2CO \end{matrix} \!\!> NBr + HBr$$

N-bromosuccinimide (NBS) is a valuable reagent or brominating olefinic compounds in the allyl position (Ziegler. 1942).

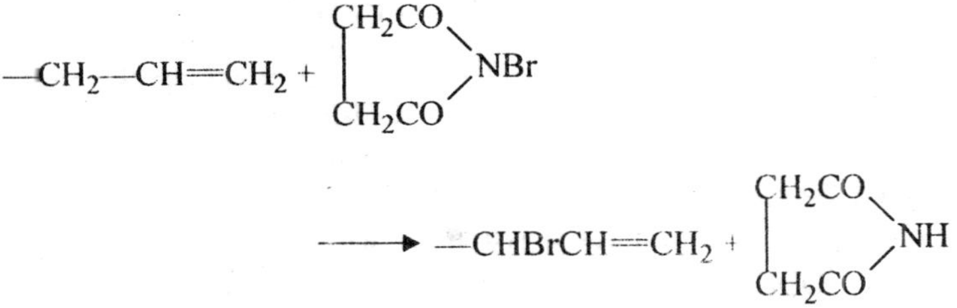

$$-CH_2-CH{=}CH_2 + \begin{matrix} CH_2CO \\ | \\ CH_2CO \end{matrix} \!\!> NBr$$

$$\longrightarrow -CHBrCH{=}CH_2 + \begin{matrix} CH_2CO \\ | \\ CH_2CO \end{matrix} \!\!> NH$$

This reaction offers a means of splitting off three carbon atoms from a compound by oxidative degradation, provided the compound contains one double bond or can have one produced, *e.g.*,

$$RCH_2CH_2CH_2OH \xrightarrow[350°C]{Al_2O_3} RCH_2CH_2CH = CH_2$$

$$\xrightarrow{\text{N-bromosuccinimide}} RCH_2CHBrCH = CH_2$$

$$\xrightarrow[\text{e tanol}]{KOH} RCH = CHCH = CH_2 \xrightarrow{\text{ozonolysis}} RCHO$$

This method has found great use in steroid chemistry.

Normally, *N*-bromosuccinimide *substitutes* alkenes in the ally position, and this reaction is believed to take place by a free-radical mechanism; it is catalysed by peroxides and is promoted by light (both are free-radical producing agents). If there are two allylic positions then two monobromo-derivatives may be obtained. Since a secondary hydrogen is substitute more readily than a primary hydrogen atom, the former gives rise to the predominan product.

The details of the mechanism are uncertain. According to Goldfinger *et al.* (1953, 1956), the function of the *N*-bromosuccinimide is to supply a *low concentration of molecular bromine.*

$$>NBr + HBr \longrightarrow >NH + Br_2$$

$$Br_2 \rightleftharpoons Br\cdot$$

$$Br\cdot + >CHCH{=}CH_2 \longrightarrow >\dot{C}CH{=}CH_2 + HBr$$

$$Br_2 + >\dot{C}CH{=}CH_2 \longrightarrow CBrCH{=}CH_2 + Br\cdot$$

N-bromosuccinimide also brominates unsaturated esters, *e.g.*,

$$CH_3CH{=}CHCO_2CH_3 \xrightarrow{>N{-}Br} BrCH_2CH{=}CH_2 + Br\cdot$$

In addition to substitution, Af-bromosuccinimide may also produce addition compounds, but these are usually formed only in small amount. Braude *et al.* (1952), however, have shown that the addition reaction is catalysed by tetraethylammonium salts, *e.g.*, *N*-bromosuccinimide and cyclohexene give 3-bromocyclohexene (Ziegler *et al.*, 1942), but in the presence of, *e.g.*, tetraethylammonium bromide, 1,2-dibromocvclohexane is the main product.

The addition reaction probably involves heterolytic fission

$$>N{-}Br + >C{=}C< \longrightarrow >N^- + \overset{+}{Br}\ \left[>C{=}C<\right] \xrightarrow{>N{-}Br}$$

$$BrC{-}CBr + >N^- + >N^+ \longrightarrow 2>N^- \xrightarrow[\text{donor}]{\text{hydrogen}} 2>NH$$

Succinyl chloride (butanedioyl chloride). This is readily prepared by the action of phosphorus pentachloride on succinic acid. It was formerly believed that succinyl chloride existed in two forms, the chain isomer, (I), and the ring isomer, (II). However, all recent work indicates structure

(1), *e.g.*, Schmid (1965, 1966), from NMR studies, showed that the molecule is symmetrical, and from inferred studies that a y-lactone oxo group (in II) is absent.

$$(CH_2COCl)_2 \quad (I) \qquad \begin{array}{l} CH_2CO \\ \;|\quad\;\; \diagdown \\ \;|\qquad O \\ \;|\quad\;\; \diagup \\ CH_2CO \end{array} \quad (II)$$

Isomeric with is succinic acid is *methylamine acid* or *isosuccinic acid*, m.p. 130°C. It may be readily prepared from sodiomalonic ester and methyl iodide, or as follows:

$$CH_3C_2CO_2H \xrightarrow{Br_2/P} CH_3CHBrCO_2H$$

$$\xrightarrow{KCN} CH_3CH(CN)CO_2H \xrightarrow{H^+} CH_3CH(CO_2H)_2$$

Oxalacetic ester is the diethyl ester of ketosuccinic acid, and may be prepared by the Claisen condensation between ethyl oxalate and ethyl acetate. It is a colourless liquid which can be distilled under reduced pressure (b.p. 132°C/24 mm), but at atmospheric pressure it eliminates a molecule of carbon monoxide to form malonic ester:

$$C_2H_5O_2CCOH_2CO_2C_2H_5 \longrightarrow CH_2(CO_2C_2H_5)_2 + CO$$

Hydrogen of the methylene group adjacent to the carbonyl group may be replaced by alkyl groups by reactions similar to those used for acetoacetic ester, *e.g.*,

$$C_2H_5O_2CCOCH_2CO_2C_2H_5 \xrightarrow{C_2H_5ONa}$$

$$[C_2H_5O_2CCOCHCO_2C_2H_5]^- Na^+ \xrightarrow{CH_3I}$$

$$C_2H_5O_2CCOCH(CH_3)CO_2C_2H_5 \xrightarrow[(ii)\,C_2H_5I]{(i)\,C_2H_5ONa}$$

$$C_2H_5O_2CCOC(CH_3)(C_2H_5)CO_2C_2H_5$$

Oxalacetic ester and its alkyl derivatives undergo 'acid hydrolysis' when boiled with alkalis, and 'ketonic hydrolysis' when boiled with dilute sulphuric acid.

'Acid hydrolysis'

$$C_2H5O_2CCOCHRCO_2C_2H_5 \xrightarrow[(ii)\,H^+]{(i)\,NaOH}$$

$$(CO_2H)_2 + RCH_2CO_2 + 2C_2H_5OH$$

'*Ketonic hydrolysis*'

$$C_2H_5O_2CCOCH_2CO_2C_2H_5 + 2H_2O \xrightarrow{HCl} [HO_2CCOCHRCO_2H]$$

$$\longrightarrow CO_2 + RCH_2COCO_2H$$

Hydrolysis of oxalacetic ester with concentrated hydrochloric acid in the cold gives oxalacetic acid (*ketosuccinic acid*):

$$C_2H_5O_2CCOCH_2CO_2H_5 + 2H_2O \xrightarrow{HCl} HO_2CCOCH_2CO_2H + 2C_2H_5OH$$

A simpler preparation of oxalacetic acid is to oxidise maleic acid with Fenton's reagent.

Oxalacetic acid is a fairly stable substance, soluble in water, the aqueous solution giving a red colour with ferric chloride, thus indicating the existence of an enolic form. Oxalacetic acid exists in two forms, one with m.p. 155°C and the other 184°C. These two forms are actually hydroxymaleic acid, (1), and *hydroxyfumaric acid*, (II), respectively:

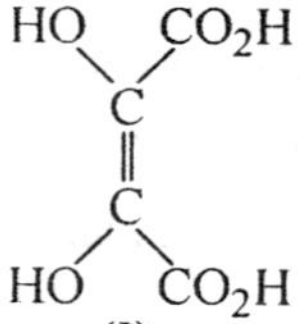

$$\begin{array}{ccc} HO & & CO_2H \\ & \diagdown\ \diagup & \\ & C & \\ & \| & \\ & C & \\ & \diagup\ \diagdown & \\ HO & & CO_2H \end{array} \qquad \qquad \begin{array}{ccc} HO_2C & & H \\ & \diagdown\ \diagup & \\ & C & \\ & \| & \\ & C & \\ & \diagup\ \diagdown & \\ H & & CO_2H \end{array}$$

(I) 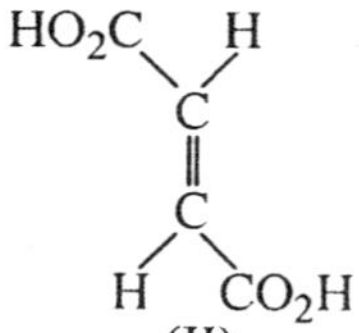 (II)

(I) is convertible into (II) by 30 per cent sulphuric acid, and the salts of (II) give (I) on treatment with dilute acid.

According to Kumler *et al.* (1962), who have examined the NMR spectrum of oxalacetic acid in deuterium oxide, at equilibrium there is 50 per cent of keto-form and 8 per cent of enol, the rest being present in some other form.

Glutaric acid (*pentanedioic acid*), m.p. 97°C (relationship to glutamic acid and tartaric acid gave rise to its name), may be prepared by a number of methods, *e.g.*:

(i) By refluxing trimethylene dicyanide with concentrated hydrochloric acid:

$$NC(CH_2)_3CN + 4H_2O \xrightarrow{HCl} HO_2C(CH_2)_3CO_2H + 2NH_3$$

(ii) By the action of methylene di-iodide on sodiomalonic ester:

$$2[CH(CO_2C_2H_5)_2]^- Na^+ + CH_2I_2 \longrightarrow {}_2NaI + CH_2[CH(CO_2C_2H5)_2]_2$$

$$\xrightarrow[\text{(ii) HCl}]{\text{(i) KOH}} CH_2[CH(CO_2H_2)_2]_2$$

$$\xrightarrow{150-200°C} HO_2C(CH_2)3CO_2H + {}_2CO_2$$

(iii) By condensing formaldehyde with malonic ester in the presence of diethylamine:

$$2(C_2H_5O_2C)_2CH_2 + CH_2O$$

$$\xrightarrow[\text{reflux}]{(C_2H_5)_2NH} (C_2H5O_2C)_2CHCH_2CH(CO_2C_2H5)_2 \quad (61\%)$$

This tetracarboxylic ester may now be treated as in (ii), or may be converted into glutaric acid directly by refluxing with concentrated hydrochloric acid:

$$(C_2H_5O_2C)_2CHCH_2CH(CO_2H_5)_2$$

$$\xrightarrow{HCl} HO_2C(CH_2)_3CO_2H \quad (76–80\ \%)$$

When heated with acetic anhydride or thionyl chloride, glutaric acid is converted into glutaric anhydride:

$$CH_2 \begin{cases} CH_2CO_2H \\ CH_2CO_2H \end{cases} \longrightarrow CH_2 \begin{cases} CH_2—CO \\ CH_2—CO \end{cases} O + H_2O$$

Succinic and glutaric acids differ from their higher homologues in that they readily form the cyclic anhydride when heated with acetic anhydride.

Adipic acid (*hexanedioic acid*), m.p. 150°C, received its name from .the fact that it was first obtained by the oxidation of fats (Latin: adeps, fat). It may be synthesised from sodiomalonic ester and ethylene dibromide, but is prepared industrially by the oxidation of cyclohexane (from the hydrogenation of benzene) by air in the presence of a catalyst, *e.g.*, cobalt acetate:

When adipic and pimelic acids are heated with acetic anhydride and the product distilled at 300°C, a cyclic ketone is obtained in each case:

Adipic acid is used for the preparation of resins (poly-esters), and is an intermediate in the manufacture of *nylon*. Nylon is used as the generic name for all synthetic fibre-forming polymeric amides having a protein-like structure. Nylon yarns and fabrics are practically non-inflammable. Nylon '66' is formed from adipic acid and hexa-methylenediamine, and polymerisation is effected by melting the mixture in the presence of various catalysts.

$$HO_2C(CH_2)_4CO_2H + H_2N(CH_2)_6NH_2 + HO_2C(CH_2)_4CO_2H + H_2N\text{—} \ldots \longrightarrow$$

$$+ HO_2C(CH_2)_4CO\text{—}[\text{—}NH(CH_2)_6NHCO(CH_2)_4CO\text{—}]_n NH(CH_2)_6NH_2$$

Nylon '6' is made from *caprolactam*, which is manufactured in several ways starting from cyclohexane:

The conversion of the oxime to caprolactam occurs via the Be Loann rearrangement, and the caprolactam is polymerised in the presence of a catalyst:

$$\text{caprolactam} \longrightarrow \text{—}[\text{—}CO(CH_2)_5NH\text{—}]_n\text{—}$$

Pimelic acid (*heptanedioic acid*), m.p. 104°C, received its name from the fact that it was obtained originally from the oxidation of fats (Greek: *pimele*, fat). It may be prepared by the hydrolysis of penta-methylene dicyanide (formed by the action of potassium cyanide on pentamethylene dibromide), or by the malonic ester synthesis, using trimethylene (*cf.* adipic acid). Pimelic acid may also be prepared by the reduction of salicylic acid with sodium and isopentanol, followed by the addition of water and then hydrochloric acid:

Subcric acid (*octanedioic acid*), $HO_2C(CH_2)_6CO_2H$, m.p. 144°C, may be synthesised by the electrolysis of potassium ethyl glutarate. It is prepared industrially by the oxidation of cork (Latin: *suber*, cork) with concentrated nitric acid.

Azelaic acid (*nonanedioic acid*), m.p. 107°C, may be synthesised from malonic ester and pentamethylene dibromide. It may be obtained .in the laboratory by refluxing castor oil with ethanolic potassium hydroxide, acidifying with sulphuric acid, and oxidising the crude ricinoleic acid so obtained with alkaline permanganate:

$$CH_3(CH_2)_5CHOHCH_2CH = CH(CH_2)_7CO_2H \xrightarrow{[O]} HO_2C(CH_2)_7CO_2H \quad (32\text{-}36\%)$$

Azelaic acid is prepared industrially by the ozonolysis folic acid.

Sebacic acid (*decanedioic acid*), m.p. 133°C, is prepared industrially by heating castor oil with sodium hydroxide.

$$CH_3(CH_2)_5CHOHCH_2CH = CH(CH_2)_7CO_2Na$$

$$\xrightarrow{NaOH} NaO_2C(CH_2)_8CO_2Na + C_6H_{13}CHOHCH_3$$

Many higher dicarboxylic acids are known.

The acid esters of the dicarboxylic acids may be prepared by refluxing a mixture of the acid with half its equivalent of the di-ester in the presence of concentrated hydrochloric acid. This is an example of acidolysis. Alternatively, the acid ester may be obtained by the half saponification of the di-ester. In either case the di-ester is separated from the acid ester by fractional distillation.

Another method of preparing a half-ester is to heat equimolecular amounts of dicarboxylic acid and alcohol: the products are unchanged acid, half-ester and di-ester, which are then separated. If the dicarboxylic acid is substituted at one a-carbon atom, then the other carboxyl group is readily esterified by the Fischer-Speier method; this is probably a question of steric hindrance.

When a cyclic anhydride is available, this is a better way to prepare the half-ester. Here again, if one α-carbon atom is substituted, it is the half-ester of the other carboxyl group that is formed.

Half-esters are useful starting materials in various syntheses, *e.g.*,

(i) $$\begin{matrix} CH_2CH_2CO_2H \\ | \\ CH_2CH_2CO_2Et \end{matrix} \xrightarrow[\text{(ii) } Br_2]{\text{(i) } SOCl_2} \begin{matrix} CH_2CHBrCOCl \\ | \\ CH_2CH_2CO_2Et \end{matrix}$$

$$\xrightarrow{ROH} \begin{matrix} CH_2CHBrCO_2R \\ | \\ CH_2CH_2CO_2Et \end{matrix}$$

(ii) $$HO_2C(CH_2)_8CO_2Et \xrightarrow[Br_2/CCl_4]{HgO} Br(CH_2)_nCO_2Et$$

(iii) $$HO_2C(CH_2)_nCO_2Et \xrightarrow{Na - EtOH} HO_2C\ (CH_2)_nCH_2OH$$

When the calcium, barium, or better still, the thorium salts of the carboxylic acids from adipic to nonadecanedioic are distilled, varying yields of cyclic ketones are obtained; *e.g.*, adipic acid gives cyclopentanone: Based on the experimental results of the ease of formation of cyclic anhydrides and cyclic ketones, is **Blanc's rule** (1905). Blanc found that dicarboxylic acids, on heating with acetic anhydride, and then distilling at 300°C (or distilling directly at 300°C), gave cyclic anhydrides or cyclic ketones according to the relative positions of the two carboxyl groups. 1,4- and 1,5-dicarboxylic acids gave cyclic anhydrides. 1,6- and 1,7- gave cyclic ketones: this is Blanc's rule. By using Blanc's rule it is possible to determine the size of rings. A double bond is introduced into the ring, and the ring opened by oxidation to the corresponding dicarboxylic acid. This is then heated with acetic anhydride and distilled. If a cyclic anhydride is obtained, the acid is either 1,4- or 1,5-; if a cyclic ketone, either 1,6- or 1,7-; if there is no change, the acid is 1,8-or more. In naturally occurring compounds the rings are usually five- or six-membered. Hence the formation of the anhydride is taken to mean a five-membered ring, and of the ketone a six-membered ring. Blanc's rule, however, is only satisfactory for simple cyclic compounds. For substituted rings, the results may be misleading, since substituents in the chain bring about ring closure much more easily; *e.g.*, substituted adipic acids may give rise to the anhydride and not to the cyclic ketone.

5

SPECIFICATION OF ABSOLUTE CONFIGURATION OF ALLENES

INTRODUCTION

This is carried out in a similar manner to that used for biphenyls. For allenes, the chiral axis passes through the double bonds.

The absolute configurations of allenes have been determined by conversion of an optically active molecule of known absolute configuration into an allene (*e.g.*, Landor, above), or by converting a chiral allene into a molecule of known absolute configuration by stereochemically unambiguous reactions.

The exception to the above procedures is the work of Mason, who determined the absolute configuration of 1,3-diphenylallene by means of electronic absorption and circular dichroism spectra. Lowe (1965) has now developed a method for predicting the sign of rotation of allenes of known absolute configuration.

Stereochemistry of the Spirans

If both double bonds in allene are replaced by ring systems, the resulting molecules are spirans. One method of naming spirans obtains the root name from the number of carbon atoms in the nucleus; this is then prefixed by the term 'spiro', and followed by numbers placed in square brackets which indicate the number of carbon atoms joined to

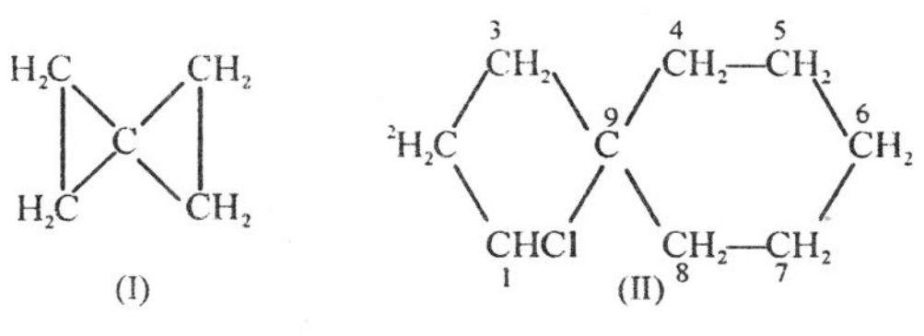

(I) (II)

Fig. 1

the 'junction' carbon atom. The positions of substituents are indicated by numbers, the numbering beginning with the smaller ring and ending on the junction carbon atom; *e.g.*, (I) is spiro-[2,2]-pentane, (II) is 1-chlorospiro-[5,3]-nonane.

Examination of these formulae shows that the two rings are perpendicular to each other, and hence suitable substitution will produce molecules with no elements of symmetry, thereby giving rise to optically active forms, *e.g.*, Mills and Nodder (1920, 1921) resolved the dilactone of benzophenone 2,2′,4,4′-tetracarboxylic acid (III). In this molecule the two shaded portions are perpendicular to each other, and consequently there are no elements of symmetry. When this compound is treated with sodium hydroxide, the lactone rings are opened to form (IV) and the optical rotation disappears.

Boescken *et al.* (1928) condensed penta-erythritol with pyruvic acid and obtained the spirocompound (V), which they resolved. Some other spiro-compounds that have been resolved are the spiro-heptane (VI) (Backer *et al.*, 1928, 1929), the spiro-hydantoin (VII) (Pope and Whitworth, 1931), and the spiroheptane (VIII) (Jansen and Pope, 1932).

$$2\ CH_3COCO_2H + C(CH_2OH)_4 \rightarrow$$

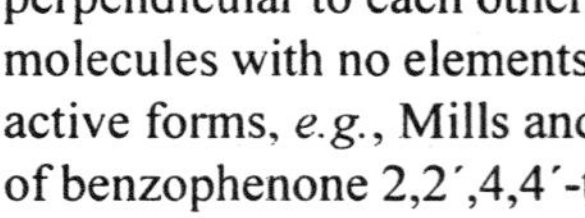

(VI) 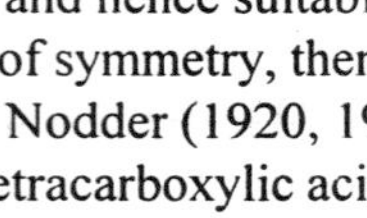 (VII) 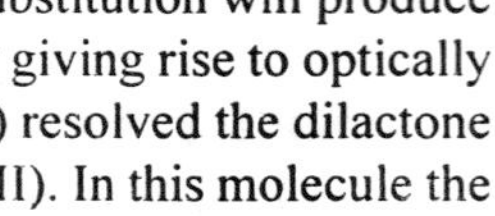 (VIII)

In all the cases so far discussed, the optical activity of the spiran is due to the asymmetry of the molecule as a whole; there is only one pair of enantiomers. If a spiro-compound also contains asymmetric carbon atoms, then the number of optically active forms is increased (above two), the actual number depending on the compound in question, *e.g.*, Sutter and Wijkman (1935) prepared the spiro-compound (IX), which contains two similar asymmetric carbon atoms (o).

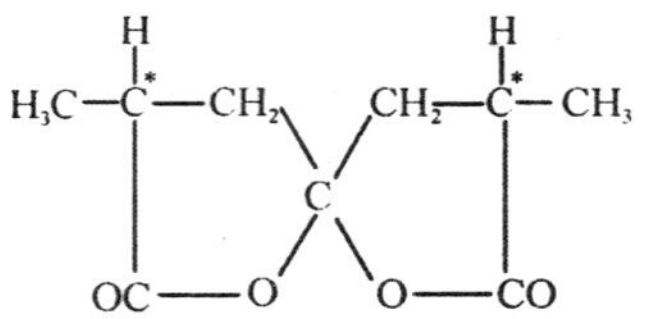

(IX) 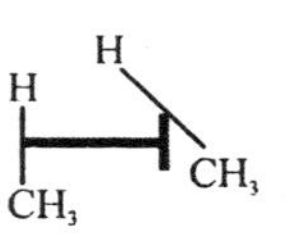 (X) 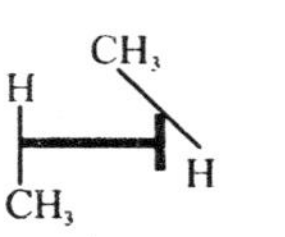 (XI) 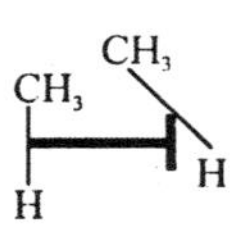 (XII)

If we imagine the left-hand ring of (IX) to be horizontal, then the righthand ring will be vertical; and if we represent them by bold horizontal and vertical lines, respectively, then there are three different geometrical isomers possible, (X), (XI) and (XII) (this can be readily demonstrated by means of models). Each of these geometrical isomers has no elements of symmetry, and so each can exist as a pair of enantiomers. Three racemic modifications were actually isolated by Sutter and Wijkman, but were not resolved.

Cram *et al.* (1954) have also prepared the following three spiro-[4,4]-nonanediols (as racemates): Various spiro-compounds have been prepared in which the spiro-atom is nitrogen, phosphorus or arsenic.

A spiran compound, acorone, has now been found in nature.

The method of specifying the absolute configuration of spirans is similar to that for allenes *e.g.*,

Spirilloxanthin (Rhodoviolascin), m.p. 218°C

This has been isolated from various sources. Karrer *et al.* (1935-1940) showed that its molecular formula was $C_{42}H_6O_{92}$ and that it contained two methoxyl groups (Zeisel method) and thirteen conjugated double bonds (hydrogenation; u.v. spectrum). Oxidation with permanganate gave bixindial and a higher dialdehyde. These workers then proposed a structure for spirilloxanthin, but were not certain of the positions of the two methoxyl groups. Weedon *et al.* (1959,1966) re-investigated this problem and assigned structure (I). They examined the NMR spectrum in the s 9-5-6-5 region. The absence of signals τ 8.35 and 8.41 indicated the absence of isopropylidene end-groups. The signal at τ 6.78 indicated two methyl groups (integration) and the signal at τ 8.02 indicated six in-chain methyls. A signal at τ 7.70 (d, J 6.8 Hz) was assigned to two methylene groups, each being adjacent to one vicinal proton, *i.e.*, $-CH_2-CH=$. The usual τ-value for such a methylene group is 8.3-8.7; the shift from this position (to 7.70) was attributed largely to the presence of an adjacent carbon atom carrying an oxygen atom (OMe; deshielding). Finally, the signal at τ 8.83, because it was exceptionally sharp, was assigned to methyl groups (four; integration) attached to saturated carbon atoms. However, since a methyl group of this type normally absorbs at τ 9.10-9.15, the downfield shift (deshielding) was attributed to the proximity of the oxygen atom (in OMe). These data led to the proposal of the structure given for spirilloxanthin (I). This structure has also been inferred from chemical evidence (Jensen, 1959), and has been confirmed by synthesis (Surmatis *et al.*, 1963). Methylheptenone was converted into the phosphorane (II) and this was condensed with crocetindial (III).

Crocetindial (III) has been prepared as follows (Isler *et al.*, 1959).

Capsorubin, m.p. 218°C

This occurs in red peppers. The early work showed that capsorubin contained four oxygen atoms, two of which were present in hydroxyl groups. The results of microhydrogenation and the similarity of the visible spectrum of capsorubin with those of β-carotene and bixindial suggested that the two remaining oxygen atoms were ketonic. This led to the proposal that capsorubin had the partial structure (I). This was supported by the fact that on heating with aqueous alkali, capsorubin gave crocetindial (II) (this is the reversal of the aldol condensation).

Further support was afforded by the observation that the infrared spectrum of capsorubin showed a band which was also present in the spectra of authentic compounds of the type (I). Also, the reduction of capsorubin with potassium borohydride produced shifts in the u.v. spectrum consistent with selective reduction of two terminal carbonyl groups to give a nonaene chromophore (Weedon *et al.*, 1958).

Thus, the problem now was the elucidation of the structures of the terminal groups (R) in capsorubin. It was originally believed that the endgroups (R) were acyclic, but Cholnoky *et al.* (1957) showed that the molecular formula of capsorubin was $C_{40}H_{50}O_4$ (and not $C_{40}H_{60}O_4$, as previously believed). This revised formula indicates that capsorubin contains two rings: D.B.E. = 41 – 28 = 13. Since capsorubin contains nine olefinic double bonds and two carbonyl double bonds, this means that two rings are present. Weedon *et al.* (1960) now examined the NMR spectrum of capsorubin. Included in the spectrum were three sharp singlets at τ 9.16, 8.80, and 8.63. These were attributed to methyl groups on saturated quaternary carbon atoms in the end-groups (R). Each signal had an intensity corresponding to two methyl groups. This suggests that the end-groups in capsorubin are identical. Weedon *et al.* (1961) then found that the Oppenauer oxidation (aluminium t-butoxide and acetone) of capsorubin gave a ictraketone, capsorubone. The infrared spectrum of this compound showed a band at 1 739 cm^{-1}, which was attributed to the carbonyl group in cyclopentanone. Hence, capsorubin contains two hydroxycyclopentane end-groups.

The NMR spectrum of capsorubone was examined in the τ 9.5-6.4 region. Four well-defined signals were shown at τ 9.02, 8.77, 8.65, and 8.02 in approximate ratios 1: 1: 1 : 2, and were attributed to methyl groups. The signal at τ 8.02 is characteristic of in-chain methyls. Since capsorubin contains four of these methyls (see (I) and (II), above), the other signals must each represent two equivalent methyls. This is understandable if the two end-groups are identical (see also above), each containing three methyls. Since the signals were sharp, all the methyls must be attached to saturated carbon atoms.

So far, it has been established that the end-groups of capsorubin are identical ; that they are cyclopentyl groups containing three methyl substituents, each on a saturated quaternary carbon atom; and that they each contain a secondary hydroxyl group (Oppenauer oxidation). It therefore follows that two methyls must constitute a gem-dimethyl group in the cyclopentane ring. Compound (III) was synthesised and examination of its NMR spectrum showed that the τ values of the methyls were upfield (0.12-0.16) with respect to those in capsorubone.

This is due to deshielding by the presence of the carbonyl group in capsorubone. Hence, the hydroxyl group in capsorubin was tentatively placed at position 3 or 4. Synthetic work then established that the hydroxyl group was at position 4. Complete synthesis of (±)-capsorubin confirmed this structure and also established that the hydroxyl group at position 4 and the keto group at position 6 were trans (as previously suggested from infrared and NMR data). The structure (and absolute stereochemistry) of capsorubin is (IV) [Weedon *et al.*, 1962].

The relative configurations of the 4-OH and 1-Me were established independently as follows. The (±)-form of (V) was synthesised and, on reduction, gave (VI)-(±)-trans and (VII)-(±)-cis. These were separated and differentiated by the fact that only (VII) formed a lactone. Each of these ((VI) and (VII); $CO_2H \rightarrow COMe$) was condensed with crocetindial, and it was shown that it was (VII) that gave a product identical with natural capsorubin. Also, the absolute configuration of C-1 has been established by correlation with that of C-1 of (+)-camphor.

Capsanthin, m.p. 175-176°C

This occurs in red peppers together with capsorubin but in far greater amount. The structure of capsanthin was actually elucidated alongside that of capsorubin (Weedon *et al.*, 1958, 1962). The early work showed that when heated with aqueous alkali capsanthin gave β-citraurin (I), a known compound. It was also found, from ultraviolet spectroscopic studies, that capsanthin contained a decaenone chromophore, and Cholnoky *et al.* (1957) showed that the molecular formula of capsanthin was

$C_{40}H_{56}O_3$. Thus, capsanthin contains two rings (cf. cap sorubin, above). The Oppenauer oxidation of capsanthin gave capsanthone, a hydroxydiketone, the infrared spectrum of which indicated the presence of a cyclopentanone ring and a hydroxyl group (Weedon *et al.*, 1960). The NMR spectrum of capsanthone was examined and some of the signals observed were the same as those for capsorubin, but only half the intensity of the latter. Thus, both compounds have a common end-group. This, as we have seen, is the 4-hydroxy-l,2,2-trimethylcyclopentyl residue. The NMR spectrum of capsanthin showed methyl signals at τ 9.02, 8.95, 8.77, 8.65, 8.3, and 8.02 in the approximate ratios 1:2:1:1:1:4 (the italicised values are the same as those shown by capsorubin). However, the signals at τ 8.95 and 8.3 are characteristic of methyls in cyclohexene rings in, *e.g.*, zeaxanthin). This is in agreement with the formation of (I) from capsanthin. Hence, from the data given, it follows that the structure of capsanthin is (II). A point to note is the failure of the Oppenauer oxidation to convert the secondary hydroxyl group in the cyclohexene end-group into the ketonic group.

Fucoxanthin, m.p. 166-167°C

The characteristic carotenoid of brown algae and diatoms; it appears to be the most abundant xanthophyll in nature. Weedon *et al.* (1964), from mass spectrometry, established the molecular formula as $C_{42}H_{58}O_6$, and the presence of two hydroxyl groups and one acetoxy-group (these groups had also been shown to be present by Jensen, 1961). Since fucoxanthin formed only a monoacetate, this suggested that one hydroxyl group was tertiary.

On reduction with lithium aluminium hydride, fucoxanthin gave a mixture of 'fucoxanthols', the u.v. spectrum of which indicated the presence of a conjugated octaene. Potassium permanganate oxidation of fucoxanthin gave dimethylmalonic acid and 2,2-dimethyl-succinic acid. Thus, fucoxanthin contains the grouping (I) and this was confirmed by examination of the i.r. spectrum, which also indicated the presence of an allenic grouping.

Partial oxidation of fucoxanthin with zinc permanganate gave four aldehydes. Structure (II) was assigned to one of these (based on molecular formula and u.v. spectrum); (III) was a known compound. The other two aldehydes had molecular formulae $C_{25}H_{34}O_4$ (τ 0.55) and $C_{27}H_{36}O_4$ (τ 0.45; d, J 7.5 Hz). These formulae were obtained by mass spectrometry.

The C_{25}- and C_{27}- aldehydes were shown to be a pentaenealdehyde and a hexaenealdehyde, respectively, by examination of their u.v. spectra before and after lithium aluminium hydride reduction (cf. capsorubin).

In addition to these aldehydes, there was also a mixture of 'allenic aldehydes', which were classified as such because their i.r. spectrum exhibited bands in the region normally associated with allenes (see also above). Further oxidation of the C_{25}- and C_{27}-aldehydes gave (II) and (III), respectively. On the other hand, complete oxidation of the C_{27}-aldehyde and the mixture of 'allenic aldehydes' gave 2,2-dimethylsuccinic acid; the former also gave dimethylmalonic acid.

The NMR spectra of fucoxanthin and the C_{27}-aldehyde showed a number of common signals, *e.g.*, τ 6.32 and 7.40; J 18 Hz, which were attributed to a methylene group adjacent to an oxo group. It was also found that the C_{27}- aldehyde formed a monoacetate, and condensed with the geranyl Wittig reagent to give a conjugated octaenone (as shown by the u.v. spectra before and after hydride reduction). On the basis of this work, Weedon *et al.* proposed (IV) as the structure of the C_{27}- aldehyde (and the C_{25}- aldehyde as its lower vinylogue). They also tentatively formulated fucoxanthin as (V). This was confirmed by later work of Weedon *et al.* (1966). The mixture of 'allenic aldehydes' was separated (chromatography) into (VI), (VII), and (VIII). All of these exhibited spectral properties (u.v., visible, i.r., NMR) in keeping with the structures assigned.

CAROTENOID ACIDS (APOCAROTENOIDS)

These are compounds which do not contain 40 carbon atoms.

Bixin, $C_{25}H_{30}O_4$

Natural bixin is a brown solid, m.p. 198°C, and is the cis-trans-form; it is readily converted into the more stable all-trans-form, m.p. 216-217°C, by iodine in benzene solution.

When boiled with potassium hydroxide solution, bixin produces one molecule of methanol and a dipotassium salt which, on acidification, gives the dibasic acid norbixin, $C_{24}H_{28}O_4$. Thus bixin is a monomethyl ester, and can be esterified to give methylbixin.

On catalytic hydrogenation, bixin is converted into perhydrobixin, $C_{25}H_{48}O_4$; thus there are 9 double bonds present in the molecule

(Liebermann *et al.*, 1915). Perhydrobixin, on hydrolysis, forms perhydronorbixin. Oxidation of bixin with permanganate produces four molecules of acetic acid (Kuhn *et al.*, 1929); hence there are four $-C(CH_3)=$ groups in the chain. Furthermore, since the parent hydrocarbon of perhydronorbixin, $C_{24}H_{46}O_4$, is $C_{22}H_{46}$ (the two carboxyl groups are regarded as substituents), the molecule is acyclic.

The thermal decomposition of bixin produces toluene, m-xylene, m-toluic acid and the methyl ester of this acid (Kuhn *et al.*, 1932). Hence the following assumptions may be made regarding the nature of the chain (cf. β-carotene).

It has been known for some time, based on infrared, ultraviolet, and visible absorption spectroscopy, that bixin had a cis-configuration about one of its double bonds. Also, the ultraviolet data excluded the possibility that the cis-isomer was one of the hindered type. Hence, the cis-isomerism cannot be at the 6,7 or the 14,15 double bond (see structure of bixin, above). Furthermore, the central double bond (10,11) was eliminated by synthesis. One suggestion was that the 2,3-double bond was involved, and the other suggestion was that it was the 4,5. Weedon *et al.* (1960) have settled the problem by the examination of the NMR spectra of natural and all-trans methylbixin. Some of the relevant data are given here. In the all-trans isomer the ends are identical and so only one signal is given by H-2 and H-2′, and one signal by H-3 and H-3′ (note that the a-proton absorbs upfield with respect to the β-proton in α, β-unsaturated oxo compounds). Since four signals were observed for the end protons in natural methylbixin, these ends must be different. However, the J values $J_{2,3}$ and $J_{2',3'}$ were the same (15.8 Hz).

This indicates that the configurations about the end double bonds (2,3 and 2′,3′) are unaffected (trans). As pointed out above, 6,7- (14,15-) and 10,11-double bonds have been excluded. This leaves the 4,5- and the 8,9-double bond as possibilities. If cis-isomerism occurred at the 8,9-double bond, it is then difficult to explain the large change in τ value for H-3. It therefore follows that the difference between the natural and all-trans methylbixin is at the 4,5-double bond. However, since bixin is unsymmetrical at its ends, the isomerism could therefore be at the 16,17-double bond (instead of 4,5). This uncertainty was settled by the examination of the NMR spectrum of cis-apo-l-norbixinal methyl ester, obtained by oxidation of natural bixin. The spectrum showed doublets at τ 4.08 and τ 2.04. Hence, the isomerism occurs at the 4,5-double bond at the ester end in bixin, *i.e.*, bixin is

Crocetin, $C_{20}H_{24}O_4$

This occurs in saffron as the digentiobioside, crocin. The structure of crocetin was elucidated by Karrer *et al.* (1928) and Kuhn *et al.* (1931). Crocetin behaves as a dicarboxylic acid and has seven double bonds (as shown by catalytic hydrogenation to perhydrocrocetin, $C_{20}H_{38}O_4$). On oxidation with chromic acid, crocetin gives 3-4 molecules of acetic acid per molecule of crocetin; thus there are 3-4 methyl side-chains. The structure of crocetin was finally shown by the degradation of perhydronorbixin, $C_{24}H_{46}O_4$, by means of the following method

$$RCH_2CO_2H \xrightarrow{Br_2-P} RCHBrCO_2H \xrightarrow{\text{hydrolysis}} RCHOHCO_2H \xrightarrow{CH_2N_2}$$

$$RCHOHCO_2CH_2 \xrightarrow{CH_3MgI} RCHOHC(OH)(CH_3)_2 \xrightarrow{(CH_3CO_2)_4Pb}$$

$$RCHO \xrightarrow{[O]} RCO_2H$$

This set of reactions was performed twice on perhydronorbixin, thereby resulting in the loss of four carbon atoms (two from each end); the product so obtained was perhydrocrocetin, $C_{20}H_{38}O_4$. On these results, crocetin is therefore:

This structure is supported by the fact that the removal of two carbon atoms from perhydrocrocetin by the above technique (one carbon atom is lost from each end) resulted in the formation of a diketone. The formation of this compound shows the presence of an a-methyl group at each end of the molecule. The structure of crocetin is further supported by the synthesis of perhydrocrocetin, and by the synthesis of crocetin diesters by Isler *et al.* (1957).

It appears, however, that part of the crocetin in crocin is present in the form of the 2,3-cis isomer. The evidence for this has been obtained from a study of the NMR spectrum of the dimethyl ester (Weedon *et al.*, 1960). This cis-form is readily converted into the all-trans form by iodine. The trans-crocetin dimethyl ester has also been synthesised by the Wittig reaction between the dialdehyde (I) and two molecules of the phosphorane (Buchta *et al.*, 1959, 1960) [structure (II), for the preparation of (I)].

BIOSYNTHESIS OF CAROTENOIDS

Biosynthetic studies of the carotenoids have been carried out, and the pathways are those for the terpenoids. Thus Braithwaite *et al.* (1957) and Grob (1957) have shown that labelled mevalonic acid is incorporated into β-carotene. Scheuer *et al.* (1959) have also shown that this acid is incorporated into lycopene.

Furthermore, Modi *et al.* (1961) have isolated mevalonic acid from carrots and Goodwin *et al.* (1967) have shown that labelled mevalonic acid is incorporated into phytoene.

The actual sequence is: isopentyl pyrophosphate (IPP; 5 carbon atoms) $\rightarrow$ geranyl pyrophosphate (10C) $\xrightarrow{\text{IPP}}$ farnesyl pyrophosphate (15C) $\xrightarrow{\text{IPP}}$ geranylgeranyl pyrophosphate (20C) $\xrightarrow{\text{2 mols}}$ phytoene (40C) $\rightarrow$ carotenoids.

Various geometrical isomers of phytoene have been isolated from natural sources, *e.g.*, the main isomer from carrot oil has the trans configuration about all unconjugated non-terminal double bonds, and a cis configuration about the central (15,15′) double bond.

On the other hand, the main isomer from diphenylamine-inhibited cultures of Flavobacterium dehydrogenases is the all-trans isomer (Weedon *et al.*, 1966-1972).

There is now a great deal of evidence to show that the route to lycopene is (the 'extra' double bonds are shown in parentheses):

$$\text{phytoene} \xrightarrow{-2H} \text{phytofluene}(11,12) \xrightarrow{-2H} \zeta-\text{carotene}(11,12;\ 11',12') \xrightarrow{-2H}$$
$$\text{neurosporene}(7,8;\ 11,12;\ 11',12') \xrightarrow{-2H} \text{lycopene}(7,8; 11,12; 11',12'; 7',8')$$

This is known as the Porter-Lincoln pathway (1950). All of these compounds occur naturally. In the laboratory it is easy to dehydrogenate phytoene by means of N-bromosuccinimide, and chromatographic separation of the products has shown the presence of the series of carotenes given above (phytoene $\rightarrow$ lycopene).

On the other hand, the mechanism of biological dehydrogenation is still uncertain. It appears that at least in some cases (atmospheric) oxygen must be present. A point of interest is that the dehydrogenation occurs successively and alternately from the 'middle' of the chain.

Lycopene can then undergo terminal cyclisation to give the carotenes. Xanthophylls are probably produced by oxygenation of the carotenes.

DIELECTRIC CONSTANT

A very rough generalisation is that ionisation of the solute increases both in amount and speed the higher the dielectric constant of the solvent.

Solvation : This factor appears to be more important than the dielectric constant. Solvation is the interaction between solvent molecules and solute molecules, and is partly accounted for by the attraction of a charge for a dipole.

If the solute has polarity, then solvent molecules will be attracted to the solute molecules. The greater the polarity of the solvent, the greater the attraction and consequently the more closely the solvent molecules will be drawn to the solute molecules. Thus more electrostatic work is done and so more energy is lost by the system, which therefore becomes more stable.

Hence, increasing the dielectric constant of the solvent increases the ionising potentiality of the solute molecules, and the higher the polarity of the solvent the more stable becomes the system due to increased solvation. Solvation, however, may also be partly due to certain chemical properties, *e.g.*, sulphur dioxide has an electrophilic centre (the sulphur atom carries a positive charge); hydroxylic solvents can form hydrogen bonds.

Some common solvents and their dielectric constants and dipole moments (in Debye units) are water (81.1; 2-3); formic acid (48.0; 1.5); nitrobenzene (35.7; 4.0); ethanol (25.8; 1.7); acetone (21.3; 3.0); acetic acid (7.1; 1-1.5); chloroform (4.6; 1.1); ether (4.3; 1.25); benzene (2.3; 0); carbon tetrachloride (2.2; 0).

There is also another problem that may arise. This is that although the solute molecules have ionised, the oppositely charged ions behave as a single unit, the pair being held together by electrostatic attraction. Such a complex is known as an ion-pair, and their recombination is known as internal return.

It has now been shown that the majority of reactions involving carbonium-ion intermediates proceed via ion-pairs rather than dissociated ions. According to some authors there are two types of ion-pairs:

(i) **Intimate or Internal Ion-pairs:** These are enclosed in a solvent cage and the ions of the pair are not separated by solvent molecules.

(ii) **Loose or External Ion-pairs:** The ions of these pairs are separated by solvent molecules but still behave as a pair. External ion-pairs may also give rise to ion-pair return (external return), but they are more susceptible to attack by other reagents than are intimate ion-pairs.

Thus, when ionisation takes place, the following steps are possible:

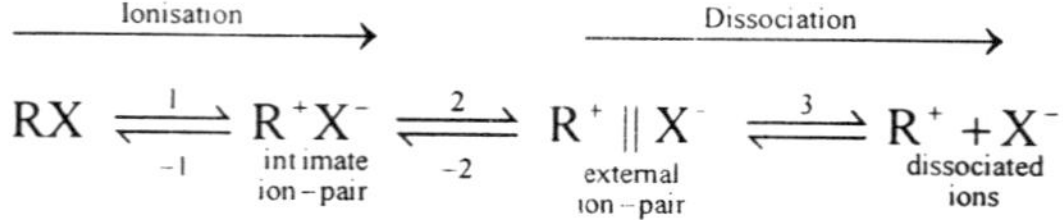

Fig. 2

N.B. : (i) –1 is internal return; (ii) –2 is external return; (iii) only equilibrium 3 is sensitive to a common ion effect; this is because an ion-pair behaves as a single particle, as has been shown by the effect on the depression of the freezing point (i = 1); (iv) the formation of ion-pairs is favoured in solvents of low polarity.

This is due to the fact that highly polar solvents, although they assist ionisation, *i.e.*, encourage the S_N1 mechanism, also cause dissociation by solvation of the ions. There is a great deal of evidence to support the formation of ion-pairs as intermediates, *e.g.*, the special salt effect. In the acetolysis of some alkyl tosylates, the rate of acetolysis is increased sharply on the addition of small amounts of lithium bromide or lithium perchlorate. On further addition of the salt, the rate drops to the normal essentially linear acceleration (which is caused by the normal salt effect). The explanation offered for the special salt effect is as follows.

$$\underset{(I)}{ROTs} \rightleftharpoons \underset{(II)}{R^+OTs^-} \rightleftharpoons \underset{(III)}{R^+O \| Ts^-} \rightleftharpoons \underset{(IV)}{R^+ + OTs^-}$$

$$(III) \xrightarrow{LiBr} \underset{(V)}{R^+ \| Br^-} \qquad (IV) \xrightarrow{LiBr} RBr$$

The solvent-separated ion-pair (III) may collapse to regenerate (II) and (I) or dissociate to give (IV). In the presence of, *e.g.*, lithium bromide, (III) but not (II) is trapped' to form (V) and so is prevented from reforming (II); (IV) reacts in the normal way to give the products. Because of the removal of (III) to form (IV) and thence the products, the return of (III) to (11) and (I) is prevented, thereby rapidly increasing the rate of formation of the product.

A point to note about intimate ion-pairs is that their geometry is very similar to that of the transition states from which they are derived. Thus, in an intimate ion-pair, the negative ion is very close to the carbonium-ion face from which it departed.

In general, S_N1 reactions will be written as direct dissociation into the two ions unless ion-pairs must be used to explain, in detail, the course of the reaction.

A number of equations have been proposed correlating rates and the nature of the solvent, but none is completely general. Hughes and Ingold (1935, 1948) proposed the following qualitative theory of solvent effects: (i) Ions and polar molecules, when dissolved in polar solvents, tend to become solvated. (ii) For a given solvent, solvation tends to increase with increasing magnitude of charge on the solute molecules or ions. (iii) For a given solute, solvation tends to increase with the increasing dipole moment of the solvent. (iv) For a given magnitude of charge, solvation decreases as the charge is spread over a larger volume. (v) The decrease in solvation due to the dispersal of charge will be less than that due to its destruction.

Since the rate-determining step in the S_N1 mechanism is ionisation, any factor assisting this ionisation will therefore facilitate S_N1 reactions. Solvents with high dipole moments are usually good ionising media and, in general, it has been found that the more polar the solvent the greater is the rate of S_N1 reactions. We have however, also to consider the problem of solvation.

$$R-X \xrightarrow{\text{slow}} \overset{\delta+}{R}---\overset{\delta-}{X} \xrightarrow{\text{fast}} R^+ + X^- \xrightarrow[\text{slow}]{OH^-} ROH$$

Increasing the polarity of the solvent will greatly increase the reaction rate, and since the transition state has a larger charge than the initial reactant molecule, the former is more solvated than the latter (rule ii). Thus the transition state is more stabilised than the reactant molecule, and solvation therefore lowers the energy of activation and so the reaction is assisted.

The rates of S_N2 reactions are also affected by the polarity of the solvent.

$$HO^- + R-X \xrightarrow{\text{slow}} \overset{\delta-}{HO} \cdots R \cdots \overset{\delta-}{X} \xrightarrow{\text{fast}} HO-R+X^-$$

A solvent with high dipole moment will solvate both the reactant ion and the transition state, but more so the former than the latter, since in the latter the charge, although unchanged in magnitude ($\delta- = -1/2$), is more dispersed than in the former (rule iv). Thus solvation tends to stabilise the reactants more than the transition state, *i.e.*, the activation energy is increased and so the reaction is retarded.

Now let us consider the Menschutkin reaction (1890):

$$R_3\ddot{N}\ R-X \rightarrow R_3\overset{\delta+}{N}\cdots R\cdots\overset{\delta-}{X} \rightarrow R_4N^+X^-$$

The charge on the transition state is greater than that on the reactant molecules; hence the former is more solvated than the latter. Thus the energy of activation is lowered and the rate of reaction thereby increased. Also, the greater the polarity of the solvent, the greater should be the solvation. The foregoing predictions have been observed experimentally.

In the following S_N2 reaction, charges decrease in the transition state,

$$HO^-\ R\overset{+}{N}R_3 \rightarrow \overset{\delta-}{H}O\text{---}R\text{---}\overset{\delta+}{N}R_3 \rightarrow HOR + R_3N$$

and hence increasing the polarity of the solvent will retard the reaction: and retardation will be greater than that in the S_N2 hydrolysis of alkyl halides (see above; only the hydroxide ion is charged in this case).

The polarity of the solvent not only affects rates of reactions, but may also change the mechanism of a reaction, *e.g.*, Olivier (1934) showed that the alkaline hydrolysis of benzyl chloride in 50 percent aqueous acetone proceeds by both the S_N2 and S_N1 mechanisms. In water as solvent, the mechanism was changed to mainly S_N1. The dipole moment of water is greater than that of aqueous acetone, and consequently the ionisation of benzyl chloride is facilitated.

Another example we shall consider is the hydrolysis of the alkyl bromides, MeBr, EtBr, isoPrBr and t-BuBr. As we have seen, Hughes, Ingold *et al.* showed that in aqueous alkaline ethanol the mechanism changed from S_N2 for MeBr and EtBr to both S_N2 and S_N1 for isoPrBr, and to S_N1 for t-BuBr. These results were explained by the + I effects of the R groups, but it also follows that the greater the ionising power of the solvent, the less will be the + I effect of an R group necessary to change the mechanism from S_N2 to S_N1.

Formic acid has been found to be an extremely powerful ionising solvent for alkyl halides, and the relative rates of hydrolysis, at 100°C, for the above series of bromides with the very weak nucleophilic reagent water, dissolved in formic acid, was found to be (Hughes *et al.*, 1937, 1940): MeBr, 1.00; EtBr, 1.71: is oPrBr, 447; t-BuBr, ca. 10^8. This continuous increase in reaction rate shows that the mechanism is mainly S_N1 (the rate increasing with the increasing + I effect of the R group). Thus both MeBr and EtBr are also hydrolysed by the S_N1 mechanism under these favourable conditions of high solvent-ionising power.

Solvents may also affect the proportions of the products in competitive reactions, *i.e.*, the attack on the same substrate by two substituting reagents in the same solution:

$$RY \xleftarrow{Y} RX \xrightarrow{Z} RZ$$

In the S_N2 mechanism there is only one reaction step, and so the overall rate and product ratio will be determined by that stage. In the S_N1 mechanism, however, the rate is determined by the rate of ionisation of RX, and the product ratio is thus determined by the competition of the fast second steps.

It therefore follows that for solvent changes, in the S_N2 mechanism the rate and product ratio will proceed in a parallel fashion, whereas in the S_N1 mechanism the rate and product ratio will be independent of each other.

A simple example that illustrates this problem is the solvolysis of benzhydryl chloride (diphenylmethyl chloride). Hammett *et al.* (1937, 1938) showed that the solvolysis of benzhydryl chloride in initially neutral aqueous ethanol gave benzhydryl ethyl ether and benzhydrol.

Hughes, Ingold *et al.* (1938) showed that if ethanol is first used as solvent and then water is progressively added, the overall rate increases, but there is very little increase in benzhydrol formation; the main effect is an increased rate of formation of benzhydryl ethyl ether. Hence the rate of the reaction and the ratio of the products are determined independently; this is consistent with the S_N1 mechanism but not with the S_N2.

It can be seen from this example that kinetic solvent effects may be used to differentiate between S_N2 and S_N1 mechanisms.

The above discussion of solvent effects has been purely qualitative, but it is also of interest to consider the problem in a quantitative way. This involves the use of the following thermochemical cycle.

Then, if the enthalpy change for each step is represented by the number (of the step) as a subscript, it follows from Hess's law that

$$\Delta H_1 = \Delta H_2 + \Delta H_3 + \Delta H_4 + \Delta H_5 + \Delta H_6 + \Delta H_7$$

(where ΔH_4 is the ionisation energy of R . and ΔH_5 is the electron affinity of X .).

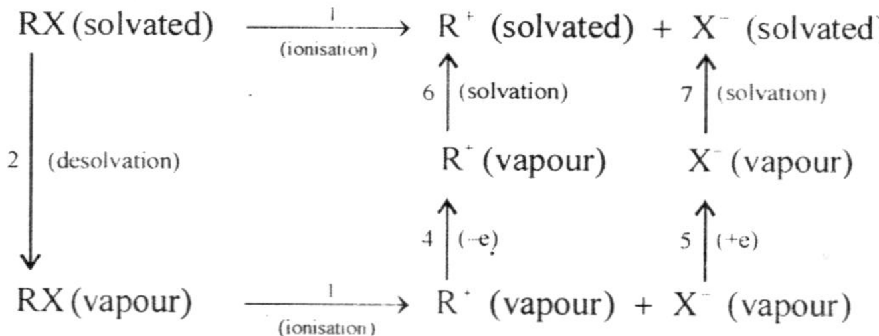

Fig. 3

All of these values except ΔH_2 have been calculated from suitable experimental data, and the value of ΔH_2 has been estimated. The values given in the following discussion are those used by Frazer and Singer (see Reading References).

In the gas phase, if the reaction is homolytic, ΔH for, step 3 in the above cycle is ΔH_3. If we use methyl chloride as our example, then ΔH_3 is +80 kcal. If, however, the reaction is heterolytic,

$$\Delta H = \Delta H_3 + \Delta H_4 + \Delta H_5 = +334.7 + 970.7 - 364.0 = 941.4 \text{ kJ}$$

It can be seen that heterolytic reaction in the gas phase is very unfavourable energetically. If the reaction is carried out in water (as solvent), then

$$\Delta H_1 = +33.7 + 334.7 + 970.7 - 364.0 - 343.1 - 288.7 = +343.1 \text{ kJ}$$

This much more favourable value is due to the large contribution of the heats of solvation (ΔH_6 and ΔH_7).

As we have already seen, in the S_N1 mechanism (with all ions and molecules solvated):

$$RX_{(aq)} \underset{}{\overset{slow}{\rightleftharpoons}} R^+_{(aq)} + X^-_{(aq)} \xrightarrow[fast]{H_2O} ROH_{(aq)} + H^+_{(aq)}$$

the rate order is t-Bu > isoPr > Et > Me. The following Table (Frazer and Singer, 1964) gives the values of AHF (heat of ionisation in kJ mol^{-1}) for various alkyl halides.

Inspection of this table shows that in going in a downward direction for a given halogen atom, ΔH_1 decreases rapidly. Moreover, calculations have shown that ΔH_1 and E (activation energy) are almost identical. Hence the rate of the S_N1 reaction will increase in the downward direction, *i.e.*, t-Bu > isoPr > Et > Me.

It has been mentioned that the 'rates of both S_N1 and S_N2 reactions for a given alkyl group is RI > RBr > RCI. The above table indicates the order (for S_N1) RBr > RCI > RI. This illustrates the point that even though a quantitative approach to a problem is always desirable, unless all the necessary values are known (reasonably) accurately, estimates may lead to some wrong conclusions. This is more the case with reactions in solution than with gas-phase reactions.

It has been mentioned above that a number of equations have been proposed correlating rates of solvolysis and the nature of the solvent. We shall now discuss this problem in a little more detail. One of the outstanding difficulties in this connection is that the structure of liquids is still uncertain. One quantitative correlation was proposed by Grunwald and Winstein (1948, 1951). This is the linear free energy equation:

$$\log (k/ko) = mY$$

The standard reaction chosen for this equation is the solvolysis of t-butyl chloride, and the standard solvent is 80 per cent aqueous ethanol (80 volumes of ethanol and 20 volumes of water). This substrate was chosen for the standard reaction because it has been well established that it undergoes solvolysis by the S_N1 mechanism. Hence, ko is the rate constant for the solvolysis of t-butyl chloride by the S_N1 mechanism in the standard solvent. k is the rate constant for the solvolysis of any particular compound in any given solvent; Y is a parameter that is a measure of the ionising power of the given solvent; m is a parameter that is a measure of the sensitivity of the rate of solvolysis of the particular compound to changes in the ionising power of the solvent. By definition, m = 1.00 for t-butyl chloride, and Y = 0.00 for the standard solvent.

Experimental results showed that the values of m were close to unity for compounds undergoing solvolysis by the SNI mechanism, but were much smaller (between 0.25 and 0.35) for compounds undergoing solvolysis by the S_N2 mechanism. This is in keeping with expectation that S_N1 reactions would be sensitive to the ionising power of the solvent, whereas S_N2 reactions would not. Thus, the determination of m could be used as a means of ascertaining whether a particular compound is undergoing solvolysis by the S_N1 or S_N2 mechanism in a given solvent.

If m and Y were truly characteristic of the nature of the compound and the solvent, respectively, the plot of log k for a particular compound against the values of Y of different solvents would be a straight line with slope m. In practice, many deviations from the equation have been observed. Winstein *et al.* (1956) modified the original equation, and other workers have also proposed various correlations. The difficulty of obtaining a good correlation between rates of solvolysis and some property of solvents is due to the fact that, apart from the problem of the structure of liquids, factors other than those considered in deriving the equations must be determined and consequently included in the equation.

THE WALDEN INVERSION (OPTICAL INVERSION)

By a series of replacement reactions, Walden (1893, 1895) transformed an optically active compound into its enantiomer. In some cases the product is 100 per cent optically pure, *i.e.*, the inversion is quantitative; in other cases the product is a mixture of the (+)- and (–)-forms in unequal amounts, *i.e.*, inversion and retention (racemisation) have taken place. The phenomenon was first discovered by Walden with the following reactions

$$\underset{\substack{(-)\text{-mallic}\\ \text{acid}\\ (I)}}{\begin{matrix} CHOHCO_2H \\ | \\ CH_2CO_2H \end{matrix}} \underset{KOH}{\overset{PCl_5}{\rightleftarrows}} \underset{\substack{(+)\text{-chlorosuccinic}\\ \text{acid}\\ (II)}}{\begin{matrix} CHClCO_2H \\ | \\ CH_2CO_2H \end{matrix}} \xrightarrow{\text{'AgOH'}} \underset{\substack{(+)\text{-mallic}\\ \text{acid}\\ (III)}}{\begin{matrix} CHOHCO_2H \\ | \\ CH_2CO_2H \end{matrix}}$$

In one, and only one, of the two reactions must there be an interchange of position between the two groups, *e.g.*, if the configuration of (I) corresponds with that of (II), the inversion of configuration must have taken place between (II) and (III). The term Walden inversion is applied to any step in a reaction in which inversion of configuration occurs.

As the above experiment stands, there is no way of telling which step is accompanied by inversion. As we have seen, change in sign of rotation does not necessarily mean that inversion of configuration has occurred.

Various methods of correlating configuration have already been described, but here we shall describe the method where bonds attached to the chiral centre are broken during the course of the reactions. This method was established by Kenyon *et al.* (1925), who carried out a series of reactions on optically active hydroxy compounds. Now it has been established that in the esterification of a monocarboxylic acid by an alcohol under ordinary conditions, the reaction proceeds by the acyl-oxygen fission mechanism thus;

$$R^1CO—OH + H—OR^2 \rightarrow R^1COOR^2 + H_2O$$

Kenyon assumed that in all reactions of this type the R^2—O bond remained intact and consequently no inversion of the alcohol is possible. The following chart shows a series of reactions carried out on ethyl (+)-lactate; Ts = tosyl group = p-toluenesulphonyl group, p-$McC_6H_4SO_2$-; the symbol → is used to represent inversion of configuration in that step. (IV) and (VI) have the same relative configurations even though the sign of rotation has changed.

Similarly, (IV) and (V) have the same relative configurations. Reaction of (V) with potassium acetate, however, produces (VII), the enantiomer of (VI).

Therefore inversion must have occurred in the formation of (VII); (V) and (VI) are produced without inversion since in these cases the C—O bond in (IV) is never broken. It should be noted here that if inversion is going to take place at all, the complete group attached to the chiral centre must be removed (in a displacement reaction).

The converse, however, is not true, *i.e.*, removal of a complete group does not invariably result in inversion.

The above series of reactions has been used as a standard, and all closely analogous reactions are assumed to behave in a similar way, *e.g.*, the action of lithium chloride on the tosylate (V) is assumed

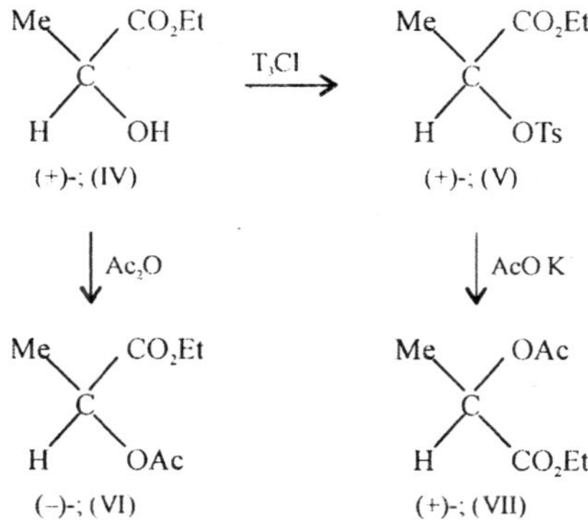

Fig. 4

to be analogous to that of potassium acetate, and the chloride produced thus has an inverted configuration

$$\underset{\text{(V)}}{\text{Me(H)C(CO}_2\text{Et)(OTs)}} \xrightarrow[\Delta]{\text{LiCl}} \text{Me(H)C(Cl)(CO}_2\text{Et)}$$

By similar procedures, Kenyon *et al.* (1929, 1930) showed that (+)-octan-2-ol and (+)-2-chloro-, 2-bromo- and 2-iodo-octane have the same relative configurations; and also that (+)-α-hydroxyethylbenzene (PhCHOHMe), (+)-a-chloro- and (+)-a-bromoethylbenzene have the same relative configurations.

MECHANISM OF THE WALDEN INVERSION

As the result of a large amount of work on the Walden inversion, it has been found that at least three factors play a part in deciding whether inversion or retention (racemisation) will occur: (i) the nature of the reagent; (ii) the nature of the substrate; (iii) the nature of the solvent. Hence it is necessary to explain these factors when dealing with the mechanism of the Walden inversion.

Many theories have been proposed, but we shall discuss only the Hughes-Ingold theory, since this is the one now accepted. According to this theory, aliphatic nucleophilic substitution reactions may take place by either the S_N2 or S_N1 mechanism.

S_N2

$$HO^- + R—X \rightarrow \overset{\delta-}{H}O\text{---}R\text{---}\overset{\delta-}{X} \rightarrow HO—R + X^-$$

Hughes *et al.* (1935) studied (a) the interchange reaction of (+)-2-iodo-octane with radioactive iodine (as NaI*) in acetone solution, and (b) the racemisation of (+)-2-iodo-octane by ordinary sodium iodide under the same conditions. These reactions were shown to take place by the S_N2 mechanism, and the rate of racemisation was shown to be twice the rate of radioactive exchange, *i.e.*, every iodide-iodide* displacement is always accompanied by inversion. (Suppose there are n molecules of optically active iodo-octane. When n/2 molecules have exchange with I* and in doing so have been inverted, racemisation is now complete although the exchange has taken place with only half of the total number of molecules.)

Thus this experiment leads to the assumption that inversion always occurs in the S_N2 mechanism. This is fully supported by other experimental work. *e.g.*, Hughes *et al.* (1936, 1938) studied the reaction of optically active a-bromoethylbenzene and a-bromopropionic acid with radioactive bromide ions, and again found that the rate of racemisation was twice the rate of exchange.

Since S_N2 reactions always occur with inversion (this is known as the stereokinetic rule for S_N2 reactions), then provided the molecularity of the S_N reaction can be determined kinetically, it is possible to correlate the configuration of the reactant with that of the product. There are four S_N2 charge-types of reaction

		Reagent	***Substrate***
1.	$Y^- + RX \rightarrow YR + X^-$	negative	neutral
2.	$Y^- + RX^+ \rightarrow YR + X$	negative	positive
3.	$Y + RX \rightarrow YR^+ + X^-$	neutral	neutral
4.	$Y + RX^+ \rightarrow YR^+ + X$	neutral	positive

The stereokinetic rule for S_N2 reactions is well established for only reactions of type 1. Hughes, Ingold *et al.* (1960) have also shown that the rule applies to type 2, *e.g.*, the reaction between a sulphonium iodide and sodium azide:

$$N_3^- + \text{(Ph)(H)(Me)C}-\overset{+}{S}Me_2 \longrightarrow N_3-C\text{(Ph)(H)(Me)} + Me_2S$$

Another example is the reaction between the acetoxyl ion and the (+)-trimethyl-x-phenylethylammonium ion to give the inverted product (Snyder *et al.*, 1949):

$$AcO^- + \text{(Ph)(H)(Me)C}-\overset{+}{N}Me_3 \longrightarrow AcO-C\text{(Ph)(H)(Me)} + Me_3N$$

Fig. 5

Hughes *et al.* (1964) have also shown that the reaction between thiourea and the dimethyl-lphenylsulphonium ion (as iodide) in methyl cyanide solution gives the thiouronium salt with substantially complete inversion; this is an example of type 4.

$$SC(NH_2)_2 + CHMePhSMe_2^+ \rightarrow CHMePhSC(NH_2)_2^+ + Me_2S$$

Examples of type 3 are also known, *e.g.*, the Menschutkin reaction.

$$R^1_3N + R^2 - Br \rightarrow R^1_3\overset{+}{N} - R^2Br^- \qquad (1)$$

Now let us consider the S_N1 mechanism.

$$R \frown N \rightarrow \overset{\delta+}{R} - \overset{\delta-}{X} \rightarrow R^+ + N^- \xrightarrow{OH^-} ROH + X^-$$

When the reaction proceeds by this mechanism, then inversion and retention (racemisation) will occur, the amount of each depending on various factors. The carbonium ion is flat (trigonal hybridisation), and hence attack by the nucleophilic reagent can take place equally well on either side, *i.e.*, equal amounts of the (+)- and (–)-forms will be produced; this is racemisation. Furthermore, on the basis that an ion-pair is formed first, the leaving group will protect its side from attack by the nucleophile, *i.e.*, inversion will occur exclusively or will predominate. Only if there is complete dissociation of the ion-pair into the individual ions can complete racemisation be expected; the shielding effect of the retiring ion is now lost. Also, racemisation will be encouraged by low concentration of the nucleophile. An example of inversion is that due to Bunton *et al.* (1955), who studied the reaction of ^{18}O-enriched water with optically active s-butanol in aqueous perchloric acid, and found that the overall rate of racemisation is twice that of the oxygen exchange. Thus every oxygen exchange causes complete inversion of configuration (cf. the iodide-iodide* exchange described above). Bunton proposed the following mechanism to explain these results:

$$EtMeCHOH + H^+ \underset{fast}{\overset{HClO_4}{\rightleftharpoons}} EtMeCH\,OH_2^+ \qquad (2)$$

$$EtMeCH\,OH_2^+ \overset{slow}{\rightleftharpoons} EtMe\overset{\delta+}{C}H --- \overset{\delta+}{O}H_2 \qquad (3)$$

$$H_2O^* + EtMe\overset{\delta+}{C}H --- \overset{\delta+}{O}H_2 \overset{fast}{\rightleftharpoons} \overset{\delta+}{H_2O^*} --- EtMeCH --- \overset{\delta+}{O}H_2$$

$$\overset{fast}{\rightleftharpoons} H_2\overset{+}{O}^* - CHMeEt + H_2O \qquad (4)$$

$$H_2\overset{+}{O}^* - CHMeEt \overset{fast}{\rightleftharpoons} HO^* - CHMeEt + H^* \qquad (5)$$

(4) occurs before the OH_2^+ has completely separated in (3), and so this side is shielded and the H_2O^* is forced to attack on the other side as shown; the result is thus inversion. The above reaction proceeds by the S_N1 mechanism since (3) is the rate-determining step (only one molecule is undergoing covalency change in this step). Had the reaction been S_N2, complete inversion would have been obtained. It was shown, however,

that the reaction rate was independent of the concentration of H_2O^*. The mechanism is therefore SN 1, since had it been S_N2, the kinetic expression would require the concentration of the H_2O^*:

$$H_2O^* + EtMeCHOH_2^+ \underset{}{\overset{slow}{\rightleftharpoons}} \overset{\delta+}{H_2O^*}---EtMeCH---\overset{\delta+}{H_2O^*}$$

$$\xrightarrow{fast} H_2\overset{+}{O^*}—CHMeEt + H_2O$$

In general, net inversion is usually observed for short-life carbonium ions. Ions of this type are produced from s- and t-alkyl derivatives. On the other hand, diarylmethyl carbonium ions have a long life (due to spreading of charge). Thus Ar^1Ar^2CHX undergoes solvolysis by the S_N1 mechanism to give a completely racemised product. These observations offer a means of estimating the relative stabilities of carbonium ions.

The stereochemical course of S_N1 reactions may also be affected by neighbouring group participation.

THE S_Ni MECHANISM

Another important S_N reaction is the S_Ni type (substitution, nucleophilic, internal). The reaction between thionyl chloride and alcohols has been studied extensively. A well-examined example is the alcohol a-phenylethanol, PhCHOHMe; this is an arylmethanol, and according to Hughes, Ingold *et al.* (1937) the first step is the formation of a chlorosulphite. No inversion occurs at this stage (which is a four-centre reaction); in the following equation, R = PhMeCH-.

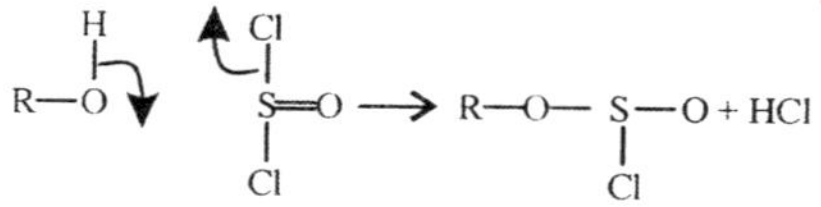

$$R—O(H) + Cl—\underset{Cl}{S}=O \longrightarrow R—O—\underset{Cl}{S}—O + HCl$$

Fig. 6

This chlorosulphite could then form a-chloroethylbenzene by one or more of the following mechanisms

(i) S_N2. This occurs with inversion.

$$R—O—\overset{Cl}{S}=O \xrightarrow{fast} Cl^- + R—O—\overset{+}{S}=O$$

$$\xrightarrow{slow} \overset{\delta-}{Cl}---R---\overset{\delta+}{OSO} \xrightarrow{fast} Cl—R + SO_2$$

(ii) S_N1. This occurs with inversion and retention (racemisation).

$$R-O-\overset{\overset{\displaystyle Cl}{|}}{S}=O \xrightarrow{\text{slow}} R^+ + O^- - \overset{\overset{\displaystyle Cl}{|}}{S}=O \xrightarrow{\text{fast}} RCl + SO_2$$

The second stage may possibly be

$$O^- - \overset{\overset{\displaystyle Cl}{|}}{S}=O \xrightarrow{\text{fast}} SO_2 + Cl^- \xrightarrow[\text{fast}]{R^+} RCl$$

(iii) S_Ni. This occurs with retention (the reaction is effectively a four-centre type).

R—O—S(=O)—Cl → [R···O···S=O, ···Cl] → RCl + SO_2

Fig. 7

In practice, the α-chloroethylbenzene obtained has almost complete retention of configuration, and consequently the mechanism must be S_Ni.

There appears to be some doubt that the S_Ni mechanism is a one-step process. It has been suggested that the S_Ni mechanism; like the S_N1 mechanism can proceed via ion-pairs, the amount of ionisation depending largely on the nature of the solvent, *e.g.*, the rate of decomposition of chlorosulphites increases with increasing polarity of the solvent. Thus, S_N1 and S_Ni mechanisms may be regarded as extremes, the latter operating when the product has retention of configuration. On this basis, the S_Ni may be formulated as follows:

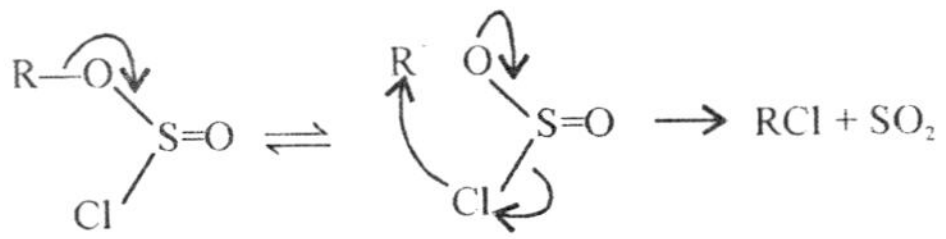

R—O—S(=O)—Cl ⇌ R O S=O Cl → RCl + SO_2

Because of the geometry of the ion-pair, the chlorine atom is forced to attack the carbonium ion from the sane side as the original R—O bond, with consequent retention of configuration.

The formation of the carbonium ion (ion-pair or otherwise) in the S_N1 mechanism is supported by much experimental work, *e.g.*, the chlorosulphite of 3-methylbutan-2-ol, on heating, gives t-amyl chloride

(Lee *et al.*, 1961). This is readily explained by the formation of initially a secondary carbonium ion which then rapidly rearranges to the more stable tertiary carbonium ion.

$$Me_2CH-CHMe-OSOCl \rightarrow [Me-\underset{\substack{|\\Me}}{CH}-\overset{+}{C}H-Me]\overset{-}{O}SOCl \rightarrow$$

$$[Me-\underset{\substack{|\\Me}}{\overset{+}{C}}-CH_2Me]\overset{-}{O}SOCl \rightarrow Me_2CClCH_2Me + SO_2$$

Fig. 8

When α-phenylethanol and thionyl chloride react-in the presence of pyridine, the α-chloroethylbenzene obtained has the inverted configuration (Hughes, Ingold *et al.*, 1937). The explanation offered is that the S_N2 mechanism is operating, the substrate now being a pyridine complex:

$$ROSOCl + C_5H_5N \rightarrow Cl^- + ROSO\overset{+}{N}C_5H_5$$

$$\rightarrow \overset{\delta-}{Cl}---R---\overset{\delta-}{O}SO\overset{+}{N}C_5H_5 \rightarrow Cl-R + SO_2 + C_5H_5N$$

Another example of the S_Ni mechanism (*i.e.*, with retention of configuration) is the decomposition of alkyl chloroformates (Kenyon *et al.*, 1933).

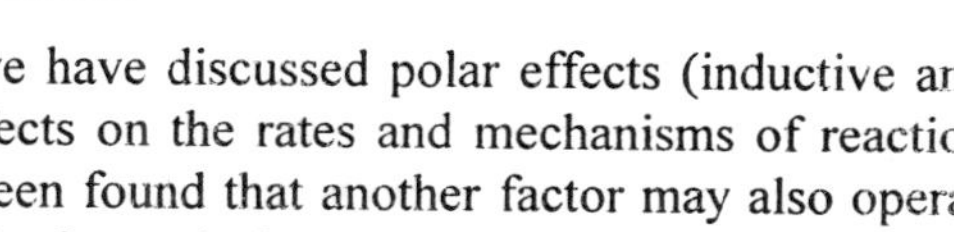

$$R-O-C(=O)-Cl \rightarrow R\ O-S(=O)-Cl \rightarrow RCl + CO_2$$

NEIGHBOURING GROUPS IN NUCLEOPHILIC SUBSTITUTIONS

So far, we have discussed polar effects (inductive and resonance) and steric effects on the rates and mechanisms of reactions. In recent years it has been found that another factor may also operate in various reactions. This factor is known as neighhouring .group participation. Here we have a group attached to the carbon atom adjacent to the carbon atom where nucleophilic substitution occurs and, during the course of the reaction, becomes bonded or partially bonded to the reaction centre to form a non-classical or bridged ion. Thus the rate and/or the stereochemistry of a reaction may be affected. When a reaction is accelerated by neighbouring group participation, that reaction is said to be anchimerically assisted (Winstein *et al.*, 1953). For anchimeric assistance to occur, the neighbouring group, which behaves as a nucleophilic reagent, must be suitably placed stereochemically with respect to the group that is ejected; this is the trans-configuration. Neighbouring group participation is also of great importance in the 1,2-shifts. As we shall see below, neighbouring group participation may also involve a group further removed than the carbon atom adjacent to the reactive centre.

In order to measure anchimeric assistance (kinetic acceleration), it is necessary to be able to estimate reaction rates that would be obtained had there been no neighbouring group participation. This can frequently be done with cyclic structures, since the geometry of the molecule often restricts the possibility of neighbouring group participation to the trans-isomer, the unassisted rate therefore corresponding to that for the cis-isomer. Acyclic structures are much more difficult to study because it is not easy to separate anchimeric assistance from polar and steric effects.

Because of the difficulties in estimating reaction rates in the absence of neighbouring group participation, it is necessary to have a reasonably large increase in rate to justify the conclusion that the rate increase is due to anchimeric assistance only. It appears that a five-fold increase has been accepted as a minimum, but obviously the case is stronger if much larger values are obtained.

In addition to anchimeric assistance as evidence for the formation of non-classical carbonium ions, other criteria have' been used, *e.g.*, stereochemistry of the reaction and structure of the products. In the latter case, the product differs from that which would be expected in the absence of neighbouring group participation.

If we accept the existence of bridged ions, the question to be answered is why should such ions be formed in preference to classical carbonium ions in any particular reaction. One reasonable answer is that when several intermediates are possible, the most stable one is the one likely to be formed. Since charge is more diffuse in the bridged ion than the classical ion, the former would be expected to be more stable than the latter.

Whether a group Z can enter into neighbouring group participation depends on the nature of Z, and for a given Z it usually depends on the size of the ring that can be formed (by n.g.p.). The following sections illustrate some examples of neighbouring group participation.

Neighbouring Carboxylate Anion

Hughes, Ingold *et al.* (1937) studied the following reaction of methyl D-a-bromopropionate

$$McCHBrCO_2Me \rightarrow McCH(OMe)CO_2Me$$

With concentrated methanolic sodium methoxide, the reaction was shown to be S_N2, and the product was L-methoxy ester (100 per cent inversion). Under these conditions, the nucleophilic reagent is the methoxide ion, and the reaction is first order with respect to both methoxide ion and ester. When the ester was subjected to methanolysis, *i.e.*, methanol was the solvent (no methoxide ion now present), the product was again L-methoxy ester (100 per cent inversion). The reaction was now first order (*i.e.*, pseudo first order), but still S_N2, the nucleophilic reagent being the solvent molecules of methanol. When the sodium salt of D-α-bromopropionic acid was hydrolysed in dilute sodium hydroxide solution, the mechanism was shown to be S_N1, and the product was now D-α-hydroxypropionate anion (100 per cent retention). In concentrated sodium hydroxide solution, however, the mechanism was S_N2 (due to the high concentration of the hydroxide ion), and the product was L-α-hydroxypropionate anion (100 per cent inversion).

The explanation for retention is uncertain, but a favoured theory is that an a-lactone is formed first (Kenyon *et al.*, 1936), with expulsion of the bromide ion. Thus, the side opposite to that of the expelled bromide ion is protected from attack by the hydroxide ion, which is consequently forced to attack from the same side as that of the expelled bromide ion, thereby leading to retention of configuration.

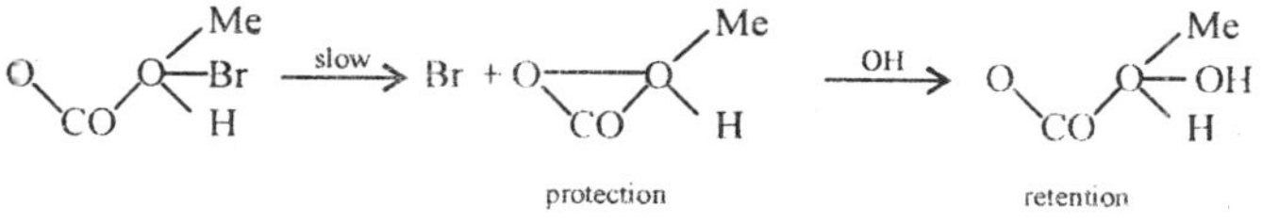

Fig. 9

Hughes, Ingold *et al.* (1950) showed that the deamination of optically active alanine by nitrous acid gave an optically active lactic acid with retention of configuration. This is also explained by neighbouring group participation of the a-carboxylate anion:

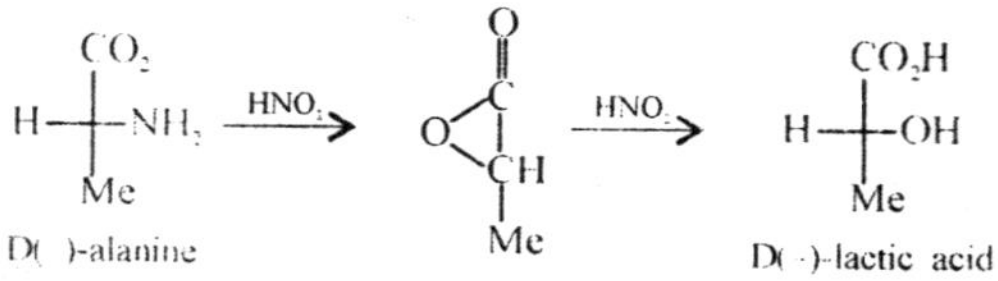

$$\text{H}-\overset{\text{CO}_2^-}{\underset{\text{Me}}{\text{C}}}-\text{NH}_2 \xrightarrow{\text{HNO}_2} \text{(α-lactone: O–C(=O)–CH(Me))} \xrightarrow{\text{HNO}_2} \text{H}-\overset{\text{CO}_2\text{H}}{\underset{\text{Me}}{\text{C}}}-\text{OH}$$

D( )-alanine D( )-lactic acid

A good example of the effect of ring-size in neighbouring group participation involving the carboxylate ion is shown by the anions of the bromocarboxylic acids:

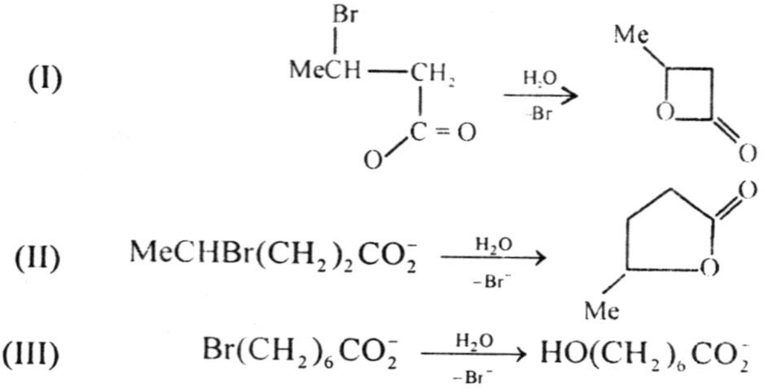

(I) $MeCH(Br)-CH_2-C(=O)O^- \xrightarrow[-Br^-]{H_2O}$ (β-lactone with Me)

(II) $MeCHBr(CH_2)_2CO_2^- \xrightarrow[-Br^-]{H_2O}$ (γ-lactone with Me)

(III) $Br(CH_2)_6CO_2^- \xrightarrow[-Br^-]{H_2O} HO(CH_2)_6CO_2^-$

Fig. 10

Lactones containing a four- to seven-membered ring can be formed under these experimental conditions, but not eight-membered rings (and larger ones). Whether a three-membered α-lactone ring is actually formed for α-bromopropionic acid is still uncertain, but such a lactone has been isolated for 2-butyl-2-bromohexanoic acid. In this case, presumably steric effects of the large butyl groups prevent attack of the hydroxide ion. At the same time, these large groups will tend to stabilise the small ring (through 'squeezing' the angle O–$CHBu_2$–CO).

Neighbouring Halogen Atoms

Brominium (bromonium) ions were first proposed by Roberts and Kimball (1937) as intermediates in the addition of bromine to alkenes. The existence of this cyclic brominium ion has been demonstrated by Winstein and Lucas (1939), who found that the action of fuming hydrobromic acid on (–)-threo-3-bromobutan-2-ol gave (±)-2,3-dibromobutane.

If no neighbouring group participation of bromine occurred in the above reactions, then if the reaction were S_N2; complete inversion would have occurred only at C_1.

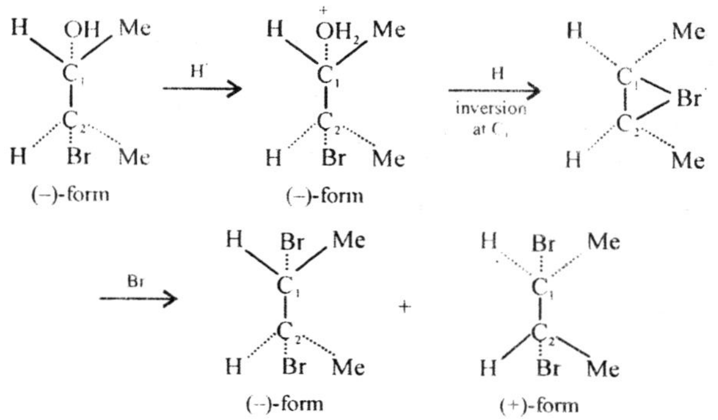

H OH Me / C_1 / C_2 / H Br Me (–)-form $\xrightarrow{H^+}$ H $\overset{+}{OH_2}$ Me / C_1 / C_2 / H Br Me (–)-form $\xrightarrow[\text{inversion at } C_1]{H}$ H Me / C_1 / Br^+ / C_2 / H Me

$\xrightarrow{Br^-}$ H Br Me / C_1 / C_2 / H Br Me (–)-form + H Br Me / C_1 / C_2 / H Br Me (+)-form

Fig. 11

If the reaction were the ordinary S_N1, the C_1 would have been a classical carbonium ion (flat), and so inversion and retention (racemisation) would have occurred only at C_1. Since either retention or inversion occurs at both C_1 and C_2, the results are explained by neighbouring group participation of the bromine atom.

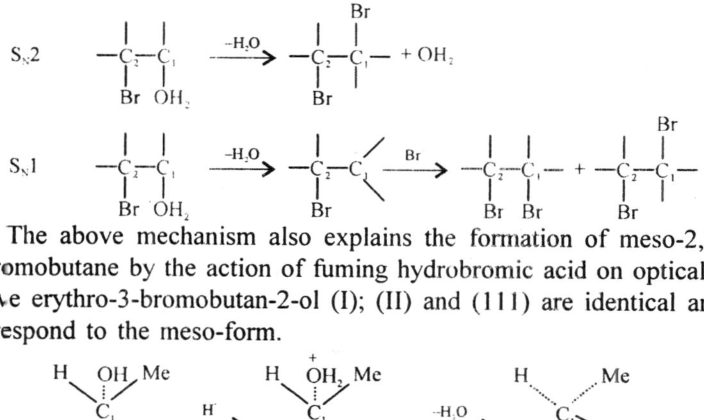

S_N2 $-C_2-C_1$ (Br, OH_2) $\xrightarrow{-H_2O}$ $-C_2-C_1-$ (Br, Br) + OH_2

S_N1 $-C_2-C_1$ (Br, OH_2) $\xrightarrow{-H_2O}$ $-C_2-C_1^+$ (Br) $\xrightarrow{Br^-}$ $-C_2-C_1-$ (Br, Br) + $-C_2-C_1-$ (Br, Br)

The above mechanism also explains the formation of meso-2,3-dibromobutane by the action of fuming hydrobromic acid on optically active erythro-3-bromobutan-2-ol (I); (II) and (111) are identical and correspond to the meso-form.

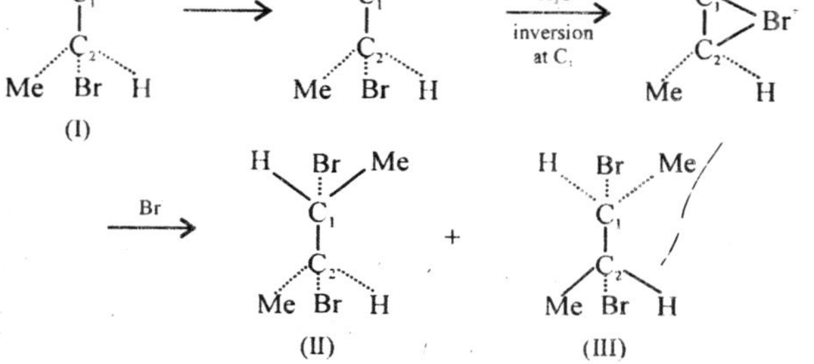

H OH Me / C_1 / C_2 / Me Br H (I) $\xrightarrow{H^+}$ H $\overset{+}{OH_2}$ Me / C_1 / C_2 / Me Br H $\xrightarrow[\text{inversion at } C_1]{-H_2O}$ H Me / C_1 / Br^+ / C_2 / Me H

$\xrightarrow{Br^-}$ H Br Me / C_1 / C_2 / Me Br H (II) + H Br Me / C_1 / C_2 / Me Br H (III)

Fig. 12

There is evidence that all the halogen atoms can form cyclic ions and offer anchimeric assistance, *e.g.*, Winstein *et al.* (1948, 1951) studied the acetolysis of cis- and trans-2-halogeno-cyclohexyl brosylates (*i.e.*, p-bromobenzenesulphonates; this group is often written as OBs):

In the absence of neighbouring group participation, the rates would be expected to be about the same. If participation occurs, then this is readily possible in the trans-isomer (I a,2a) by attack of X at the rear of the ejected OBs^- ion, but this is not so for the cis-isomer. The rate ratios observed were trans/cis: X = I, $2.7 \times 10^6/1$; X = Br, 800/1; X = Cl, 3.8/1.

NEIGHBOURING HYDROXYL OR ALKOXYL GROUP

Hydroxyl groups may enter into neighbouring group participation, one way being via the alkoxide ion. Thus, Bartlett (1935) showed that alkali converts trans-2-chlorocyclohexanol into cyclohexene oxide, and proposed a mechanism in which an alkoxide ion is formed first and this then ring-closes with ejection of the chloride ion: Bergvist (1948) showed that this reaction proceeds more than 100 times as fast as that when the cis-compound is used. Here again, the trans-form permits ready attack at the rear of the chloride ion whereas the cis-isomer does not. The fact that the cis-form does react may be explained by assuming that the reaction proceeds via cis-elimination of the chlorine atom. This would require a distorted (*i.e.*, highly strained) transition state, and consequently the activation energy for this path would be greater than that for trans-elimination. Another example is the conversion of sugars into epoxy-sugars. The hydroxyl group itself may also participate as a neighbouring group, the most important example being the case with chlorohydrins of the type $Cl(CH_2)_nOH$. In these compounds, anchimeric assistance is greatest when n = 4 (to give a five-membered ring), far less for n = 5 (to give a six-membered ring), and is absent for other values of n. Thus, tetramethylene chlorohydrin in water is converted into tetrahydrofuran about 103 times as fast as ethylene chlorohydrin is converted into ethylene epoxide. This is readily explained on the basis' that five-membered rings are very stable.

Alkoxyl groups behave similarly to hydroxyl groups in neighbouring group participation, *i.e.*, show anchimeric assistance in the formation of five-membered rings, and to a far less extent in the formation of six-membered rings. Thus, the acetolysis of 4-methoxybutyl brosylate is about 650 times as fast as that of n-butyl brosylate. It should be noted, however, that there appears to be some doubt about this explanation.

Neighbouring Acetoxyl Group

Winstein et at. (1942, 1943) showed that a neighbouring acetoxyl group leads to the formation of an acetoxonium ion. trans-2-Acetoxycyclohexyl brosylate (I) forms trans- l,2-diacetoxycyclohexane (II) when treated with silver acetate, and the same product (II) is obtained when the starting material is trans-2-acetoxycyclohexyl bromide (III).

The authors believe that the course of the reaction, based on the stereochemical evidence, proceeds through the same acetoxonium ion (IV). This mechanism is supported by the fact that in each case, when the reaction was carried out in the presence of a small amount of water, the product was now the monoacetate of cis-cyclohexane-l,2-diol,(V); some diacetate of this cis-diol was also obtained.

Further support for the formation of (IV) is afforded by the fact that the cis-isomers of (1) and (III) undergo the same reactions but at much slower rates. The formation of the intermediate (Va) is supported by the fact that when the solvolysis of (I) is carried out in ethanol, (VI) is obtained (Winstein *et al.*, 1943).

Neighbouring Phenyl Group

One type of molecular rearrangement is the 1,2-shift, and this may involve neighbouring group participation. An example is the acetolysis of the tosylates (p-toluenesulphonates) of 3-phenylbutan-2-ol (Cram, 1949). If the mechanism were S_N2, then each tosylate would produce an inverted active acetate, inversion occurring only at the C–OTs carbon atom, the result being that the threo-tosylate would give the erythro-acetate and the erythrotosylate the threo-acetate. If the mechanism were SN1, ionisation of the tosyloxy group would produce the classical carbonium ion and the expected result would be racemisation (and retention) at the C–OAc carbon atom, the net result being that both threo- and erythro-tosylate yield the same product.

Cram found that the L-threo-tosylate gave the racemic threo-acetate, whereas the L-ervthro tosylate produced the almost optically pure L-erythro-acetate. Hence the mechanism cannot be S_N2 or S_N1. These results, however, can readily be explained on the basis of the formation of an intermediate bridged carbonium ion, the phenonium cation.

This phenonium cation is symmetrical, and since attack by an acetic acid molecule can be expected to occur with equal probability at carbons 1 and 2, products (1) and (2) will be formed in equal amounts. As these are mirror images, the result is therefore the racemic modification. The phenonium cation from the L-erythro-tosylate is also attacked at carbons 1 and 2 with equal probability, and gives the sane optically active product [(1) = (2)] which has the same configuration as the starting material.

The above explanation is completely satisfactory for the results given, but is no longer so when other experimental work is also considered. In acetic acid, the L-threo-tosylate racemises at a rate which is considerably faster than the rate of formation of p-toluenesulphonic acid, and it was estimated that 80 per cent of the tosylate is racemised before its conversion into the acetate (Winstein *et al.*, 1952). It was therefore suggested that an ion-pair is formed, and this can undergo racemisation through internal return prior to the formation of the postulated phenonium ion which leads to the products. Not only are acetates formed in this reaction, but so are alkenes (25-35 per cent), and when the structures of these alkenes are considered, the results are not explained on the basis of a phenonium ion intermediate. Cram (1952) therefore proposed that a classical cation is formed as well as the bridged ion.

The mechanism was then further elaborated to explain the formation of 4 per cent of erti thro-acetate from threo-tosylate. (This formation of erythro- from the threo-compound has been termed 'leakage' from the threo to the erythro system.) Brown *et al.* (1965) have examined this reaction and have provided evidence to show that the results can be explained in terms of a rapidly equilibrating pair of classical ions: In this case, the phenonium ion may be regarded as the transition state formed from either classical carbonium ion and not as the intermediate proposed above.

Another example involving the phenyl group is the ethanolysis of the conjugate base of 2-(p-hydroxyphenyl)ethyl bromide (I), which is about 106 times as fast as that of the corresponding p-methoxy-compound (II) (Winstein *et al.*, 1963). The oxygen atom in (I) is much more electron-releasing than that in (II), and so the intermediate (Ia, which has been isolated) is much more stable than (IIa). Hence, reaction involving the formation of (Ia) requires a lower energy of activation than that involving the formation of (IIa), and so the former proceeds faster than the latter. This, of course, is based on the argument that the formation of (Ia) or (IIa) is the rate-determining step.

SOME OTHER EXAMPLES OF NEIGHBOURING GROUP PARTICIPATION

Many nucleophilic groups can enter into neighbouring group participation but, as we have seen, the extent depends on the structure of the molecule. The sulphur compound (I) undergoes hydrolysis in aqueous dioxan about 10^4 times

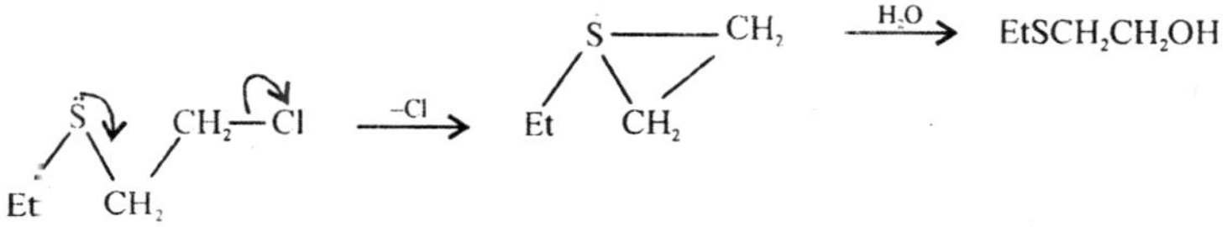

$$\underset{\text{(I)}}{Et\ddot{S}CH_2CH_2-Cl} \xrightarrow{-Cl} Et\overset{+}{S}(CH_2CH_2) \xrightarrow{H_2O} EtSCH_2CH_2OH$$

Fig. 13

as fast as that of the oxygen analogue. As we have seen, alkoxyl groups exhibit neighbouring group participation only when the formation of a five- (and six-) membered ring is involved. Similarly, the alkaline hydrolysis of (III) proceeds much faster than that of n-butyl bromide (to n-butanol); the five-membered ring is stable. On the other hand, the alkaline hydrolysis of (IV) produces the rearranged product (V). The course of this reaction is readily explained on the basis that a cyclic intermediate is formed and that this undergoes

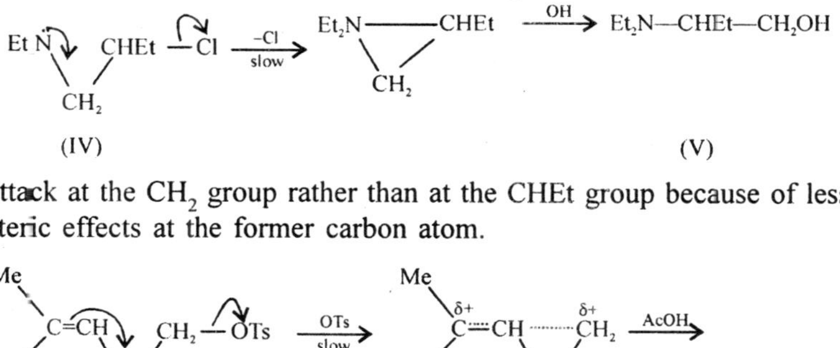

$$\underset{\text{(IV)}}{Et_2\ddot{N}CH_2CHEt-Cl} \xrightarrow[\text{slow}]{-Cl} Et_2\overset{+}{N}(CH_2CHEt) \xrightarrow{OH^-} \underset{\text{(V)}}{Et_2N-CHEt-CH_2OH}$$

attack at the CH_2 group rather than at the CHEt group because of less steric effects at the former carbon atom.

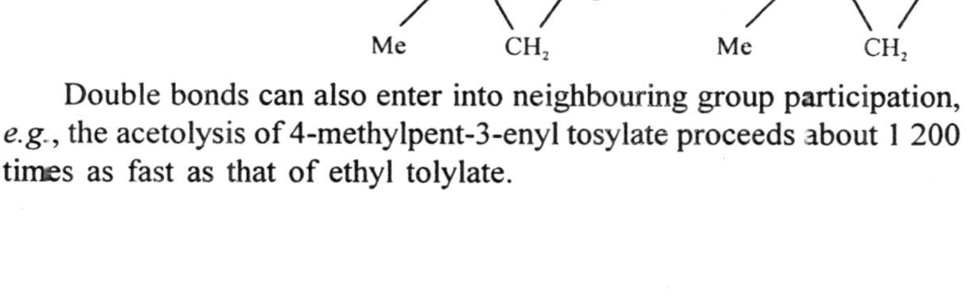

$$Me_2C{=}CH-CH_2-CH_2-OTs \xrightarrow[\text{slow}]{OTs^-} Me_2\overset{\delta+}{C}{=\!=}CH\cdots\overset{\delta+}{C}H_2(CH_2) \xrightarrow{AcOH} Me_2C{=}CH-CH_2-CH_2-OAc + H_2C(Me)C-CH(CH_2)CH_2$$

Double bonds can also enter into neighbouring group participation, *e.g.*, the acetolysis of 4-methylpent-3-enyl tosylate proceeds about 1 200 times as fast as that of ethyl tolylate.

The products are 2-cyclopropylpropene and 4-methylpent-3-enyl acetate

The intermediate ion (VI) is the homoallylic cation

Partial Asymmetric Synthesis

Partial asymmetric synthesis may be defined as a method for preparing optically active compounds from symmetric compounds by the intermediate use of optically active compounds, but without the necessity of resolution (Marckwald, 1904). In ordinary laboratory syntheses, a symmetric compound always produces the racemic modification.

This definition included the term `partial' in order to distinguish this type of asymmetric synthesis from another type in which a `physical reagent' was used instead of optically active compounds. This physical reagent was circularly polarised light, and in this case the process was labelled an 'absolute asymmetric synthesis'. It is now common practice to regard asymmetric synthesis as a special case of stereoselectivity in which a prochiral unit (centre or face) is converted into a chiral centre and results in unequal amounts of stereoisomers (with concomitant optical activity of the product). In this 'definition', no distinction is drawn between 'chemical' and 'physical' methods. Even so, the term 'absolute' is still often used to specify the use of a physical 'reagent'.

The first asymmetric synthesis was carried out by Marckwald (1904), who prepared an active (–)-valeric acid (laevorotatory to the extent of about 10 per cent of the pure compound) by heating the half-brucine salt of ethylmethylmalonic acid at 170°C.

(I) and (II) are diastereoisomers; so are (III) and (IV). (V) and (VI) are enantiomers, and since the mixture is optically active, they must be present in unequal amounts. Marckwald believed this was due to the different rates of decomposition of diastereoisomers (I) and (II), but according to Eisenlohr and Meier (1938), the half-brucine salts (I) and (II) are not present in equal amounts in the solid form (as thought by Marckwald). These authors suggested that as the less soluble diastereoisomer

H_3C, CO_2H, C_2H_5, CO_2H on C $\xrightarrow{(-)\text{-brucine}}$ H_3C, CO_2H [(–)-brucine], C_2H_5, CO_2H on C (I) + H_3C, CO_2H, C_2H_5, CO_2H [(–)-brucine] on C (II) $\xrightarrow{170°C}$

H_3C, CO_2H [(−)-brucine], C_2H_5, H — C (III) + H_3C, H, C_2H_5, CO_2H [(−)-brucine] — C (IV) $\xrightarrow{HCl}$ H_3C, CO_2H, C_2H_5, H — C (V) + H_3C, H, C_2H_5, CO_2H — C (VI)

crystallised out (during evaporation of the solution), some of the more soluble diastereoisomer spontaneously changed into the less soluble diastereoisomer to restore the equilibrium between the two; thus the final result was a mixture of the half-brucine salt containing a larger proportion of the less soluble diastereoisomer. If this be the explanation, then we are dealing with an example of asymmetric transformation and not of asymmetric synthesis. Further work, however, has shown that Marckwald had indeed carried out an asymmetric synthesis. Kenyon and Ross (1951) decarboxylated optically active ethyl hydrogen ethylmethylmalonate (VII) and obtained an optically inactive product, ethyl (±)-a-methylbutyrate (VIII).

H_3C, CO_2H, C_2H_5, $CO_2C_2H_5$ — C (VII) active $\longrightarrow$ CO_2 + H_3C, H, C_2H_5, $CO_2C_2H_5$ — C (VIII) inactive

These authors (1952) then decarboxylated the cinchonidine salt of (VII) and still obtained the optically inactive product (VIII).

H_3C, CO_2H (cinchonidine), C_2H_5, $CO_2C_2H_5$ — C $\longrightarrow$ H_3C, H, C_2H_5, $CO_2C_2H_5$ — C (VIII) inactive + CO_2 + cinchonidine

Kenyon and Ross suggested the following explanation to account for their own experiments and for those of Marckwald. Decarboxylation of diastereoisomers (I) and (II) takes place via the formation of the same carbanion (Ia), and decarboxylation of (VII) and its cinchonidine salt via (VIIa).

Combination of carbanion (Ia) with a proton will produce diastereoisomers (III) and (IV) in different amounts, since, in general, diastereoisomers are formed at different rates. Since the carbanion may be represented as a resonance hybrid of (Ia) and (Ib), this carbanion is essentially flat. Also,

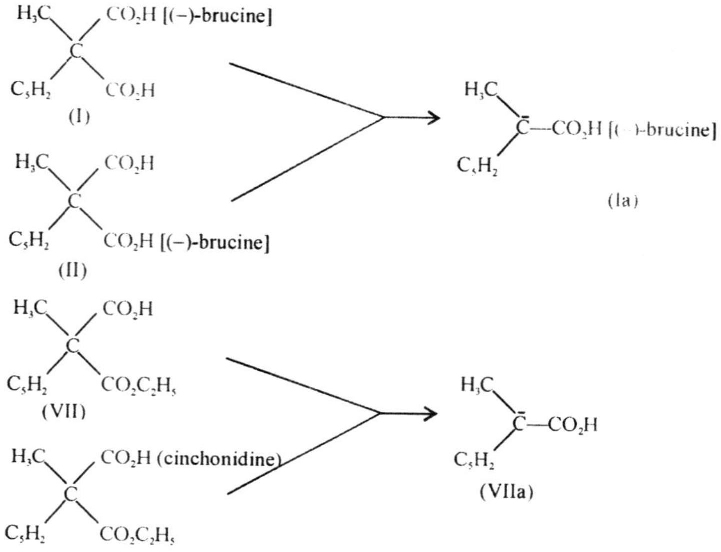

Fig. 14

because of the presence of the chiral centre in (–)-brucine, the two faces of the carbanion are diastereotopic. Since (VIIa) may also be represented as a resonance hybrid of (VIIa) and (VIIb), the molecule is essentially flat and has enantiotopic faces

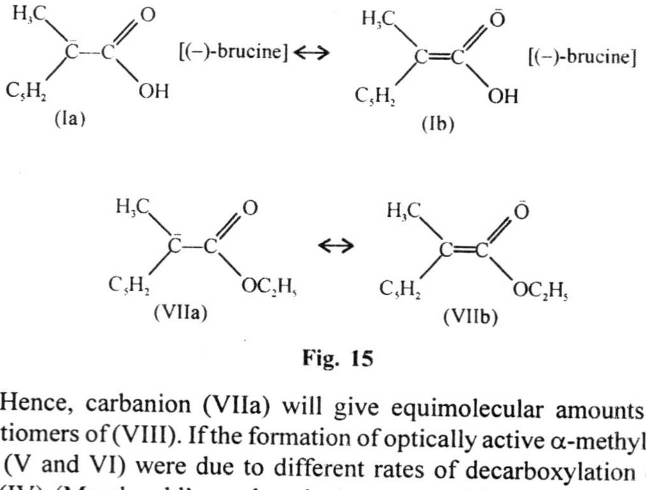

Fig. 15

Hence, carbanion (VIIa) will give equimolecular amounts of the enantiomers of (VIII). If the formation of optically active α-methylbutyric acid (V and VI) were due to different rates of decarboxylation of (III) and (IV) (Marckwald's explanation) or to partial asymmetric transformation during crystallisation (Eisenlohr and Meier's explanation), then these effects are nullified if Kenyon's explanation is correct, since the intermediate carbanion is the same for both diastereoisomers.

Thus, if the asymmetric transformation theory were correct, then decarboxylation of the dibrucine salt of ethylmethylmalonic acid to α-methylbutyric acid should give an optically inactive product, since only one type of crystal is now possible (asymmetric transformation is now impossible).

$$\underset{C_2H_5}{\overset{H_3C}{}}\!\!>\!C\!<\!\!\underset{CO_2H\ [(-)\text{-brucine}]}{\overset{CO_2H\ [(-)\text{-brucine}]}{}} \longrightarrow \underset{C_2H_5}{\overset{H_3C}{}}\!\!>\!\bar{C}\text{—}CO_2H\ [(-)\text{-brucine}]$$

(Ia)

Fig. 16

On the other hand, if the carbanion (Ia) is an intermediate in this decomposition, it is still possible to obtain an optically active product. Kenyon and Ross did, in fact, obtain a laevorotatory product.

McKenzie (1904) carried out a number of asymmetric syntheses by reduction of the keto group in various keto-esters in which the ester group contained a chiral group, *e.g.*, benzoylformic acid was esterified with (–)-menthol, the ester reduced with aluminium amalgam, and the resulting product hydrolysed; the mandelic acid so obtained was slightly laevorotatory.

$$C_6H_5COCO_2H + (-)\text{-}C_{10}H_{19}OH \rightarrow C_6H_5COCO_2C_{10}H_{19} + H_2O \xrightarrow{Al/Hg}$$

$$C_6H_5CHOHCO_2C_{10}H_{19} \xrightarrow{H_2O} \underset{(-)\text{-rotation}}{C_6H_5CHOHCO_2C} + (-)\text{-}C_{10}H_{19}OH$$

Similarly, the pyruvates of (–)-menthol, (–)-pentyl alcohol and (–)-borneol gave an optically active lactic acid (slightly laevorotatory) on reduction.

$$CH_3COCO_3R(-) \xrightarrow{[H]} CH_3CHOHCO_2R(-) \xrightarrow{H_2O}$$

$$\underset{(-)\text{-rotation}}{CH_2CHOHCO_2H} + (-)\text{-ROH}$$

McKenzie (1904) also obtained similar results with Grignard reagents, *e.g.*, the (–)-menthyl ester of benzoylformic acid and methylmagnesium iodide gave a slightly laevorotatory atrolactic acid.

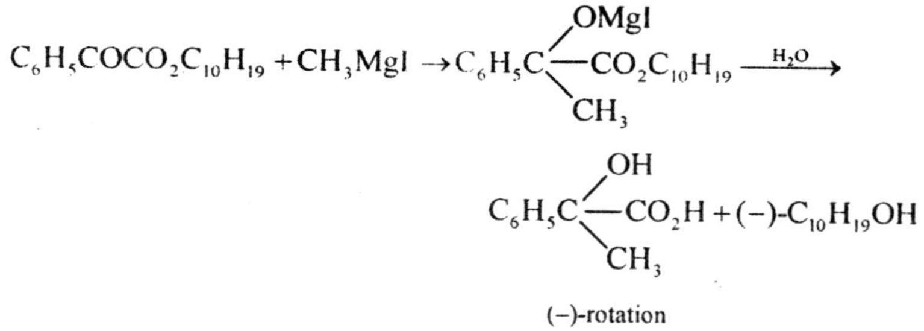

$$C_6H_5COCO_2C_{10}H_{19} + CH_3MgI \rightarrow C_6H_5C(OMgI)(CH_3)-CO_2C_{10}H_{19} \xrightarrow{H_2O}$$

$$C_6H_5C(OH)(CH_3)-CO_2H + (-)\text{-}C_{10}H_{19}OH$$

(–)-rotation

Fig. 17

Another example of asymmetric synthesis involving the use of a Grignard reagent is the reduction of 3,3-dimethylbutan-2-one into a dextrorotatory 3,3-dimethylbutan-2-ol by means of (+)-2 methylbutylmagnesium chloride (Mosher *et al.*, 1950). The authors explained the result by

$$(CH_3)_3CCOCH_3 \xrightarrow{(+)-CH_3CH_2CH(CH_3)CH_2MgCl} \underset{(+)-\text{rotation}}{(CH_3)_3CCHOHCH_3}$$

postulating the formation ofa cyclic transition state in which the preferred configuration has the methyl group.of the optically active Grignard reagent on the same side of the ring as the larger of the two groups in the ketone. This configuration of the T.S. is preferred because steric effects are less than in

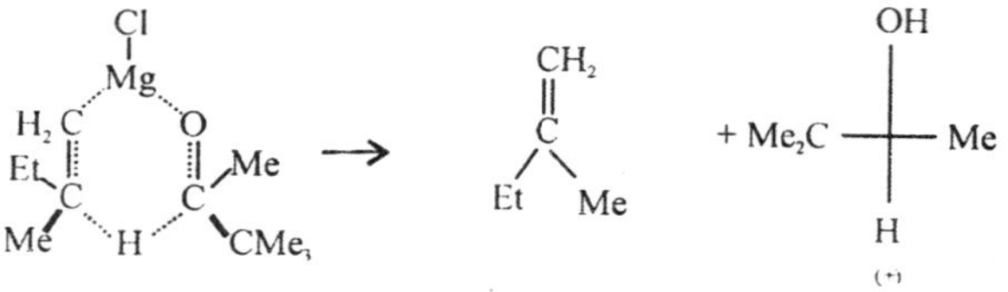

Cl–Mg cyclic T.S. (H_2C, Et, Me, C, H, O=C, Me, CMe_3) $\rightarrow$ $CH_2=C(Et)(Me)$ + $Me_3C-C(OH)(H)-Me$ (+)

the alternative T.S., and so the energy content of the former is lower than that of the latter. Mosher used a variety of ketones and obtained a predominance of one enantiomer, but the optical purity was always low (most values were between ~ 4 and 10 per cent). This suggests that the energy differences between the two transition states may be small. Mosher *et al.* (1961, 1966), however, have concluded that the size of the groups is not the only factor that determines the stereo selectivity of these asymmetric, reductions.

In fact, although the actual size of a group is fixed, it appears that its effective size depends on the environment and on the mechanism of the reaction. This may lead to difficulties in predicting the steric course of a particular reaction.

In the above examples, the reduction involves a hydrogen atom transfer from the chiral β-carbon atom of the Grignard reagent. Morrison (1967) has shown that isopropyl phenyl ketone is reduced by, *e.g.*, $PhMeCHCH_2CH_2MgCl$, in which the β-carbon atom is not chiral, to give predominantly one enantiomer of $PhCHOHPr^i$. The two hydrogen atoms in the Grignard reagent are diastereotopic (by internal comparison) and so produce diastereoisomeric transition states with the ketone, thereby resulting in the formation of excess of one enantiomer over the other.

Doering *et al.* (1950) have carried out the reduction of ketones by means of the MeerweinPonndorf-Verley method. The reduction of methyl isohexyl ketone with (+)-butan-2-ol in the presence of aluminium 2-butoxide gave (+)-methylisohexylmethanol. Here again the results are explained by the formation of a cyclic transition state with the isohexyl group predominantly on the same side as the methyl group of the butoxide. In both reactions, transfer of a hydride ion takes place.

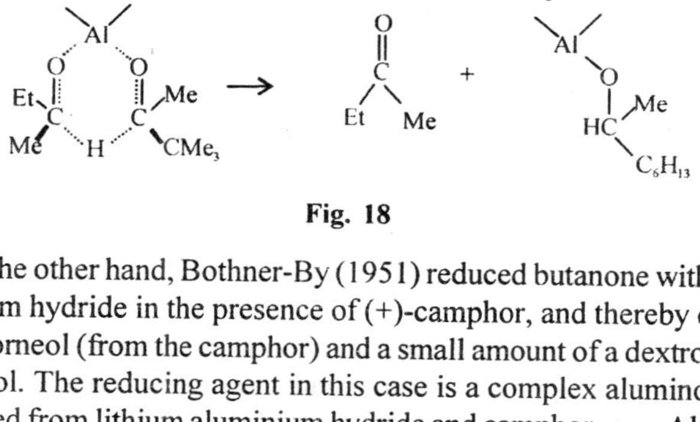

Al; O; O; Et; C; Me; Me; H; C; CMe_3 → O; C; Et; Me + Al; O; Me; HC; C_6H_{13}

Fig. 18

On the other hand, Bothner-By (1951) reduced butanone with lithium aluminium hydride in the presence of (+)-camphor, and thereby obtained (+)-isoborneol (from the camphor) and a small amount of a dextrorotatory butan-2-ol. The reducing agent in this case is a complex aluminohydride ion formed from lithium aluminium hydride and camphor, *e.g.*, $Al(OR)H_3^-$

$$CH_3COC_2H_5 \xrightarrow[(+)-\text{camphor}]{LiAlH_4} \underset{(+)-\text{rotation}}{CH_3CHOHC_2H_3}$$

Turner *et al.* (1949) carried out a Reformatsky reaction (see Vol. I) using acetophenone, (–)menthyl bromoacetate and zinc, and obtained a dextrorotatory β-hydroxy-β-phenylbutyric acid.

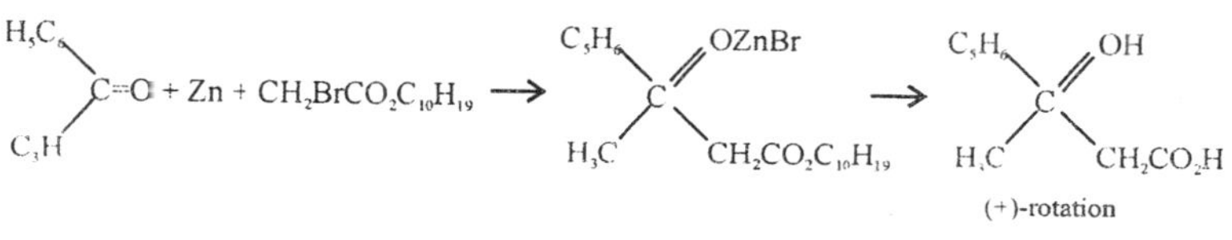

H_5C_6, C_3H >C=O + Zn + $CH_2BrCO_2C_{10}H_{19}$ → C_5H_6, OZnBr, C, H_3C, $CH_2CO_2C_{10}H_{19}$ → C_5H_6, OH, C, H_3C, CH_2CO_2H (+)-rotation

Reid *et al.* (1962) have also used aldehydes in the Reformatsky reaction, *e.g.*, benzaldehyde gave a laevorotatory β-hydroxy-β-phenylpropionic acid. Alkenes have been oxidised with optically active peroxy-acids to give optically active epoxides, *e.g.*, Montanari *et al.* (1969) used (+)-peroxycamphoric acid:

$$\text{Ph(H)C=C(Me)H} \xrightarrow{(+)\text{-RCO}_3\text{H}} \text{Ph(H)C—C(Me)H (epoxide)}$$

(+)-

Asymmetric hydrogenation has also been effected with the use of a rhodium complex containing, *e.g.*, (+)- or (–)-PhCHMeNHCHO as ligands (McQuillin *et al.*, 1969).

$$\text{PhCMe=CHCO2Me} \xrightarrow[H_2]{(+)\text{-cat.}} (+)\text{-PhCHMeCH}_2\text{CO}_2\text{Me}$$

Prelog *et al.* have studied, by means of conformational analysis, the steric course of the addition of Grignard reagents to benzoylformic (phenylglyoxylic) esters of chiral alcohols. If the letters S, M and L refer respectively to small, medium and large groups attached to the carbinol carbon atom of the chiral alcohol, then the general reaction may be written:

$$C_6H_5COCO_2CSML \xrightarrow{RMgX} C_6H_5CR(OH)CO_2CSML$$

$$\xrightarrow{\text{hydrol}} C_6H_5CR(OH)CO_2H$$

Prelog *et al.* found that the configuration of the asymmetric carbon atom in the stereoisomer that predominated in this reaction could be correlated with that of the carbinol carbon of the alcohol. The authors considered that the most stable conformation of the ester was the one in which the two oxo groups are planar and trans to each other. It was also believed that the atoms O–C also lie in this plane, *i.e.*, the fragment shown is essentially planar. The problem now was the rotation of the

$$\text{Ph—C(=O)—C(=O)—O—C}$$

group CSML about the single O–C bond. There are three energetically favourable conformations, and they are those in which, in turn, S, M, and

L will (i) lie in the plane; (ii) in front of the plane; (iii) behind the plane, of the rest of the molecule (*i.e.*, the fragment shown). Prelog originally proposed (1953) that the most populated conformation was that in which group S lay in this plane and the groups M and L skew (IXa and Xb; see Chart, in which the thick lines represent groups in front, broken lines groups behind, and ordinary lines groups in the plane). Prelog, however, later (1956) proposed that the most populated conformation was that in which group L lay in the plane and the groups S and M skew (IXb and Xc; see Chart).

The third energetically favourable conformation in which the M group lies in the plane is also shown in the Chart (IXc and Xa). The enantiomeric alcohols (IX) and (X) each give rise to the three energetically favourable conformations for the corresponding ester (see Chart).

Another assumption made by Prelog (1953) was that the Grignard reagent attacks the oxo group (of PhCO) from the less hindered side. With methylmagnesium iodide as the Grignard reagent, the resulting α-hydroxy-acid is atrolactic acid (IXd or Xd). The direction of attack is indicated in the Chart by a thick arrow when attack is from the front face and a broken arrow when attack is from the back. This may be illustrated as shown, with (IXb) as the example.

Ph–C(=O)–C(=O)–O–C(M)(S)(L) (IXb) —(i) MeMgI, (ii) H_3O^+→ Ph(Me)C(HO)–C(=O)–O–C(M)(S)(L) —(i) aq KOH, (ii) H_3O^+→ Me–C(CO$_2$H)(OH)(Ph) (IXd)

Examination of the Chart shows that the predominant enantiomer of atrolactic acid produced is related to the configuration of the chiral alcohol. (IX), via the most populated conformation (IXh), produces (IXd), and (X), via the most populated conformation (Xc), produces (Xd), the enantiomer of (IXd). It might be noted that even if each set of conformations were present in equal populations, the same results are achieved since two of each set of three produce the same atrolactic acid. Hence, all chiral alcohols of the type CSML(OH) which are stereochemically related, *i.e.*, the groups S, M, and L have the same spatial relationship, will produce the same enantiomer of atrolactic acid in excess. Also, if the sign of rotation of the atrolactic acid produced in excess is measured, the stereochemical relationships of the chiral alcohols can be determined. Now, (–)-menthol, (–)-borneol, and (–)-octan-2-ol are all configurationally related to L(–)-glyceraldehyde and all give (–)-atro lactic acid as the predominant enantiomer. The absolute configurations of these alcohols have been established; they are as shown.All have the same configuration for groups S, M, and L, and all are alcohols of the type (IX), *e.g.*,

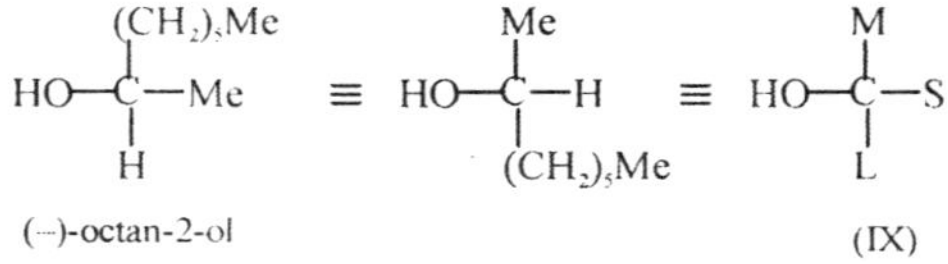

(–)-octan-2-ol (IX)

Fig. 19

It therefore follows that (–)-atrolactic acid is (IXd). Hence, if the configuration of the chiral alcohol is known, it is possible to deduce that of the enantiomer of atrolactic produced in excess, and vice versa. This method of correlating configurations has been called the 'atrolactic acid method'. Furthermore, since the absolute configuration of (–)-atrolactic acid is that of (IXd) [determined as described above, and confirmed by other methods], (–)-atrolactic acid is the (R)(–)- and (+)-atrolactic acid is the (S)(+)-acid. Hence, the absolute configurations of chiral alcohols can be correlated with the sign of rotation of the atrolactic acid produced in excess.

It also follows from the above discussion that if the keto-acid is pyruvic acid and the Grignard reagent is phenylmagnesium bromide then, for a given chiral alcohol, the atrolactic acid is predicted to have the opposite sign of that produced by benzoylformic acid and methylmagnesium iodide. The Ph and Me groups are now interchanged; this is illustrated as shown.

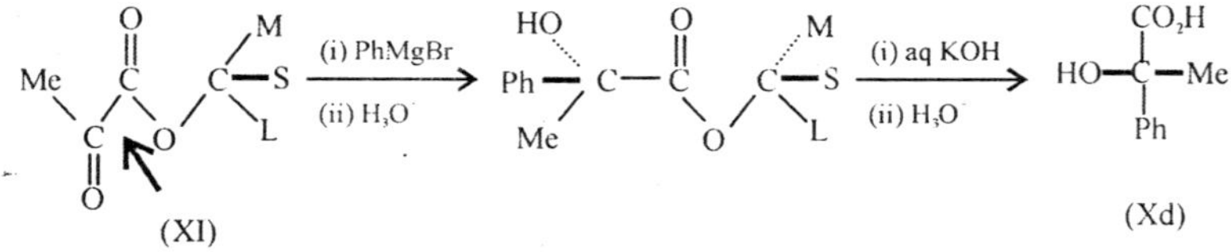

(XI) (i) PhMgBr (ii) H_2O (i) aq KOH (ii) H_2O (Xd)

Fig. 20

(XI) corresponds to (IXb), but now atrolactic acid (Xd) is obtained in excess. These predictions have been verified by Prelog *et al.*

An alternative method for the determination of the absolute configuration of a chiral alcohol is that due to Horeau (1961-1964). The alcohol is acetylated in pyridine with an excess of (±)-phenylethylacetic anhydride, $(PhCHEtCO)_2O$, then the excess of the anhydride is hydrolysed and the optical activity of the phenylethylacetic acid is measured. Since one enantiomer of the anhydride reacts with the alcohol faster than the other, the other enantiomer will be in excess in the recovered acid (cf. the kinetic method of resolution. From the rotation of this acid, the configuration of the alcohol may be deduced from the following empirical rule: The alcohol with the absolute configuration (X) gives, by application of Prelog's method, excess of (+)-atrolactic acid (Xd) and, by Horeau's method, excess of (–)-phenylethylacetic acid (Xe).

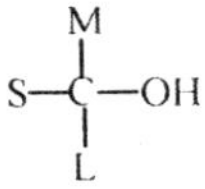

```
      M              CO2H            CO2H
      |               |               |
  S—C—OH        HO—C—Me         H—C—Et
      |               |               |
      L              Ph              Ph
     (X)            (Xd)            (Xe)
```

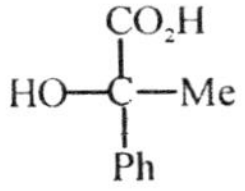

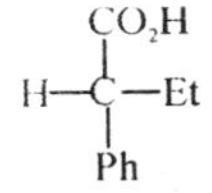

Horeau's method is more convenient to carry out than Prelog's method, and has the added advantage in being highly stereoselective.

Cram *et al.* (1952) have also dealt with asymmetric syntheses in which the molecule contains a chiral centre that belongs to the molecule, *i.e.*, remains in the molecule, *e.g.*, the Kiliani reaction in

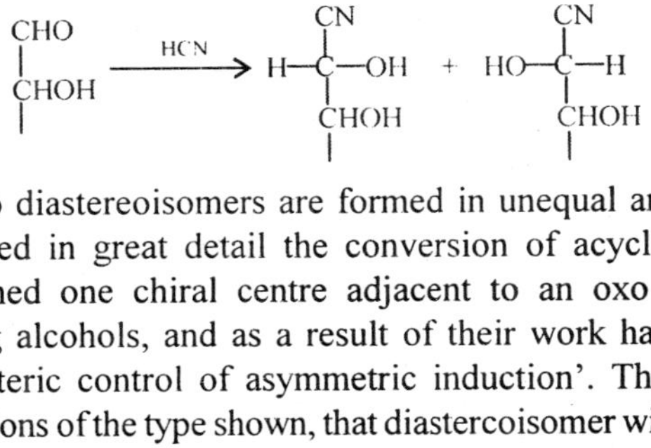

```
                        CN                 CN
CHO                     |                  |
 |       HCN        H—C—OH    +     HO—C—H
CHOH  ———————→          |                  |
 |                     CHOH               CHOH
                        |                  |
```

which the two diastereoisomers are formed in unequal amounts. These workers studied in great detail the conversion of acyclic compounds which contained one chiral centre adjacent to an oxo group to the corresponding alcohols, and as a result of their work have formulated the rule of 'steric control of asymmetric induction'. This is: 'in non-catalytic reactions of the type shown, that diastercoisomer will predominate which would be formed by the approach of the entering group from the least hindered side of the double bond when the rotational conformation of the C–C bond is such that the double bond is flanked by the two least bulky groups attached to the adjacent asymmetric centre.' Thus, using the Newman projection formulae (Z = MgX):

According to Cram's rule, the product from reaction (a) should predominate. An example that demonstrates this is the reaction between α-phenylpropionaldehyde (S = H, M = Me, L = Ph) and methylmagnesium bromide (R^1 = Me); two products can be formed, viz., (XII) (the ervthrocompound) and (XIII) (the threo-compound):

The ervthro-compound (XII) is the major product.

Some exceptions to the predictions have been observed, particularly when group L is very large. Other models have been proposed.

Karabatsos (1967), using certain assumptions, has been able to predict semiquantitatively from the Cram model (non-catalytic type) the stereospecificity of the product.

As pointed out above, the rule will not apply to catalytic reduction, and it also does not apply to compounds in which the asymmetric carbon atom is joined to a group capable of complexing with the reagent, *e.g.*, OH, NH_2. In this case, a different model is to be used (Cram *et al.*, 1959). Such a model is a rigid cyclic one and the predominant product is as shown:

The influence of enzymes on the steric course of reactions has also been investigated, *e.g.*, Rosenthaler (1908) found that emulsin converted benzaldehyde and hydrogen cyanide into dextrorotatory mandelonitrile which was almost optically pure. It has been found that in most enzymic reactions the product is almost 100 per cent of one or other enantiomer. Enzymes are proteins and are optically active, but since they are so 'one-sided' in their action, it appears likely that the mechanism of the reactions in which they are involved differs from that of asymmetric syntheses where enzymes are not used. It has been suggested that enzymes are the cause of the formation of optically active compounds in plants. Although this is largely true, the real problem is: How were the optically active enzymes themselves produced?

Ferreira's work, however, shows that optically active compounds may possibly be produced in living matter by activation of a racemic modification. This theory appears to be superior to that of the formation of optically active compounds by the action of naturally polarised light.

Asymmetric syntheses involving elimination reactions have also been carried out to prepare optically active alkenes, *e.g.*, Goldberg *et al.* (1966) subjected (-)-cis-4-methylcyclohexyl hydratropate to pyrolysis and obtained (-)-4-methylcyclohexene (in very low optical purity).

Now let us consider reactions in which the products are not chiral but have been formed by two identical groups in the substrate reacting at different rates. These are highly stereoselective, and since they involve enzymes, may be regarded as a special class of asymmetric synthesis. A classical example of this type is the oxidation of citric acid (XIV) to α-ketoglutaric acid (XV) by means of an enzyme:

$$HO_2CCH_2-\overset{\displaystyle H}{\underset{\displaystyle CO_2H}{\overset{|}{\underset{|}{C}}}}-CH_2CO_2H \xrightarrow{\text{enzyme}} HO_2CCH_2CH_2COCO_2H$$

(XIV) (XV)

According to this equation, only the right-hand side methylene group is oxidised; the two groups are enantiotopic and react at different rates with a chiral reagent. This was demonstrated by the use of deuterium-labelled citric acid (XIVa) to give (XVa).

$$HO_2CCD_2-\overset{\displaystyle H}{\underset{\displaystyle CO_2H}{\overset{|}{\underset{|}{C}}}}-CH_2CO_2H \xrightarrow{\text{enzyme}} HO_2CCD_2CH_2COCO_2H$$

(XIVa) (XVa)

Absolute Asymmetric Synthesis

As we have seen, right- and left-circularly polarised light is unequally absorbed by enantiomers, provided the light has a wavelength in the neighbourhood of the characteristic absorption bands of the compound.

It has been suggested that circularly polarised light produced the first natural active compounds, and to support this theory, racemic modifications have been irradiated with circularly polarised light and attempts made to isolate one enantiomer. There was very little success in this direction until W. Kuhn and Braun (1929) claimed to have obtained a small rotation in the case of ethyl a-bromopropionate. The racemic modification of this compound was irradiated with right- and left-circularly polarised light (of wavelength 2 800Å), and the product was found to have a rotation of + or –0.05°, respectively. Thus we have the possibility of preparing optically active products from inactive substances without the intermediate use of optically active reagents (cf. Ferreira's work). This type of synthesis is known as an absolute asymmetric synthesis; it is also known as an absolute asymmetric decomposition (or destruction), since one enantiomer is decomposed (or destroyed) preferentially. On the other hand, there is another type of absolute asymmetric synthesis in which a chiral physical 'reagent' converts an achiral substrate into a chiral product.

From 1930 onward, more conclusive evidence for absolute asymmetric decompositions has been obtained, *e.g.*, W. Kuhn and Knopf (1930) irradiated (±)-α-azidopropionic dimethylamide, $CH_3CHN_3CON\ (CH_3)_2$, with right-circularly polarised light and obtained an undecomposed product with a rotation of +0.78°; with left-circularly polarised light, the undecomposed product had a rotation of −1.04°. Thus the (−)- or (+)-form is decomposed (photochemically) by right- or left-circularly polarised light, respectively. Similarly, Mitchell (1930) irradiated humulene nitrosite with right- and left-circularly polarised red light, and obtained slightly optically active products.

Davis and Heggie (1935) found that the addition of bromine to 2,4,6-trinitrostilbene in a beam of right-circularly polarised light gave a dextrorotatory product; this is an example of an absolute asymmetric synthesis.

Small (+)-rotations were also observed when a mixture of ethyl fumarate and anhydrous hydrogen peroxide in ethereal solution was irradiated with right-circularly polarised light (Davis *et. al.*, 1945).

As mentioned above, there has been much speculation about how optically active compounds are produced in nature. A difficulty of the 'circularly polarised light' theory is that, up to the present, no satisfactory answer has been offered to explain the source of such light in nature. On the other hand, Lee *et al.* (1956) showed that electrons emitted in β-decay are polarised, and if they slow down they lose some of their energy by emission of γ-radiation which is circularly polarised. Garay (1968) has irradiated D- and L-tyrosine with β-particles derived from strontium-90 Y and showed that the D-enantiomer was decomposed more quickly. He failed to 'activate' racemic DL-tyrosine, but suggested that β-particle bombardment may he the source of optically active compounds in nature.

Molecular Overcrowding

All the cases discussed so far owe their asymmetry to restricted rotation about a single bond. There is, however, another way in which steric factors may produce molecular asymmetry. It has been found that, in general, non-bonded carbon atoms cannot approach closer to each other than about 3.0Å. Thus, if the geometry of the molecule is such as to produce 'intramolecular overcrowding', the molecule becomes distorted. An example of this type is 4,5,8-trimethyl-l-phenanthrylacetic acid (I). The phenanthrene nucleus is planar and substituents lie in this plane. If, however, there are fairly large groups in positions 4 and 5, then there will not be enough room to accommodate both groups in the plane of the nucleus.

This leads to strain being produced by intramolecular overcrowding, and the strain may be relieved by the bending of the substituents out of the plane of the nucleus, or by the bending (buckling) of the aromatic rings, or by both. Thus the molecule will not be planar and consequently will be asymmetric and therefore (theoretically) resolvable. Newman *et al.* (1940, 1947) have actually resolved it, and have also resolved (II). Bell *et al.* (1949) resolved (III), and it was these authors who introduced the term 'intramolecular overcrowding'.

Theilacker *et al.* (1953) resolved (IV), a heterocyclic analogue of phenanthrene. All of these compounds were found to have low optical stability, but Newman *et al.* (1955, 1956) have prepared (V) and (VI hexahelicene) which, so far, are the most optically stable compounds of the intramolecular overcrowding type.

It will be noticed that in (VI) the only way in which out-of-plane distortion can occur is through buckling of the molecule. The simplest molecule exhibiting overcrowding and consequent out-ofplane buckling of the molecule is 3,4-benzophenanthrene (VII); this has been shown to be nonplanar by X-ray analysis (Schmidt *et al.*, 1954). Similarly, Robertson *et al.* (1954) have shown that (VIII) exhibits out-of-plane buckling.

Another point to note in connection with out-of-plane buckling is that the buckling is distributed over all the rings in such a manner as to cause the minimum distortion in any one ring.

This distortion, which enables non-bonded carbon atoms to avoid being closer together than 3.0Å (marked with dots in VII and VIII), forces some of the other carbon atoms to adopt an almost tetrahedral valency arrangement (the original hybridisation is trigonal), and this affects the physical and chemical properties of the molecule, *e.g.*, Coulson *et al.* (1955) have calculated that the deformation in (VIII) produces a loss of resonance energy of about 75 kJ mol^{-1}.

We may now summarise the problem of molecular overcrowding as follows. A molecule is said to be overcrowded if, when the standard values are assigned to the bond lengths and bond angles, at least one pair of non-bonded atoms are closer to each other than the sum of their accepted van der Waals radii. If these atoms were to remain very close to each other and the rest of the molecule remained unchanged, there would be a large steric strain. If, however, the molecule were deformed (buckled) so that the overcrowded atoms became separated to the sum of their van der Waals radii there would now be a considerable strain energy in the molecule. Since a large amount of energy is required to stretch bonds and much less energy is required to bend them in overcrowded molecules the geometry of the molecule adjusts itself so that the energy is a minimum by mainly bending various bonds, with the bond lengths not much changed from the unstrained analogue. Thus overcrowded molecules are non-coplanar and have the form of a segment of a helix.

Mason *et al.* (1965) have measured the circular dichroism spectra of (–)-(VIIa) and (+)-(VIIb) (these are derivatives of VII), and analysis of these spectra led the authors to conclude that (–)-(VIIa) has the M-(minus, left-handed) and (+)-(VIIb) the P-(plus, right-handed) helical configuration viewed in the direction perpendicular to the mean molecular plane.

Helical molecules are those in which the arrangement of the atoms or groups is an imaginary helix. Such molecules are optically active due to the presence of helical dissymmetry or helicity. Thus, helicity is a particular type of chirality.

Just as benzene rings may suffer distortion, so can a molecule which owes its planarity to the presence of a double bond. Such an example is dianthronylidene (IX). The carbon atoms marked with dots are overcrowded (the distance between each pair is 2.9Å), and the strain is relieved by a rotation of about 40° around the olefinic double bond (Schmidt *et al.*, 1954).

Even in such simple molecules as tiglic acid (X) the two methyl groups give rise to molecular overcrowding with the result that the β-methyl group appears to be displaced from the molecular plane, thereby relieving overcrowding which is also partly relieved by small distortions in bond angles. These results were obtained by Robertson *et al.* (1959) from X-ray studies, and these authors also showed similar distortions in angelic acid (XI).

In polynuclear aromatic hydrocarbons in which the strain tends to be overcome by out-of-plane displacements of substituents and out-of-plane ring buckling, these effects cause changes in the ultraviolet spectra, but it is not yet possible to formulate any correlating rules. NMR studies by Reid (1957) have shown a shift for the hydrogen atoms in positions 4 and 5 in phenanthrene itself.

A similar phenomenon has been detected by Brownstein (1958) in 2-halogenobiphenyls, and the explanation offered is that the shift is due to the steric effect between the 2-halogen and the 2′-hydrogen atom.

Although molecular overcrowding is normally confined in the polynuclear type to systems containing three or more rings, nevertheless various substituted benzenes may also exhibit out-of-plane displacements of the substituents.

Electron-diffraction studies of polyhalogenobenzenes suggest that such molecules are non-planar (Hassel *et al.*, 1947), whereas X-ray studies indicate that in the solid state such molecules are very closely or even exactly planar (Tulinsky *et al.*, 1958; Gafner *et al.*, 1960). Ferguson *et al.* (1959, 1961) have examined, by X-ray analysis, polysubstituted benzenes containing not more than one halogen atom, *e.g.*, o-chloro- and bromobenzoic acid, and 2-chloro5-nitrobenzoic acid.

In all three molecules the steric strain is relieved by small out-of-plane displacements of the exocyclic valency bonds in addition to the larger in-plane displacements of these bonds away from one another. Ferguson *et al.* (1962) have also shown that in 2-chloro-5-nitrobenzoic acid the carboxyl group is twisted further out of the benzene plane than in o-chlorobenzoic acid.

RACEMISATION OF BIPHENYL COMPOUNDS

Since the optical activity of biphenyl compounds arises from restricted rotation, it might be expected that racemisation of these compounds would not be possible. In practice, it has been found that many optically active biphenyl compounds can be racemised under suitable conditions, *e.g.*, boiling in solution. The general theory of these racemisations is that heating increases the amplitude of the vibrations of the substituent groups in the 2,2′,6,6′-positions, and also the amplitude of vibration of the two benzene rings with respect to each other, thereby permitting the substituent groups to slip by one another. Thus the nuclei pass through a common plane and hence the probability is that the final product will contain an equimolecular amount of the (+)- and (–)-forms. Westheimer (1946-1950) has assumed, in addition to the above bond-stretchings, that the angles α, β and γ are deformed, and also the benzene rings themselves are deformed during racemisation. Westheimer obtained good agreement between the estimated and measured activation energies of racemisation of some di-orthosubstituted biphenyls. His calculations were based on known values of van der Waals radii and stretching and bending-force constants of the various bonds. In the absence of the assumption of deformations, the agreement between estimated and calculated values was very poor. At the same time, these calculations offer very strong evidence for the obstacle theory.

These ideas may be represented pictorially by the energy profile. The enantiomers (A) and (A') have the same energy content, and as they approach the planar configuration (B) the energy content increases due to steric repulsion.

The larger the groups a and b, the higher will be the energy barrier, *i.e.*, the slower will be the rate of racemisation.

There is, however, a further point that requires consideration. There are two possible planar configurations for the transition state, (B) and (C). As we have seen earlier, the general principle is that crossed steric interactions between groups of different size are less than the sum of the steric interactions between groups of equal size. Hence, the transition state for racemisation is more likely to be (B) than (C).

2,2′,6,6′-Tetrasubstituted biphenyl compounds may be classified under three headings according to the nature of the substituent groups.

(i) *Non-resolvable:* These contain any of the following groups: hydrogen, methoxyl or fluorine. The volumes (effective volumes) of these groups are too small to prevent rotation about the single bond. Thus, 2,2′-difluoro-6,6′-dimethoxybiphenyl-3,3′-dicarboxylic acid (I) is non-resolvable.

(ii) *Resolvable, but Easily Raceinised:* These must contain at least two amino-groups, or two carboxyl groups, or one amino- and one carboxyl group; the remaining groups may be any of those given in (i) [but not hydrogen]. Thus, 6,6′-difluorodiphenic acid (II) is resolvable, and is readily racemised.

(iii) *Not Racemisable at All:* Biphenyl compounds which fall in this group are those which contain at least two nitro-groups; the other groups can be any of those given in (i)-but not hydrogenand (ii). Thus 2,2′-difluoro-6,6′-dinitrobiphenyl (III) is resolvable, and cannot be racemised.

The order of steric hindrance produced by various groups appears to be :

$$Br \gg Me > Cl > NO_2 > CO_2H \gg OMe > F$$

This order corresponds roughly to the order of the van der Waals radii of the groups.

In addition to the size of the groups in the ortho-positions, the nature and position of other substituent groups also play a part in the rate of racemisation, *e.g.*, the rate of racemisation of (IV) is much slower than that of (V) (Adams *et al.*, 1932, 1934). Thus the nitro-group in position 3′ has a much greater stabilising influence than in position 5′. The reason for this is uncertain, but one possible explanation is as follows. In (VI), the methyl group of the methoxyl group is probably in the configuration shown. In (VII), the nitro-group in the 3′-position would tend to force the methyl group away, the resulting configuration being somewhat as shown in (VII); in this condition there would be greater interference between the methoxyl group and the two groups in the other benzene ring. This buttressing effect of groups is in the order

$$NO_2 > Br > Cl > Me$$

It is interesting to note that this order is very different from that in which these. groups produce steric hindrance in the o-position. It has also been found that groups in the 4- and/or 4′-positions affect the rate of racemisation. The reason for this is not clear. It has been proposed that the effect of a 4 - (and/or 4′-) nitro-group in retarding the racemisation of 6-nitrodiphenic acid is entirely due to a change in entropy of activation (Harris *et al.*, 1957).

Adams *et al.* (1954, 1957) have examined the rate of racemisation of (VIII). The rate is increased when R is an electron-attracting group such as NO_2 or CN, and is decreased when R is an electronreleasing group such as Me or OMe. These results were explained as follows. With, *e.g.*. R = NO_2, (IX) contributes to the resonance hybrid as well as (VIII). The resonance hybrid therefore has increased C=N double bond character and consequently it is now easier for the molecule to pass through a planar transition state. With, *e.g.*, R = Me, the C–N bond acquires far less double bond character than in its absence, and so it is more difficult for the molecule to pass through a planar transition state.

Adams *et al.* (1957, 1961) also examined the optical stability of compounds of type (X); they found that the half-life was in the following order for R: Me < Et < i-Pr < t-Bu. If the effect of R were due merely to the inductive effect, then the unexpected value for t-Bu cannot be explained on this basis. The authors have proposed the following explanation.

The t-Bu group, because of its large bulk, displaces the adjacent Me groups out of the plane of the benzene ring, thereby causing molecular overcrowding; this decreases the interference to rotation about the N–C (ring) bond. A molecular model of this compound showed such an interference. According to Bryan *et al.* (1960), it is possible that steric repulsion also operates to cause considerable angle distortion.

The racemisation of 2,2′-bridged biphenyls is, in general, easy to effect, since many of them are optically labile. Furthermore, the optical stability of a bridged compound is generally considerably lower than that of the corresponding unbridged biphenyl. If, however, the bridge contains double bonds, it will have additional rigidity and so will be less optically labile, *e.g.*, (XIII) is more optically stable than (X), (XI) and (XII).

Racemisation of biphenyls is usually effected by heating in a suitable solvent, but Mislow *et al.* (1963) have racemised (XIa) by irradiation (in ether solution) with ultraviolet light or by heating in the dark above 200°C. On the other hand, the ketone (XIb) underwent simultaneous racemisatior and decarbonylation to (XIc) when irradiated under the same conditions.

Mislow *et al.* (1963) have also compared the rates of racemisation of (XII) and (XIII), and found that the deutero compound (XIII) racemised 1.13 times as fast as (XII). This is an example of the secondary isotope effect and has also been referred to as an inverse isotope effect. These results may be explained by the fact that deuterium has a smaller van der Waals radius than hydrogen, and in keeping with this explanation is that Mislow *et al.* (1964) were unable to detect any optical activity in (XIV) (which was mixed with other deuteromethyl compounds: CD_3,CD_2H; CD_2H,CD_2H; CD_3, CDH_2).

EVIDENCE FOR THE OBSTACLE THEORY

Evidence for the obstacle theory, *i.e.*, interference of groups, amounts to proving that the two benzene rings in optically active biphenyl compounds are not coplanar. A direct chemical proof for the noncoplanar configuration was given by Meisenheimer *et al.* (1927). The method was to unite the 'obstacle groups' in optically active biphenyl compounds, thereby forming five- or six-membered rings. Meisenheimer started with 2,2′-diamino-6,6′-dimethylbiphenyl, resolved it and then carried out the following reactions on one of the enantiomers:

In all the optically active compounds, the rings cannot be coplanar, since if they were, the molecules would possess a centre or plane of symmetry.

If the dilactam, however, is not planar, then it would possess no elements of symmetry, and consequently would be optically active. If the dilactam is planar, then it has a centre of symmetry, and consequently cannot be optically active. This compound was, in fact, not optically active, and so must be planar. This planarity is readily explained in terms of resonance; all bonds have double bond character and so the molecule is planar (cf.. 2,2′-bridged biphenyls).

According to Dhar (1932), X-ray analysis studies have shown that in the solid state the biphenyl molecule is planar. On the other hand, according to Robertson (1961), who also examined crystalline biphenyl by X-ray analysis, the molecule is not strictly planar.

This non-planarity has been attributed to steric repulsion between the o-hydrogen atoms. Gas phase electron-diffraction studies indicated that the two rings are inclined at about 45° to one another (Brockway *et al.*, 1944; Bastiansen, 1949). In the solid state, crystal forces presumably tend to keep the biphenyl molecule almost planar.

Ultraviolet spectra measurements of biphenyl compounds have shown that the two rings in o-substituted biphenyls are not coplanar. The ultraviolet spectrum of biphenyl, λ_{max} 248 (19 000) rim, is different from that of benzene, λ_{max} 198 (8 000) nm. This shift to the longer wavelength can be explained on the basis that biphenyl is a resonance hybrid, one of the contributing structures being the extended conjugated form (II). Thus the interannular bond will have some partial double bond character and consequently the molecule will tend to be planar.

When o-positions are occupied, the coplanarity is prevented and consequently the spectrum will be different, the maximum absorption shifting to shorter wavelengths. Thus, 2-methylbiphenyl has λ_{max} 236 (10 000) nm and 2,2′-dimethylbiphenyl Amax 224 (~700) nm. Also, Pickett *et al.* (1936) have shown that the ultraviolet absorption spectra of bimesityl (III),λ_{max} 267 (545) nm, and mesitylene (IV), λ_{max} 266 (260) nm, were almost identical but different from the spectrum of biphenyl.

In (III) there is steric inhibition of resonance, and so coplanarity is prevented. Thus E for (III) is approximately double that of (IV), but 'Lax remains essentially unchanged.

In 2,2′-bridged biphenyls in which the bridge is saturated it is possible to calculate the angle of twist between the two benzene rings from normal bond angles and bond lengths. These values depend on the type of bridge, *i.e.*, the number of atoms and their nature (C, S, O. etc.). The ultraviolet spectra of these compounds have confirmed that the two benzene rings are not coplanar (when the bridge is of the appropriate type).

Other evidence for the obstacle theory is the behaviour of Chichibabin's hydrocarbon. Wheland *et al.* (1952) examined this compound by means of the ESR method and found that it existed in the form of the free diradical to the extent of 4-5 per cent. When, however, there were four chlorine atoms in the four o-positions, the amount of free diradical was very much greater. In the latter, the coplanarity of the two rings is now prevented.

Westheimer's calculations of the activation energies of racemisation of some biphenyl compounds are also evidence for the obstacle theory.

Stereochemistry of the Allenes

Allenes are compounds which have the general structure (I).

ahC=C=Cde   abC=C=Cah
(I)   (II)

Examination of the space formula of compounds of this type shows that the molecule and its mirror image are not superimposable. The $\sigma-\pi$ way of writing (I). The two end carbon atoms are in a state of trigonal hybridisation, and the centre carbon atom is in the digonal state. Thus the centre carbon atom forms two π-bonds which are perpendicular to each other the π_x-bond is perpendicular to the plane of the paper, and the π_y-bond is in the plane of the paper.

In the trigonal state, the π-bond is perpendicular to the plane containing the three σ-bonds consequently the groups a and b lie in the plane of the paper, and the groups d and e in the plane perpendicular to the plane of the paper. This molecule does not possess a plane or centre of symmetry: this is also true for molecule (II). Thus (I) and (II) will be resolvable.

The resolvability of allenes was predicted by van't Hoff in 1875, but experimental verification was not obtained until 1935, when Mills and Maitland carried out a catalytic asymmetric dehydration on 1,3-dil-naphthyl-l,3-diphenylprop-2-enol (III), to give the dinaphthyl-diphenylallene (IV).

When the dehydration was carried out with an optically inactive dehydrating catalyst, *e.g.*, p-toluenesulphonic acid, the racemic modification of the allene derivative was obtained. When, however, the alcohol (III) was boiled with 1 per cent benzene solution of (+)-camphorsulphonic acid, a dextrorotatory allene was obtained.

Similarly, (–)-camphorsulphonic acid gave a laevorotatory allene. Another asymmetric synthesis of an allene is that of 1,3-diphenylallene. Jacobs *et al.* (1957) rearranged 1,3-diphenylpropyne by adsorption on alumina impregnated with brucine or quinine; the former gave the (-)-allene and the latter the (+)-allene:

$$PhC \equiv CCH_2Ph \xrightarrow[\text{alkaloid}]{Al_2O_3} PhCH{=}C{=}CHPh$$

The first successful resolution of an allene derivative was carried out by Kohler *et al.*, also in 1935. Lapworth and Wechsler (1910) prepared 3-1-naphthyl-l,3-diphenylallene-l-carboxylic acid (V) but failed

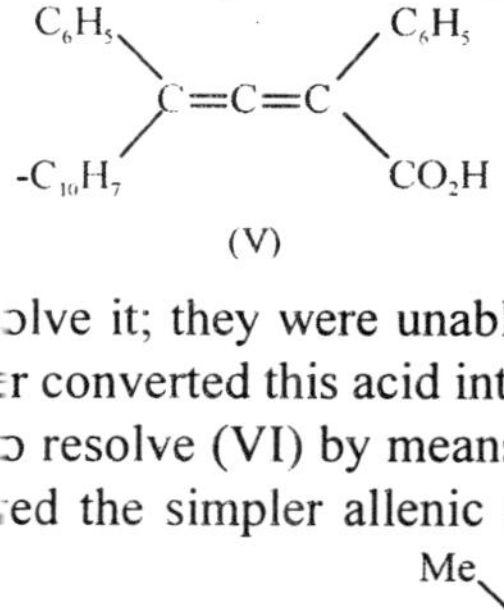

$$(C_6H_5)(\text{-}C_{10}H_7)C{=}C{=}C(C_6H_5)(CO_2H)$$

(V)

$$(C_6H_5)(1\text{-}C_{10}H_7)C{=}C{=}C(C_6H_5)(COOCH_2CO_2H)$$

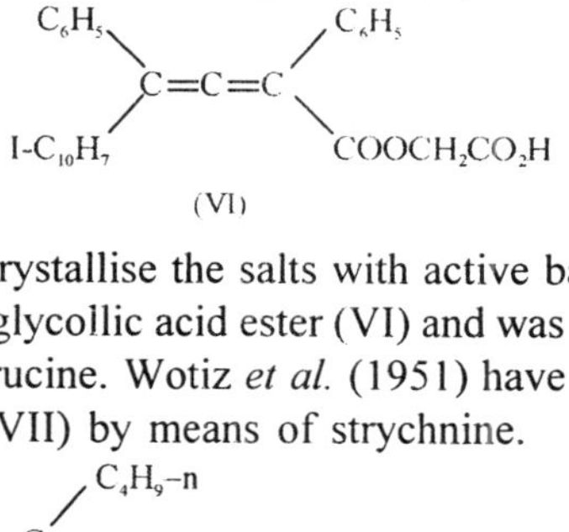

(VI)

to resolve it; they were unable to crystallise the salts with active bases. Kohler converted this acid into the glycollic acid ester (VI) and was then able to resolve (VI) by means of brucine. Wotiz *et al.* (1951) have also resolved the simpler allenic acid (VII) by means of strychnine.

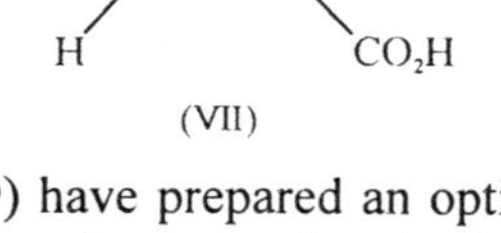

$$(Me)(H)C{=}C{=}C(C_4H_9\text{-}n)(CO_2H)$$

(VII)

Landor *et al.* (1959) have prepared an optically active allene by a method which correlates it stereochemically with a tetrahedrally asymmetric alcohol. An optically active acetylenic alcohol, on treatment with thionyl chloride, gave an optically active allene; the mechanism is possibly S_Ni'.

Landor *et al.* (1962) have also deduced the absolute configuration of the (4)-chloride by first determining the absolute configuration of the (+)-alcohol; the (R)-(–)-alcohol gave the (S)-(–)allene.

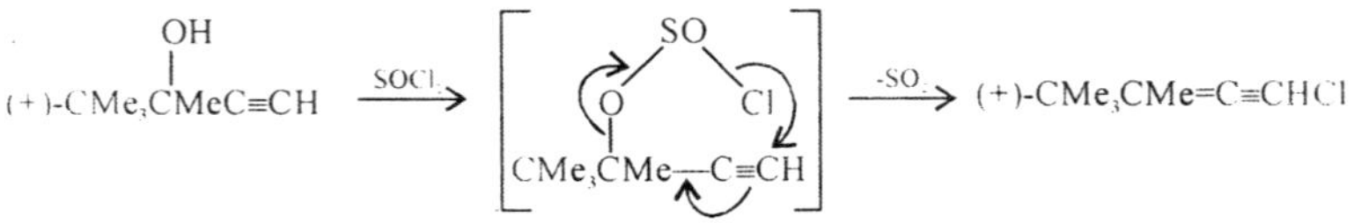

$$(+)\text{-}CMe_3\overset{OH}{\overset{|}{C}}MeC\equiv CH \xrightarrow{SOCl_2} \left[\begin{array}{c} SO \\ O \quad Cl \\ CMe_3CMe-C\equiv CH \end{array} \right] \xrightarrow{-SO_2} (+)\text{-}CMe_3CMe=C=CHCl$$

Fig. 21

It has been previously pointed out that if the number of double bonds in the cumulene is odd the molecule exhibits geometrical isomerism, but if even, then it exhibits optical isomerism. The allenes discussed above contain two double bonds, but more recently, Nakagawa *et al.* (1961) have prepared the following cumulene with four double bonds, 1,5-di-t-butyl-l,5-dip-chlorophenylpentatetraene, in optically active forms.

Although allenes were not successfully resolved until 1935, compounds with a similar configuration were resolved as early as 1909. In this year, Pope *et al.* resolved 4-methylcyclohexylidene-1-acetic acid (VIII); in this compound one of the double bonds of allene has been replaced by a six-membered ring, and the general shape of the allene molecule is retained.

It is interesting to note, in connection with allenes, that the antibiotic mycomycin and other natural poly-acetylenes have been shown to contain the allene grouping. Mycomycin is optically active, and owes its optical activity to the presence of this grouping. Celmer and Solomons (1953) have shown that the structure of mycomycin is:

$$CH\equiv CC\equiv CCH=C=CHCH=CHCH=CHCH_2CO_2H.$$

Polynuclear Hydrocarbons

The polynuclear hydrocarbons may be divided into two groups, those in which the rings are isolated, *e.g.*, biphenyl, diphenylmethane, etc.; and those in which two or more rings are fused together in the o-positions, *e.g.*, naphthalene, anthracene, etc.

Isolated systems : Biphenyl (diphenyl), m.p. 71°C, occurs in small quantities in coal-tar. It may be prepared:

(i) **By Fittig's reaction** (1863). This is carried out by treating bromobenzene with sodium in ethereal solution:

$$2C_6H_5Br + 2Na \longrightarrow C_6H_5C_6H_5 + 2NaBr \qquad (20\text{–}30\%)$$

(ii) When bromobenzene in ethanolic solution is made alkaline and refluxed with hydrazine in the presence of a palladium catalyst on a calcium carbonate support, biphenyl is obtained (Busch *et al.*, 1936):

$$2C_6H_5Br \longrightarrow C_6H_5C_6H_5 + 2HBr \qquad (80\%)$$

(iii) (a) By warming benzenediazonium sulphate in ethanol with copper powder (see also the Gomberg reaction):

$$2C_6H_5N_2^+ HSO_4^- \xrightarrow[C_2H_5OH]{Cu} C_6H_5C_6H_5 + 2N_2 + 2H_2SO_4 \ (25\%)$$

The yield has been improved (to 50%) by warming a mixture of aniline, pentyl nitrite and benzene and finally refluxing (Cadogan, 1962). This method may also be used for preparing biphenyl derivatives.

(iv) By the **Ullmann biaryl synthesis** (1903). Iodobenzene is heated with copper powder in a sealed tube:

$$2C_6H_5I + 2Cu \longrightarrow C_6H_5C_6H_5 + 2CuI \qquad (80\%).$$

(v) Biphenyl is formed by reaction between phenylmagnesium bromide and bromobenzene in the presence of a small amount of cobaltous chloride.

$$C_6H_5MgBr + C_6H_5Br \xrightarrow{CoCl_2} C_6H_5C_6H_5 + MgBr_2 \ (g.)$$

Industrially, biphenyl is prepared by passing benzene vapour through a red-hot tube, preferably packed with pumice (600-800°C):

A more recent method is to mix benzene vapour (preheated to 650°C) with superheated steam (1000–1100°C), and passing the mixture into steel vessels coated internally with a film of Fe_3O_4.

Biphenyl is increasing in use as a heat transfer medium; chlorinated biphenyls are used as plasticisers.

Biphenyl undergoes the usual nuclear substitution reactions, the phenyl group being o- and p-orienting (one phenyl group behaving as an electron-releasing group and the other as an electron-acceptor).

The first substituent enters mainly the 4-position, and to a lesser extent the 2-position. Whether this group is electron-releasing or withdrawing, introduction of a second substituent usually takes place in the *unsubstituted* ring in the p-and o-positions, *e.g.*, on nitration, the main product is 4-nitrobiphenyl, together with a small amount of 2-nitrobiphenyl.

On the other hand, if the stabilities of the carbonium ion intermediates are examined, it will be seen that there is greater dispersal of charge in the -(or o-) ion than in the m-. Consequently, the former is more stable and so this orientation is favoured.

When oxidised with chromic acid, biphenyl forms a small amount of benzoic acid, most of the hydrocarbon being oxidised completely to carbon dioxide and water.

On the other hand ozonolysis of biphenyl at –20°C gives an 86 per cent yield of benzoic acid (Copeland *et al.*, 1961).

Benzidine (4,4-*diaminobiphenyl*) may be prepared by the benzidine rearrangement; hydrazobenzene, on warming with hydrochloric acid, rearranges to benzidine. Benzidine is a colourless solid, m.p. 127°C. Its hydrochloride is soluble, but its sulphate is sparingly soluble in water. It is very important commercially since it is used in the preparation of azo-dyes, *e.g.*, Congo red.

o-Tolidine (4,4'-diammo-3,3'-*dmethylbiphenyl*), m.p. 129°C, and dianisidine (4,4'-diamino-3'-dimethoxybiphenyl), m.p. 138°C, are manufactured on a large scale as intermediates in the preparation of azo-dyes. If at least three of the positions 2, 2', 6 and 6' are occupied by sufficiently large groups, free rotation about the single bond joining the two phenyl groups is no longer possible.

Provided each ring has no vertical plane of symmetry, this restricted rotation gives rise to optical activity due to the molecule being *asymmetric as a whole*; *e.g.*, 6-*nitrobiphenyl*-2,2'-*dicarboxylic acid* (1), and 6,6'-*diamino*-2,2'-*dimethylbiphenyl* (II), have been resolved.

If the substituent group is large enough, then only two groups in the o- and o'-positions will cause restricted rotation, *e.g.*, *biphenyl-2,2-disulphonic* acid (III), has -been shown to be optically active.

The cause of the restricted rotation is due mainly to the stearic effects of the groups in the o- and o'-positions.

This type of stereoisomerism, arising from restricted rotation about a single bond and where the stereoisomers can be isolated, is called **atropisomerism**; the isomers are known as *atropisomers*.

Since the optical activity of biphenyl compounds arises from restricted rotation, it might be expected that racemisation would not be possible. In practice, however, many optically active biphenyls can be racemised, *e.g.*, by boiling in solution.

The general theory of these racemisations is that heating increases the amplitude of vibrations of the substituents and also the amplitude of vibration of the interannular bond. In addition to these bond stretchings, the various valency angles between ring and substituents are also deformed. Thus the nuclei will pass through a common plane, and hence, when the molecule returns to its restricted state, it will do so equally in the (+)- and (–)-forms, *i.e.*, the molecules are racemised. A number of polyphenyls and some of their derivatives have been prepared, *e.g.*, *terphenyl*, *quaterphenyl*, *quinquephenyl* and *sexiphenyl*.

Diphenylmethane, m.p. 26°C. Some related compounds of diphenylmethane have already been discussed, *e.g.*, benzophenone. Diphenylmethane may be prepared:

(i) By the Friedel-Crafts condensation between benzyl chloride and benzene:

$$C_6H_5CH_2Cl + C_6H_6 \xrightarrow{AlCl_3} C_6H_5CH_2C_6H_5 + HCl$$

(50%)

(ii) By the condensation between one molecule of formaldehyde and two molecules of benzene in the presence of concentrated sulphuric acid:

$$2C_6H_6 + CH_2O \xrightarrow{H_2SO_4} C_6H_5CH_2C_5H_5 + H_2O$$

(iii) By heating benzophenone with hydriodic acid and red phosphorus at 150°C under pressure, by the Wolff-Kishner reduction, or by LAH–$AlCl_3$.

$$C_6H_5COC_6H_5 \xrightarrow{HI/P} C_6H_5CH_2C_6H_5 \text{ (95\%)}$$

(iv) By the following Grignard reaction:

$$C_6H_5MgBr + C_6H_5CH_2Cl \xrightarrow[60-70\,°C]{PhH} C_6H_5CH_2C_6H_5 \text{ (ex)}$$

The reactions of diphenylmethane are similar to those of biphenyl, *e.g.*, on nitration, the nitro-group enters mainly the 4-positions; the second nitro-group then enters the 4'-position (the benzyl group is o-p-orienting).

Since the hydrogen of the methylene group is very active (due to each ring being electron-attracting), bromination of diphenyl-methane gives rise to substitution at the methylene carbon atom, and not in the ring, to form, *e.g.*, diphenylmethyl bromide:

$$C_6H_5CH_2C_6H_5 + Br_2 \longrightarrow C_6H_5CHBrC_6H_5 + HBr$$

Diphenylmethyl bromide is readily hydrolysed to alkali to *benzhydrol (diphenlcarbinol:)*. When oxidised with chromic acid, diphenylmethane forms benzophenone. When its vapour is passed through a red-hot tube, diphenylmethane forms fluorene:

Triphenylmethane (tritane) may be prepared by the following methods:

$$2C_6H_5 + C_6H_5\,CHCl_2 \xrightarrow{AlCl_3} (C_6H_5\,)_3CH + 2HCl$$

$$3C_6H_5 + CHCl_3 \xrightarrow{AlCl_3} (C_6H_5\,)_3CH + 3HCl\ (33\%)$$

$$C_6H_5\,CHO + 2C_6H_6 \xrightarrow{AnCl_2} (C_6H_5\,)_3CH + H_2O$$

Triphenylmethane, m.p. 93°C, is the parent substance of the triphenylmethane dyes. When brominated, it forms tripheylmethyl bromide, $(C_6H_5\,)_3CBr$ (cf. diphenylmethane), and may be oxidised directly to triphenylcarbinol. Triphenylmethane is acidic (more so than diphenylmethane), the conjugate base being stabilised by resonance. Thus, when treated with sodium in ethereal solution (or in liquid ammonia), triphenylmethane forms triphenyl-methylsodium:

$$(C_6H_5)_3CH + Na \longrightarrow (C_6H_5)_3\ \bar{C}Na^+ + \frac{1}{2}Na$$

Triphenylmethylsodium is also formed when triphenylmethyl chloride is treated with sodium amalgam.

Derivatives of triphenylmethane are best prepared synthetically.

Triphenylmethyl chloride (trityl chloride), m.p. 112°C, may be prepared from triphenyl-carbinol and acetyl chloride (see below), or by the Friedel-Crafts condensation between benzene and carbon tetrachloride; tetraphenylmethane is not formed:

$$3C_6H_6 + CCl_4 \xrightarrow{AlCl_3} (C_6H_5)_3CCl + 3HCl\ (68\text{-}84\%)$$

The halogen atom is extremely reactive, *e.g.*, when boiled with water, triphenylmethyl chloride forms the corresponding alcohol, and with ethanol it forms the ethyl ether:

$$(C_6H_5)_3COC_2H_5 \xleftarrow{C_2H_5OH} (C_6H_5)_3CCl \xrightarrow{H_2O} (C_6H_5)_3COH$$

Triphenylmethyl chloride may be converted into triphenylmethane as follows:

$$(C_6H_5)_3CCl \xrightarrow{Mg} (C_6H_5)_3CMgCl \xrightarrow{H_2O} (C_6H_5)_3COH$$

This reaction is particularly interesting since the formation of the Grignard reagent from triphenylmethyl bromide was shown to occur via a free-radical mechanism (Gomberg and Bachmann, 1930). These authors showed that when half the total amount of magnesium had reacted, the solution contained no Grignard reagent:

$$2(C_6H_5)_3CBr \longrightarrow 2(C_6H_5)_3C\cdot + MgBr$$

$$Mg + MgBr_2 \rightleftharpoons 2\cdot MgBr$$

$$(C_6H_5)C\cdot + \cdot MgBr \longrightarrow (C_6H_5)_3CMgBr$$

When triphenylmethyl chloride is dissolved in liquid sulphur dioxide, it ionises:

$$Ph_3CCl \rightleftharpoons Ph_3C^+ + Cl^-$$

Triphenylcarbinol, m.p. 165°C; may be prepared by the oxidation of triphenylmethane with chromium trioxide:

$$(C_6H_5)_3CH \xrightarrow{CrO_3} (C_6H_5)_3COH\ (85\%)$$

It may also be prepared from benzophenone or ethyl benzoate as follows:

$$(C_6H_5)_2CO \xrightarrow[(ii)\,H_2O^+]{(i)\,C_6H_5MgBr} (C_6H_5)_3COH \xleftarrow[(ii)\,H_3O^+]{(i)\,2C_6H_5MgBr} C_6H_5CO_2C_2H_5$$

Triphenylcarbinol reacts extremely rapidly with hydrochloric acid or acetyl chloride to form triphenylmethyl chloride (cf. tertiary alcohols):

$$(C_6H_5)_3COH + CH_3COCl \longrightarrow (C_6H_5)_3CCl + CH_3CO_2H\ (93–95\%)$$

It also reacts rapidly with alcohols in the presence of a trace of acid or iodine to form ethers, *e.g.*,

$$(C_6H_5)_3COH \overset{H^+}{\rightleftharpoons} (C_6H_5)_3C—OH_2^+ \rightleftharpoons H_2O + (C_6H_5)_3C^+$$

$$\xrightarrow{C_2H_5OH} (C_6H_5)_3COC_2H_5 + H^+$$

Triphenylcarbinol is reduced to triphenylmethane by formic acid or LAH—$AlCl_3$, and condenses with aniline hydrochloride to form p-aminotetraphenylmethane.

Tetraphenylmethane, m.p. 282°C, may be prepared as follows:

$$(C_6H_5)_3CCl + C_6H_5MgBr \longrightarrow (C_6H_5)_4C + MgClBr\ \ (v.p.)$$

It may also be prepared by diazotising p-aminotetraphenylmethane in ethanolic solution and then boiling (see also triphenylcarbinol, above):

$$(C_6H_5)_3COH + C_6H_5NH_2{\cdot}HCl \longrightarrow (C_6H_5)_3CC_6H_4NH_2{\cdot}HCl$$

$$\xrightarrow[C_2H_5OH]{NaNO_2/H_2SO_4} (CH_5)_4C$$

Tetraphenylmethane *cannot* be prepared by reaction between benzene and carbon tetrachloride in the presence of aluminium chloride (see triphenylmethyl chloride, above).

The reason for this is not certain, but it is likely that the steric effect plays a large part in hindering the Reaction

When triphenylmethyl chloride, in benzene solution, is treated with silver *in the absence of* air, and the yellow benzene solution then evaporated, the residue is a white solid, m.p. 145-147°C.

This was believed to be *hexaphenylethane*, $(C_6H_5)_6C_2$, the yellow colour being due to its dissociation into triphenylmethyl free radicals (Gomberg, 1900). Lankamp *et al.*, (1968) and Guthrie *et al.*, (1969), however, have now shown that the dimer of triphenylmethyl radical is not hexaphenylethane but is **1-diphenylmethylene-4-tritylcyclohexa-2,5-diene**:

The stability of this free radical is attributed to resonance, a large number of resonating structures contributing to the resonance hybrid:

Triphenylmethyl free radical reacts immediately with a number of reagents to form triphenyl-methyl derivatives, *e.g.*, with oxygen, it forms a colourless peroxide:

$$(C_6H_5)_3C\cdot \xrightarrow{O_2} (C_6H_5)_3C—O—O\cdot$$

$$\xrightarrow{(C_6H_5)_6C_2} (C_6H_5)_3C—O—O—C(C_6H_5)_3 +$$

$$(C_6H_5)_3C\cdot$$

With iodine it forms triphenylmethyl iodide, and with nitric oxide, nitrosotriphenylmethane (the formation of the latter compound appears to be uncertain):

$$2(C_6H_5)_3C\cdot + I_2 \rightleftharpoons 2(C_6H_5)_3CI$$

$$(C_6H_5)_3C\cdot + NO \longrightarrow (C_6H_5)_3CNO$$

It also combines with sodium to form triphenylmethylsodium, a brick-red solid and an electrical conductor:

$$(C_6H_5)3C\cdot + Na \longrightarrow (C_6H_5)_3\ \bar{C}Na^+$$

In addition to the above reactions, triphenylmethyl free radical can act as a powerful reducing agent, and will reduce the salts of silver, gold and mercury to the metals, *e.g.*,

$$(C_6H_5)_3C\cdot + AgCl \longrightarrow (C_6H_5)_3CCl + Ag$$

The triphenylmethyl free radical is an example of a long-life free radical. The problem here is why does 'hexaphenylethane' dissociate so readily (as compared to ethane)? The answer is believed to be as follows.

The free radical is stabilised by resonance, which therefore acts as a 'driving force' in the dissociation. At the same time, steric factors also operate. In 'hexaphenylethane' (sp^3 hybridisation) there is steric strain due to crowding, and this is relieved by dissociation to form a planar or almost planar free radical (sp^2 hybridisation; resonance requires complete or almost complete planarity). That the steric effect is operating is supported by the fact that methyl groups in the o-positions increase the dissociation. Further support for the steric effect is the case of schitschibabin's hydrocarbon: This compound contains about 4.5 per cent free radical, but when all four o-positions are occupied by chlorine the free-radical content is almost 100 per cent. In the chlorinated compound coplanarity of the two benzene rings is prevented, *i.e.*, this is an example of steric inhibition of resonance. When dissolved in liquid sulphur dioxide, the solution shows a high conductivity (Walden, 1903), and this was explained by assuming that the 'hexaphenylethane' dissociates into positive and negative ions:

$$(C_6H_5)_6C_2 \rightleftharpoons (C_6H_5)_3C^+ + (C_6H_5)_3C^-$$

The existence of the triphenylmethyl carbonium ion has definitely been established by Dauben *et al.*, (1960), who prepared the crystalline perchlorate, $Ph_3C^+ClO_4^-$, and the fluoroborate, $Ph_3C^+BF_4^-$.

These carbonium ions have been shown to have the property of abstracting hydride ions from hydrocarbons of low nucleophilic power:

$$Ph_3C^+ClO_4^- + RH \rightleftharpoons Ph_3CH + R^+ClO_4^-$$

This property thus affords a means of preparing new carbonium salts.

Bibenzyl (1,2-*diphenylethane*), m.p. 52°C, may be prepared by the Friedel-Crafts condensation between benzene and ethylene dichloride:

$$2C_6H_6 + ClCH_2CH_2Cl \xrightarrow{AlCl_3} C_6H_5CH_2CH_2C_6H_5 + 2HCl$$

Another method of preparation is by the action of sodium or copper on benzyl bromide:

$$2C_6H_5CH_2Br \xrightarrow{Na} C_6H_5CH_2CH_2C_6H_5$$

Bibenzyl is oxidised by permanganate or chromic acid to benzoic acid.

trans-stilbene (trans-1, 2-diphenylethylene) may be prepared:

(i) By treating benzylmagnesium bromide with benzaldehyde and dehydrating the product, benzylphenylmethanol:

$$C_6H_5CHO + C_6H_5CH_2MgBr \longrightarrow C_6H_5CH\begin{cases} OMgBr \\ CH_2C_6H_5 \end{cases} \xrightarrow{H^+}$$

$$C_6H_5CHOHCH_2C_6H_5 \xrightarrow[(heat)]{-H_2O} C_6H_5CH{=}CHC_6H_5$$

(ii) By reducing benzoin with amalgamated zinc and an ethanolic solution of hydrogen chloride:

$$C_6H_5CHOHCOC_6H_5 \longrightarrow C_6H_5CH = CHC_6H_5$$ (53-57%)

(iii) By heating α-phenylcinnamic acid in quinoline in the presence of copper chromite

$$C_6H_5CHO + C_6H_5CH_2CO_2K \xrightarrow{(CH_3CO)_2O}$$

$$C_6H_5CH = C(C_6H_5)CO2H \longrightarrow C_6H_5CH = CHC_6H_5$$

(iv) Stilbene may also be prepared by the Meerwein Reaction

trans-Stilbene, m.p. 124°C, is the stable isomer; cis-stilbene (isostilbene), b.p. 145° (13mm), is the unstable isomer;

$$\begin{array}{c} C_6H_5 \quad H \\ \diagdown \; \diagup \\ C \\ \| \\ C \\ \diagup \; \diagdown \\ H \quad C_6H_5 \end{array} \qquad \qquad \begin{array}{c} C_6H_5 \quad H \\ \diagdown \; \diagup \\ C \\ \| \\ C \\ \diagup \; \diagdown \\ C_6H_5 \quad H \end{array}$$

trans-stilbene cis-stilbene

trans-Stilbene is reduced by sodium and ethanol to bibenzyl. It adds on bromine to form stilbene dibromide which, on heating with ethanolic potassium hydroxide, forms **tolan** (*diphenylacetylene*), m.p. 62°C:

$$C_6H_5CH{=}\ CHC_6H_5 \xrightarrow{Br_2} C_6H_5CHBrCHBrC_6H_5$$

$$\xrightarrow{KOH} C_6H_5C{\equiv}CC_6H_5$$

Stilbene dibromide reacts with silver acetate to form two isomeric diacetates. These, on hydrolysis, form hydrobenzoin and isohydrobenzoin :

$$C_6H_5CH(OCOCH_3)CH(OCOCH_3)C_6H_5 \xrightarrow{H_2O} C_6H_5CHOHCHOHC_6H_5$$

cis-Stilbene may be prepared by reducing diphenylacetylene with zinc dust and ethanol, or by irradiating trans-stilbene with ultra-violet light.

$$C_6H_5C \equiv CC_6H_5 \xrightarrow{e;\,H^+} C_6H_5CH = CHC_6H_5$$

cis-Stilbene is readily converted into trans-stilbene under the catalytic influence of traces of hydrogen bromide and peroxides.

Benzoin, m.p. 137°C, is usually prepared by the benzoin condensation, this reaction being carried out by refluxing benzaldehyde with aqueous ethanolic potassium cyanide:

$$2C_6H_5CHO \xrightarrow{KCN} C_6H_5CHOHCOC_6H_5\ (83\%)$$

The generally accepted mechanism is

$$C_6H_5\overset{\overset{O}{\|}}{\underset{\underset{H}{|}}{C}} + :CN^- \rightleftharpoons \left[C_6H_5\overset{\overset{O^-}{|}}{\underset{\underset{H}{|}}{C}}{-}CN\right] \rightleftharpoons \left[C_6H_5{-}\overset{\overset{OH}{|}}{\underset{..}{C}}{-}CN\right] \underset{}{\overset{C_6H_5CHO}{\rightleftharpoons}}$$

$$\left[C_6H_5\overset{\overset{OH}{|}}{\underset{\underset{CN}{|}}{C}}{-}\overset{\overset{O^-}{|}}{\underset{\underset{CN}{|}}{C}}{-}C_6H_5\right] \rightleftharpoons \left[C_6H_5\overset{\overset{O^-}{|}}{\underset{\underset{CN}{|}}{C}}{-}\overset{\overset{OH}{|}}{\underset{\underset{H}{|}}{C}}{-}C_6H_5\right]$$

$$\rightleftharpoons C_6H_5\overset{\overset{O}{\|}}{C}{-}\overset{\overset{OH}{|}}{\underset{\underset{H}{|}}{C}}C_6H_5 + :CN^-$$

Fig. 22

When a mixture of aldehydes is treated with aqueous ethanolic potassium cyanide, 'mixed' benzoins (as well as the 'single' benzoins) are obtained:

$$Ar^1CHO + Ar^2CHO \xrightarrow{KCN} Ar^1CHOHCOAr^2 + Ar^2CHOHCOAr^1$$

Furthermore, the reversibility of the benzoin condensation is indicated by the fact that when the benzoin $Ar^1CHOHCOAr^1$ is heated with the aldehyde Ar2CHO in the presence of potassium cyanide, a mixed benzoin is obtained (Buck and Ide, 1933):

Aliphatic aldehydes do not undergo the benzoin condensation.

Benzoins (and acyloins) tautomerise to *enediols*:

$$\text{—CHOHCO—} \rightleftharpoons \text{HO—}\overset{|}{C}\text{=}\overset{|}{C}\text{—OH}$$

There is much evidence to support this, *e.g.*, acetylation produces the enediol diacetate. The enediol is usually less stable than its corresponding a-hydroxyketone; in certain cases the former has been isolated.

Benzoin is reduced to **deoxybenzoin** (*desoxybenzoin*), m.p. 60°C, which may also be prepared by the following Friedel-Crafts reaction:

$$C_6H_5CH_2OCl + C_6H_6 \xrightarrow{AlCl_3} C_6H_5CH_2COC_6H_5 + HCl\ (82\text{-}83\%)$$

The prefix *deoxy-* (formerly *desoxy-*) indicates the replacement of a hydroxyl group by hydrogen. Deoxybenzoin contains an active methylene group; *e.g.*, it forms a sodio-derivative, condenses with aldehydes in the presence of piperidine, and adds on to a, unsaturated carbonyl compounds (cf. ethyl malonate).

Benzoin is an α-hydroxyketone; hence it reduces Fehling's solution and forms an osazone.

When reduced by sodium amalgam, benzoin forms mainly **hydrobenzoin**, and a small amount of **isohydrobenzoin**:

$$C_6H_5CHOHCOC_6H_5 \xrightarrow{e\,;\,H^+} C_6H_5CHOHCHOHC_6H_5$$

This formula (tolylene glycol) contains two identical asymmetric carbon atoms, and hence the compound can exist in the dextro, laevo and meso-forms (cf. tartaric acid). The racemic modification is known as hydrobenzoin (m.p. 139°C), and the meso-form as isohydrobenzoin (m.p. 121°C). Both forms give benzoic acid on oxidation.

Benzoin is oxidised by chromic acid to a mixture of benzaldehyde and benzoic acid, and by nitric acid to *benzil:*

$$C_6H_5CHOHCOC_6H_5 \xrightarrow[HNO_3]{[O]} C_6H_5COCOC_6H_5 \ (96\%)$$

Purer benzil is obtained by oxidation with copper sulphate in aqueous pyridine (86% yield).

Benzil is a yellow crystalline solid, m.p. 95°C. It behaves as a typical α-diketone, *e.g.*, it is oxidised by hydrogen peroxide in acetic acid to benzoic acid, forms a monoxime and dioxime etc.

Two monoximes of benzil are known: α-(cis-), m.p. 134°C, and β-(trans), m.p. 113°C. Three dioximes are possible and all are known: α-(cis-cis-), m.p. 237°C, β-(trans-trans-), m.p. 207°C, and γ-or *amphi-(cis-trans-)*, m.p. 166°C.

Benzil condenses with o-phenylenediamine to form 2,3-*diphenylquinoxaline*: The nature of the reduction products of benzil depends on the reducing agent used.

Thus reduction with sodium amalgam gives hydrobenzoin (and some isohydrobenzoin); amalgamated zinc and hydrogen chloride in ethanol, stilbene; sodium hyposulphite in ethanol benzoin; amalgamated tin and hydrogen chloride in ethanol deoxybenzoin: and catalytic reduction using nickel at 230°C, bibenzyl.

When heated with ethanolic potassium hydroxide, benzil undergoes the **benzilic acid rearrangement:**

$$C_6H_5COCOC_6H_5 + KOH \longrightarrow (C_6H_5)_2C(OH)CO_2K \ (77\text{-}79\%)$$

The rearrangement has been shown to be first order with respect to both benzil and hydroxide ion (Westheimer, 1936), and it has also been shown that when benzil is heated for a very short time with methanolic sodium hydroxide in water containing ^{18}O, the benzil recovered contained ^{18}O (Roberts and Urey, 1938).

This is in keeping with the assumption that the first step is the rapid *reversible* addition of hydroxide ion to benzil.

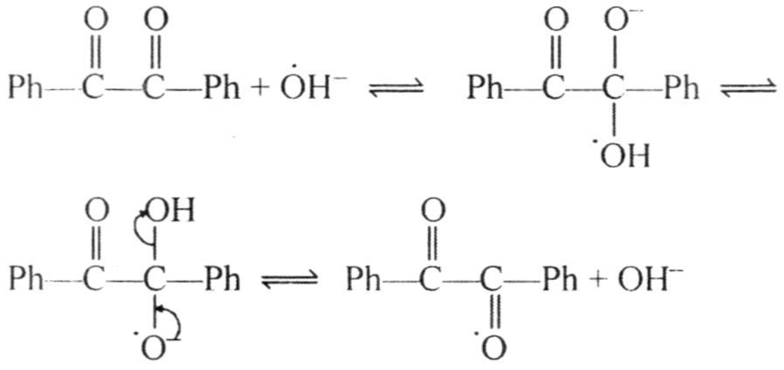

$$\mathrm{Ph{-}\overset{O}{\overset{\|}{C}}{-}\overset{O}{\overset{\|}{C}}{-}Ph + \dot{O}H^{-} \rightleftharpoons Ph{-}\overset{O}{\overset{\|}{C}}{-}\underset{\dot{O}H}{\overset{O^{-}}{C}}{-}Ph \rightleftharpoons}$$

$$\mathrm{Ph{-}\overset{O}{\overset{\|}{C}}{-}\underset{\dot{O}^{-}}{\overset{OH}{C}}{-}Ph \rightleftharpoons Ph{-}\overset{O}{\overset{\|}{C}}{-}\underset{\dot{O}}{\underset{\|}{C}}{-}Ph + OH^{-}}$$

All of these results are in keeping with the mechanism proposed by Ingold in 1928:

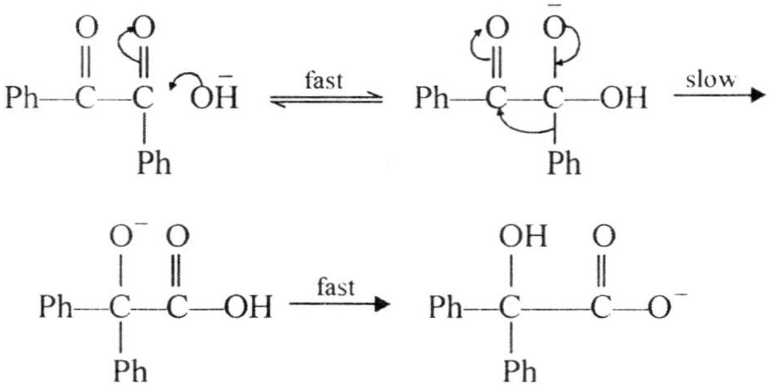

$$\mathrm{Ph{-}\overset{O}{\overset{\|}{C}}{-}\underset{Ph}{\overset{O}{\overset{\|}{C}}}\ \bar{O}H \xrightleftharpoons{\text{fast}} Ph{-}\overset{O}{\overset{\|}{C}}{-}\underset{Ph}{\overset{\bar{O}}{C}}{-}OH \xrightarrow{\text{slow}}}$$

$$\mathrm{Ph{-}\underset{Ph}{\overset{O^{-}}{C}}{-}\overset{O}{\overset{\|}{C}}{-}OH \xrightarrow{\text{fast}} Ph{-}\underset{Ph}{\overset{OH}{C}}{-}\overset{O}{\overset{\|}{C}}{-}O^{-}}$$

Fig. 23

This is an example of the 1,2-shift, and since a phenyl group is involved, it is possible that a phenonium ion is formed as an intermediate. A prediction that can be made from the above mechanism is that when the two aryl groups are different, the one that is more electron-releasing (due to the presence of a suitable substituent, *e.g.*, p-MeO) will tend to neutralise the positive charge on the carbon atom to which it is attached when the C=O bond is polarised. Thus, it will be the *other* CO group that will link with the hydroxide ion, and consequently it will be the aryl group attached to this 'other' CO group that migrates preferentially. This prediction is borne out in practice, *e.g.*, in p-methoxybenzil, the methoxyphenyl group migration was about 32 per cent. (Roberts *et al.*, 1951), and Eastham, *et al.*, (1958), using labelled p-methylbenzil, showed that the phenyl group migrated almost exclusively.

Furthermore, it has also been shown that the hydroxide ion is *not* a specific catalyst; sodium methoxide and potassium t-butoxide produce the corresponding esters of benzilic acid (Doering *et al.*, 1956). Formation of these esters is strong evidence that the first step in the rearrangement is the addition of the nucleophilic reagent.

$$\mathrm{Ph{-}C({=}O){-}C({=}O)(Ph)} + \mathrm{OMe^-} \longrightarrow \mathrm{Ph{-}C(O^-)(Ph){-}C({=}O){-}OMe}$$

$$\xrightarrow{H_2O} \mathrm{Ph_2C(OH)CO_2Me}$$

Benzilic acid (m.p. 150°C) forms diphenylacetic acid when refluxed in acetic acid solution with hydriodic acid and red phosphorus:

$$(C_6H_5)_2C(OH)CO_2H \xrightarrow{HI/P} (C_6H_5)_2CHCO_2H + H_2O \ (94\text{–}97\%)$$

When oxidised with chromic acid, benzilic acid forms benzophenone.

BENZENE ANALYSIS

Analysis and molecular-weight determinations show that the molecular formula of benzene is C_6H_6. The corresponding alkane is hexane, C_6H_{14}, and since the number of hydrogen atoms in benzene is eight less, it is to be expected that benzene would exhibit marked 'unsaturated reactions'. This is found to be so in practice, *e.g.*,

(i) Benzene adds on halogen, the maximum number of halogen atoms being six.

(ii) Benzene may be catalytically hydrogenated to cyclohexane, the maximum number of hydrogen atoms added is six.

(iii) Benzene forms a *triozonide,* $C_6H_6(O_3)_3$.

All these reactions indicate that benzene contains *three* double bonds. Further examination, however, shows that these double bonds behave in a most remarkable manner in comparison with double bonds in aliphatic compounds, *e.g.*,

(iv) Alkaline permanganate has no action on benzene in the cold, but on prolonged boiling, benzene is broken down into carbon dioxide and water.

(v) In the absence of sunlight (and preferably in the presence of a halogen carrier), benzene undergoes substitution when treated with halogen *(cf.* however, isobutene,

(vi) Halogen acids do *not* add on to benzene.

Reactions (i) – (vi) lead to the conclusion that benzene contains three double bonds, but that these double bonds are different from aliphatic double bonds. This difference gives rise to 'aromatic properties' (unusual degree of saturation, stability, etc., see also later).

Kekule (1865) was the first to suggest a ring structure for benzene. He proposed formula (I), and believed *(but did not prove)* that this formula satisfied the following points.

(i) That benzene contains three double bonds.

(ii) That all the six hydrogen atoms in benzene are equivalent; consequently there is only one possible mono-substituted derivative, and there are three possible disubstitution products of benzene.

Kekule's theory stimulated a large amount, of research into the structure of aromatic compounds, a notable result being achieved by Ladenburg in 1874, when he *proved experimentally* that all the six hydrogen atoms in benzene were equivalent.

Kekule's formula, however, did not explain the peculiar behaviour of the three double bonds in benzene. Claus (1867) therefore introduced his diagonal formula, (II), to overcome this difficulty. This formula also appeared to suggest a reason for the simultaneous formation of *o*-and p-compounds (see later). In 1882, Claus modified his formula, now postulating that the *para-bonds* were not like ordinary bonds, but could be easily ruptured.

Dewar (1867) suggested a number of formulae for benzene, one being (III). This is not perfectly symmetrical and therefore was unacceptable (but see later). Ladenburg (1869) attacked Kekule's formula on the grounds that it should *give four* disubstituted derivatives.

Thus there should be *two ortho-derivatives,* 1,2- and 1,6-; these have never been observed in practice. Ladenburg therefore proposed his prism formula, (IV). The six carbon atoms are at the corners of a regular prism, the edges of which denote linkages. This formula does not contain any double bonds, and therefore does not account for the addition products of benzene.

V. Meyer (1870) opposed Ladenburg; he pointed out that the 1,2- and 1,6-derivatives differed only in the position of double bonds and believed that this difference would be too slight to be noticeable. V. Meyer thus supported Kekule's formula.

Kekule (1872) felt that too much importance was being attached to the possible difference between the 1,2- and 1,6- positions, and pointed out that the difficulty arose from the inadequate representation of molecules by structural formulas. According to Kekule, the carbon atoms in benzene were continually in a state of vibration, and due to this vibration, each C — C pair had a single bond half of the time and a double bond the other half. This amounts to an oscillation between the two forms (V) and (VI), each molecule spending half its time in (V) and the other half in (VI). Thus neither (V) nor (VI) represents the benzene molecule satisfactorily; benzene is a 'combination' of the two, and so all bonds will be identical (neither single nor double), and hence there is no real difference between 1,2- and 1,6-disubstituted benzenes.

Baeyer (from about 1884-92) carried out a detailed investigation of benzene and some of its derivatives in order to decide between the formulae of Kekule and Ladenburg. One of the first things Baeyer proved was that cyclohexane and hexahydrobenzene were identical, thus establishing the ring structure of benzene (which Kekule had assumed).

Baeyer observed that as soon as one double bond was removed from benzene, the 'saturation' proper ties were lost, and that dihydrobenzene (cyclohexadiene) behaved as would be expected of an alkadiene. Baeyer showed that no *para* linkage was apparently present and hence rejected Claus' diagonal formula; he also rejected Dewar's formula for similar reasons. Baeyer also showed, by investigation of various benzene derivatives, that Ladenburg's prism formula was untenable and concluded that benzene contained six carbon atoms in a ring, but did not accept Kekule's formula.

He adopted a suggestion of Armstrong (1887) and proposed what is known as the Armstrong-Baeyer centric formula (VII). According to this, the fourth valency of each carbon atom is represented as directed towards the centre of the ring but not actually linked to its opposite neighbour as in the Claus formula. This centric bond is not real but potential; by mutual action the power of each is rendered latent and there is a condition of equilibrium. Such a centric formula is unknown in aliphatic chemistry, and thus this formula could account for 'aromatic properties'.

The centric formula, however, is unsatisfactory for several reasons, *e.g.*, it did not explain the stability of the ring or the behaviour of the polynuclear hydrocarbons. The outcome of Baeyer's work was that 'aromatic properties' depend on the peculiar symmetrical arrangement of the fourth valency of each carbon atom in the ring.

Physico-chemical methods were also applied to the benzene problem, *e.g.*, Thomsen (1880) and Stohmann (1893) came to the conclusion that the heat of combustion of benzene was incompatible with the existence of three double bonds. On the other hand, Bruhl (1880) believed that the value of the refractive index of benzene proved their existence. These methods are particularly interesting in that they are some of the first examples of the application of physico-chemical methods to the elucidation of the structure of organic compounds.

In 1899, Thiele applied his theory of partial valency to the benzene problem, and suggested formula (VIII). This formula dispenses with Kekule's oscillation hypothesis, and in this formula there is no real difference between 'single' and 'double' bonds, and so accounts for there being no difference between the 1,2- and 1,6- positions. At the same time it also accounts for the 'saturation' of benzene.

If Thiele's formula is correct, *i.e.*, 'aromatic character" is due to the symmetrical conjugation of the benzene ring, then cyclo-octatetraene should also exhibit 'aromatic character'. Willstatter prepared this compound with a view to testing Thiele's formula for benzene, and found that it had typical unsaturated properties, whereas, owing to its symmetrical conjugation (structure IX), it would have been expected to resemble benzene.

Structural representation of aromatic compounds : According to the *Chemical Society,* the Kekule type of structure should, in general, be used for benzenoid compounds. On the other hand, large circles representing *six delocalised n-electrons in cyclic systems* (with or without positive or negative signs as appropriate) may be used for certain types of compounds. Cyclic systems having more or fewer than *six* delocalised n-electrons may be represented by circles comprised of broken lines.

AROMATICITY (AROMATIC CHARACTER)

Originally, the term *aromaticity* was used to describe all compounds that had the properties of benzene, and was confined, in consequence, only to compounds which contained benzene rings or a condensed system of benzene rings. The reason for this description was that compounds which had 'aromatic character' exhibited properties which were very much different from those of the analogous aliphatic and alicyclic compounds, *e.g.*, the ease of substitution (although the benzene ring was 'unsaturated'), the stability of the benzene ring, the weaker basic properties of aromatic amines, the acidic properties of phenols (as compared with alcohols), etc. These differences in chemical properties led chemists to seek an explanation, and in consequence, to define *aromaticity*.

CLASSIFICATION OF MONOCYCLIC SYSTEMS

Monocyclic systems have been classified according to the number of carbon atoms in the ring: **small rings,** 3 – 4; **common rings,** 5 – 7; **medium rings,** 8 – 11; **large rings,** 12 –. As we shall see, many chemical properties depend on the class of the cycloalkane, and these differences in behaviour have been explained largely in terms of steric strain.

Baeyer's strain theory. Baeyer (1885) was the first to point out that the angle subtended by the corners and centre of a regular tetrahedron –109° 28′ – lies between the values of the angles in a regular pentagon (108°) and a regular hexagon (120°). On this observation was based the **Baeyer strain theory.** According to the strain theory, the valency angle can be altered from this normal value (109° 28′), but when altered, a strain is set up in the molecule, and the greater the deviation from the normal angle, the greater is the strain. Thus, according to Baeyer, five- and six-membered rings form most readily, and are the most stable because they involve the least strain (or distortion) from the normal valency angle. If all the rings are *assumed* to be planar, the distortions for each ring-size can be readily calculated, *e.g.*, in cyclopropane the three carbon atoms each occupy a corner of an equilateral triangle. Since the angles of an equilateral triangle are 60°, the distortion in cyclopropane will be 1/2 (109° 28′ – 60°) = + 24° 44′. It should be noted that the distortion of the bond angle has been assumed to be *equally* shared between the two bonds. A list of distortions, and it can be seen that these decrease rapidly from ethylene (counting this as a 'two-membered' ring) to cyclopentane and then increase more slowly. Thus, the smaller rings (including ethylene) suffer large distortion and this was believed to be the cause of their enhanced reactivity.

Baeyer's strain theory is based on a mechanical concept of valency and on the assumption that *all* the rings are planar. This has led to conclusions which have now been shown to be wrong. In the first place, physical methods have shown that all rings, except cyclopropane, are not planar (cyclopropane is a three-point system and consequently *must* be planar).

In the second place, quantum mechanical calculations do not permit very large distortions of bond angles. If cyclopropane were an equilateral triangle with bonds directed *along* the lines joining the carbon atoms, then the ring valency angle of each carbon atom would be 60°. This value is impossible, since the carbon valency angle can never be less than 90° (when they are pure p-orbitals).

Furthermore, mixture *of p* with s-orbitals *opens* the valency angle. According to Coulson *et al.* (1949), calculation has shown that the smallest carbon valency angle that one can reasonably expect to have is 104°. Coulson has therefore suggested that in cyclopropane, the carbon hybridised orbitals are not pointing towards one another in the same straight line, and consequently there is a loss of overlap.

It is this loss of overlap that gives rise to instability, the cyclopropane molecule being in a state of 'strain' due to 'bent' bonds. Applying this argument to cyclobutane, we see that this molecule also has 'bent' bonds, but loss of overlap is less in this case than for cyclopropane, and so the former will be more stable than the latter.

Since strain affects chemical stability, measurement of the latter will indicate the presence of the former. Chemical stability may be measured in various ways, *e.g.*, by the heat of formation, heat of combustion, dipole moment, absorption spectra, etc. One of the most convenient to work with, in respect to hydrocarbons is the heat of combustion; this gives a measure **of thermochemical stability** and is the *total* strain, *i.e,,* the sum of the *Baeyer strain* (bond angle deformation only) and other sources of strain (steric repulsions, etc.). If the strain in a ring changes with the size of the ring, then this should be observed by changes in the heat of combustion.

The *total strain* in the molecule may be calculated as follows.

(number of C atoms in ring) × (observed heat of combustion/CH_2 – observed heat of combustion CH_2 for n-alkanes)

According to the results stability increases up to the six-membered ring, then decreases from seven to eleven-membered rings, and from the twelve-membered onwards attains the stability of the six-membered ring.

General Properties of the Cycloalkanes

In many chemical respects the cycloalkanes are similar to the alkanes, but in others they are different in that the lower members form addition products with *ring fission.* Chlorine and bromine, in the presence of a catalyst, form halogenocycloalkanes, but with cyclopropane, the product is a mixture of 1,1-, 1,2-, and 1,3-dihalogenopropanes. Concentrated hydrogen iodide has no action on cycloalkanes except cyclopropane (at room temperature), which gives w-propyl iodide, and cyclobutane (on heating) which gives n-butyl iodide. Hydrogen in the presence of nickel at elevated temperatures does not attack cycloalkanes except cyclopropane (at 80°C to give propane) and cyclobutane (at 200°C to give «-butane).

Depending on the nature of the compound and on the reagents used, various rings may be opened or changed in size.

The infra-red absorption region of the CH_2 group (C – H stretch) in cycloalkanes is the same as that for alkanes except in the case of cyclopropane where the region is 3 080 – 3040cm^{-1} (v).

On the other hand, the absorption region of the carbonyl group (C = O stretch) in cycloalkanones depends on the size of the ring: 4-ring, 1780–1760 cm^{-1} (s), 5–ring, 1750 – 1740 cm^{-1} (s), 6- and 7 -rings, 1720–1700 cm^{-1} (s).

Thus the first two rings can be distinguished from acyclic ketones, but not the last two. However, the molecular formula will help here.

THE BIPHENYL MOLECULE

Kaufler (1907) proposed the 'butterfly' formula (I) in order to account for the chemical behaviour of various biphenyl derivatives, *e.g.*, Michler and Zimmermann (1881) had condensed benzidine with carbonyl chloride and obtained a product to which Kaufler assigned structure (II). According to Kaufler, the co-axial structure (III) was impossible, since the two amino-groups are too far apart to react simultaneously with carbonyl chloride; it should be noted that this simultaneous reaction at both ends was assumed by Kaufler. Simultaneous reaction, however, is reasonable (according to Kaufler) on the folded structure (II). Now Schultz (1880) had prepared a dinitrodiphenic acid by the nitration of diphenic acid, and Schmidt *et al.* (1903), from their work on this acid, believed it to be 6,6′-dinitrodiphenic acid (IV); these workers, it should be noted, did not synthesise the acid. In 1921, however, Kenner *et al.* synthesised 6,6′-dinitrodiphenic acid by means of the Ullmann reaction on the ethyl ester of 2-chloro-3-nitrobenzoic acid, and hydrolysing the product. This acid (V) (written with the two benzene rings co-axial), did not have the same melting point as Schultz's acid, and so Kenner, believing that his and Schultz's acid were both 6,6′-dinitrodiphenic acid, suggested that the two were stereoisomers.

Then Christie and Kenner (1922) showed that Kenner's acid was resolvable, and pointed out that this could be explained on the Kaufler formula (IV), since this structure has no elements of symmetry. These authors, however, also pointed out that the optical activity could also be accounted for by the co-axial structure (V), provided that the two benzene rings do not lie in one plane.

Kaufler's formula, as we have seen, was based on the assumption that the two amino-groups in benzidine react simultaneously with various reagents. Re-investigation of these reactions showed that this was not the case, *e.g.*, Turner and Le Fevre (1926) found that the compound produced from benzidine and carbonyl chloride was not as originally formulated (see II or III), but had a free aminogroup, *i.e.*, the compound was

$$[NH_2C_6H,C_6H,NH]_2CO$$

Hence Kaufler's reason for his butterfly formula is incorrect, and although it does not necessarily follow that the formula is incorrect, nevertheless Turner's work weakened Kaufler's claim. One of the strongest bits of chemical evidence for rejecting Kaufler's formula is that of Barber and Smiles (1928).

These workers prepared the three dimercaptobiphenyls, (VI), (VII) and (VIII), and oxidised each one. Only one of them, the 2,2′-derivative (VI), gave the intramolecular disulphide (diphenylene disulphide, IX). On the Kaufler formula, all three dithiols would be expected to give the intramolecular disulphides, since the two thiol groups are equally distant in all three compounds.

Physico-chemical methods have also been used to determine the configuration of the biphenyl molecule, *e.g.*, the crystal structure of 4,4′-biphenyl derivatives shows a centre of symmetry; this is only possible for the co-axial formula. Dipole moment measurements also confirm this configuration, *e.g.*, the dipole moment of 4,4′-dichlorobiphenyl is zero; this again is only possible if the two benzene rings are co-axial.

OPTICAL ACTIVITY OF BIPHENYL COMPOUNDS

Christie and Kenner's work has been extended by other workers, who showed that compounds in which at least three of the four ortho-positions in biphenyl are occupied by certain groups could be resolved. It was then soon found that two conditions were necessary for biphenyl compounds to exhibit optical activity

(i) Neither ring must have a vertical plane of symmetry. Thus (I) is not resolvable, but (II) is.

(ii) The substituents in the ortho-positions must have a large size, *e.g.*, the following compounds were resolved: 6-nitrodiphenic acid, 6,6′-dinitrodiphenic acid, 6,6′-dichlorodiphenic acid, 2,2′diamino-6,6′-dimethylbiphenyl.

The earlier work showed that three groups had to be present in the ortho-positions. This gave rise to the theory that the groups in these positions impinged on one another when free rotation was attempted, *i.e.*, the steric effect prevented free rotation. This theory of restricted rotation about the single bond joining the two benzene rings (in the co-axial formula) was suggested simultaneously in 1926 by Turner and Le Fevre, Bell and Kenyon, and Mills.

Consider molecule (III) and its mirror image (IV). Provided that the groups A, B and C are large enough to 'interfere mechanically', *i.e.*, to behave as 'obstacles', then free rotation about the single bond is restricted. Thus the two benzene rings cannot be coplanar and consequently (IV) is not superimposable on (III), *i.e.*, (III) and (IV) are enantiomers. In molecule (III) there is no chiral centre; it is the molecule as a whole which is chiral, due to the restricted rotation.

In biphenyl the two benzene rings are co-axial, and in optically active biphenyl derivatives the rings are inclined to each other due to the steric and repulsive effects of the groups in the orthopositions. The actual angle of inclination of the two rings depends on the nature of the substituent groups, but it appears to be usually in the vicinity of 90°, *i.e.*, the rings tend to be approximately perpendicular to each other. Thus, in order to exhibit optical activity, the substituent groups in the ortho-positions must be large enough to prevent the two rings from becoming coplanar, in which case the molecule would possess a plane or a centre of symmetry, *e.g*, diphenic acid is not optically active. In configuration (V) the molecule has a plane of symmetry, and in configuration (VI) a centre of symmetry;

of these two, (VI) is the more likely because of the repulsion between the two carboxyl groups.

If restricted rotation in biphenyl compounds is due entirely to the spatial effect, then theoretically we have only to calculate the size of the group in order to ascertain whether the groups will impinge and thereby give rise to optical activity. The 'size' of the group, however, must be calculated from van der Waals radii the results are in good agreement with experiment.

Later work has shown that if the substituent groups are large enough, then only two in the o- and o′-positions will produce restricted rotation, *e.g.*, Lesslie and Turner (1932) resolved biphenyl-2,2'disulphonic acid (VII). In this molecule the sulphonic acid group is large enough to be impeded by the ortho-hydrogen atoms. This molecule was readily racemised on heating, but Lesslie *et al.* (1962) prepared the enantiomers of 2,2′-di-t-butylbiphenyl from the corresponding optically active 6,6′-di-t-butylbiphenyl-3,3′-dicarboxylic acids. These were found to be highly optically stable (the t-butyl group is very large). Lesslie and Turner (1933) have also shown that the (+)-camphorsulphonate of 3′-bromobiphenyl-2-trimethylarsonium iodide (VIII) undergoes mutarotation. The trimethylarsonium group is large enough to be impeded by the ortho-hydrogen atoms (the bromine atom in the meta-position gives asymmetry to this ring). Attempts to isolate the active biphenyl compound failed because it racemised rapidly. This mutarotation indicates that the biphenyl is optically active and that the two enantiomers are readily interconvertible.

Since one phenyl group can rotate with respect to the other (about the interannular bond), the various positions would correspond to different conformations. In acyclic compounds, although there are preferred conformations the energy barriers are very low, and so the various conformations are interconvertible. In the case of the biphenyls, however, because of steric hindrance, the molecules have large energy barriers separating the two forms, and these barriers are large enough (75-105 kJ mol^{-1}) to produce separable rotational isomers. Such isomers are called atropisomers, and the two conditions for atropisomerism have been given above.

It has already been pointed out that diphenic acid is not optically active, and that its configuration is most probably (VI). Now calculation shows that the effective diameter of the carboxyl group is large enough to prevent configuration (V) from being planar, and consequently, if the two rings could be held more or less in this configuration, the molecule would not be coplanar and hence would be resolvable.

Such a compound, (IX), was prepared and resolved by Adams and Kornblum (1941). The two benzene rings are not coplanar and are held fairly rigid by the large methylene bridge.

Many biphenyls have been studied from the point of view of the effect of a 2,2′-bridge on the optical activity of molecules of the type (X). When n = 1, the molecule is a disubstituted fluorene. Since this molecule is flat, it is not resolvable. When n = 2, the molecule is a disubstituted 9,10-dihydrophenanthrene. Such compounds have been resolved, *e.g.*, (XVI) and (XVII) (see below). When n = 3, the molecules are resolvable and are highly optically stable. Iffland *et al.* (1956) prepared the optically active biphenyl (XI) which has two amino-groups in the 6,6′-positions. On the other hand, these authors have also prepared (XII) in optically active forms.

Mislow (1957) has also obtained the dibenzocyclo-octadiene acids (XIII) in optically active forms; both forms were highly optically labile. Similar to (XIII) is (XIV) which has been resolved by Bell (1952). Mislow *et al.* (1961) have also resolved the biphenyl derivative (XV), and in 1962, prepared the (−)-form of (XVI). Turner *et al.* (1955) had prepared (XVII), and Mislow, on comparing the optical stability of (XVI) with (XVII), found that the latter was the more stable one. In (XVI), the two methyl groups can slip past each other comparatively easily by bending out of the ring plane, but in (XVI) the second benzene ring in each naphthalene nucleus behaves as a large group and is also rigid, and consequently bending is very difficult.

Apart from the resolution of 2,2'-biphenyls, the stereochemistry of these compounds has been studied by u.v. and NMR spectroscopy and by X-ray analysis. In this way, a more detailed threedimensional structure of these molecules has been obtained, *e.g.*, Wahl, jun. *et al.* (1972) have carried out an X-ray diffraction study of (XX) and have shown that this molecule exists in the pseudochair form (XXI). (XX) had been prepared and resolved chromatographically (on cellulose acetate) by Liittringhaus *et al.* (1967).

ABSOLUTE CONFIGURATIONS OF BIPHENYLS

Since biphenyls owe their asymmetry to the molecule being asymmetric, the methods used for correlating configurations of compounds containing asymmetric carbon atoms cannot be applied. However, Mislow *et al.* (1957) used asymmetric synthesis as a means of establishing the absolute configuration of 6,6′-dinitro-2,2′-diphenic acid. Their method was chemical; assignment of absolute configuration has been obtained from a consideration of the transition states in the Meerwein-Ponndorf-Verley reduction of a dissymmetric diphenylic ketone by asymmetric alcohols of known absolute configuration. The (+)- and (–)-ketones (I) were partially reduced with (S)-(+)-methyl-t-butylmethanol in the presence of aluminium t-butoxide. The products were unchanged ketone with the (+)-form predominating, and the (+)- and (–)-alcohols, (II), with the (–)-form predominating.

Thus, as far as the unchanged ketone is concerned, this reaction is a kinetic resolution, since the ketone is now enriched in the (+)-form, and at the same time, the alcohol has become enriched in the (-)-form. Examination of models showed that, in the single conformation possible for the (S)-enantiomer (of the ketone), hydride transfer to either side of the carbonyl group is hindered by steric repulsion between the t-butyl group and a phenyl group, whereas for the (R)enantiomer, the repulsion is only between the methyl group and a phenyl group. Thus the (+)-ketone the (+)-ketone is (S)-(+) and the (–)-alcohol is (R)-(–). Furthermore, the (S)-(+)-ketone had been prepared from (–)-6,6′-dinitro-2,2′-diphenic acid (via the dimethyl ester), and so this (–)-enantiomer is the (S)-(–)-acid. This assignment of absolute configuration has been confirmed by X-ray diffraction (Akimoto *et al.*, 1968). The method of chemical correlation is reliable only if there is no change in configuration during the transformation or if the change in configuration occurs in a predictable manner. Thus, using the (S)-(–)-acid as absolute standard, Mislow *et al.* carried out a number of chemical correlations, *e.g.*,

Mislow *et al.* (1958), using the (S)-(–)-acid as the absolute standard, also correlated configurations in the biphenyl series by the quasi-racemate method. In this way these authors determined the configurations of 6,6′-dichloro- and 6,6′-dimethyl-2,2′-diphenic acid. Mislow *et al.* (1960) have also confirmed absolute configurations in the biphenyl series by the rotatory dispersion method. The authors showed that the shapes of the ORD curves depended on the configuration and conformation of the biphenyl compounds; these compounds contain inherently dissymmetric chromophores.

The specification of absolute configuration of biphenyl compounds is carried out as follows. Since biphenyls do not owe their asymmetry to the presence of asymmetric carbon atoms (which are chiral centres), the criterion now is the presence of a chiral axis (*i.e.*, an axis of asymmetry). This chiral axis may be derived from a chiral centre Z, (III), by 'extending' the point into a line AB which now passes through an elongated tetrahedron, (IV). For Z to be a chiral centre in (III), a, a′, b, and b′ must all be different, but for AB to be a chiral axis in (IV), it is sufficient that a and b be different, and a′ and b′ be different; it is not necessary that a should be different from a′, or b from b′.

To apply the sequence rule to axial chirality (or axial asymmetry), it is necessary to use an additional rule, viz. with respect to an external point on the chiral axis, groups at the near end of the axis are given precedence over groups at the far end. If (IV) is viewed from point A, then the pair a–b, being nearer A, precedes the pair a′–b′; and in (V), pair a–b precedes pair c–d. If the order of priority is a > b and a′ > b′ (for IV), and a > b and c > d (for V), then both of these models give the final order of priority shown in (VI).

If now, in accordance with the conversion rule, (VI) is viewed from the side remote from 4, then (VII) is obtained. This, by two interchanges gives (VIIa), and since the sequence 1 → 2 → 3 is clockwise, (IV) and (V) are (R)-configurations. Had (IV) and (V) been viewed from point B on the chiral axis, then pairs a-b, a′–b′ and c–d will be 1 and 2, and the pair a–b will be 3 and 4. This gives (VIII), which in turn gives (IX), and this, by two interchanges, gives (IXa), which is still the (R)-configuration.

Let us now consider some substituted biphenyls. First, the four ortho-substituent groups are inspected. If they are different, as pairs (2–6 and 2′–6′), then they are used. Thus, in (X), NO_2 = a, and CO_2H=b (priority a > h), and this molecule is therefore the (S)-form (the last diagram has 1 → 2 → 3 anticlockwise). In (XI), since the upper ring has Cl in both ortho-positions, groups H and Me are therefore selected; NO_2 = a, CO_2H = b, Me = c, and H = d. Hence, (XI) is the (S)-form (two interchanges in the last diagram give 1 → 2 → 3 in an anticlockwise direction).

OTHER EXAMPLES OF ATROPISOMERISM

In addition to the biphenyl compounds, there are many other examples where optical activity in the molecule is produced by restricted rotation about a single bond which may or may not be one that joins two rings.

The following examples are only a few out of a very large number of compounds that have been resolved.

(i) 1,1′-Binaphthyl-8,8′-dicarboxylic acid has been obtained in optically active forms by Stanley (1931). This compound gives rise to asymmetric transformation; resolution with brucine gave 100 per cent of either the (+)- or (–)-compound.

Other compounds similar to the binaphthyl which have been obtained in optically active forms are 1,1′-binaphthyl-5,5′-dicarboxylic acid (I) (Bell *et al.*, 1951), the bianthryl derivatives, (II) and (III) (Bell *et al.*, 1949), and the 4,4′- and 5,5′-biquinolyls, (IV) and (V) (Crawford *et al.*, 1952).

(ii) Mills and Elliott (1928) obtained N-benzenesulphonyl-8-nitro-l-naphthylglycine (VI) in optically active forms; these were optically unstable, undergoing asymmetric transformation with brucine. Mills and Kelham (1937) also resolved N-acetyl-N-methyl p-toluidine-3-sulphonic acid (VII) with brucine, and found that it racemised slowly on standing. In both (VI) and (VII) the optical activity arises from the restricted rotation about the C–N bond (the C being the ring carbon to which the N is attached). Asymmetry arising from the same cause is also shown by 2-acetomethylamido-4′,5-dimethyldiphenylsulphone (VIII); this was partially resolved by Buchanan *et al.* It is also interesting to note in this connection that Adams *et al.* (1950) have isolated pairs of geometrical isomers of compounds of the types (IX) and (X); here geometrical isomerism is possible because of the restricted rotation about the C–N bonds.

(iii) Luttringhaus *et al.* (1940, 1947) isolated two optically active forms of 4-bromogentisic acid decamethylene ether (XI). This belongs to the group known as 'ansa' compounds, and the methylene ring is perpendicular to the plane of the benzene ring; the two substituents, Br and CO_2H, prevent the rotation of the benzene nucleus inside the large ring. Cram *et al.* (1955) have obtained paracyclophanes in optically active forms, *e.g.*, (XII). In this molecule, the planes of the two benzene rings are approximately parallel (and the carboxyphenyl ring cannot rotate to give the enantiomer). When the bridges each contained four methylene groups, the compound could not be resolved (the carboxyphenyl ring can now rotate to give the enantiomers). On the other hand, Blomquist *et al.* (1961) have resolved the simple paracyclophane (XIII).

(iv) Terphenyl compounds can exhibit both geometrical and optical isomerism when suitable substituents are present to prevent free rotation about single bonds, *e.g.*, Shildneck and Adams (1931) obtained (XIV) in both the cis- and trans-forms. Interference of the methyl and hydroxyl groups in the ortho-positions prevents free rotation and tends to hold the two outside rings perpendicular to the centre ring. Inspection of these formulae shows that if the centre ring does not possess a vertical plane of symmetry, then optical activity is possible. Thus, Browning and Adams (1930) prepared the dibromo cis- and trans-forms of (XV) and resolved the cis-isomer; the trans-isomer is not resolvable since it has a centre of symmetry. It can be seen from the terphenyl compounds discussed that atropisomerism does not necessarily imply enantiomerism. The different forms may be related to each other as diastereoisomers, but whether optical isomerism is also exhibited depends on the substitution pattern.

(v) A very interesting case of restricted rotation about a single bond is afforded by the compound 10-m-aminobenzylideneanthrone (XVI). This was prepared by Ingram (1950), but he failed to resolve it. He did show, however, that it was optically active by the mutarotation of its camphorsulphonate salt, and by the preparation of an active hydriodide. Thus the molecule is asymmetric, and this asymmetry can only be due to the restricted rotation of the phenyl group about the C-phenyl bond, the restriction being brought about by hydrogen atoms in the ortho-positions. The two hydrogen atoms labelled $\overset{\times}{H}$ overlap in space, and consequently the benzene ring cannot lie in the same plane as the 10-methyleneanthrone skeleton. Another example is the substituted cinnamic acids (XVII) (R = Cl, Me, OMe) [Adams *et al.*, 1940, 1941]. The benzene ring and the ethylenic double bond cannot become coplanar, and it was found that the order of stability to racemisation was Cl > Me > OMe.